Protein Structure and Modeling

Natalya Kurochkina

Protein Structure and Modeling

Springer

Natalya Kurochkina
Department of Biophysics
School of Theoretical Modeling
Washington, DC, USA

ISBN 978-981-13-6600-0 ISBN 978-981-13-6601-7 (eBook)
https://doi.org/10.1007/978-981-13-6601-7

This Springer imprint is published by the registered company Springer Nature Singapore Pte Ltd.
The registered company address is: 152 Beach Road, #21-01/04 Gateway East, Singapore 189721, Singapore

Preface

Structure of proteins, nucleic acids, polysaccharides, and lipids is very important for our understanding of biological processes and life functions. Biological macromolecules are very complex objects to study. Proteins represent the largest and most extensively studied group of biopolymers. Comprising thousands of atoms, they exhibit several levels of organization, multiple isoforms, dynamic behavior, and diverse functions. With advancements in sequencing, complete genomes of many organisms are being decoded. We are able to compare healthy and diseased organisms and observe changes during development. These rapid methods bring new data on the basic building blocks of living organisms and allow their thorough comparison and elucidation of principles of organization. X-ray crystallography, NMR, electron microscopy, and other methods reveal structure of macromolecules at atomic resolution which contributes to significant progress in many areas of biomedical research, medicine, biotechnology, and design of new drugs. With growing amount of data, new approaches in biochemistry, mathematics, biophysics, structural biology, and other areas lead to development and refinement of theoretical models. Increase in accuracy and number of experimental and theoretical models helps better understand and even design new living organisms.

Proteins are remarkably regular objects in spite of being very heterogeneous and irregular at some levels. They perform functions that require precise organization. We explain their precise organization and performance by structural principles. As molecular machines they function according to concerted work of their constituent parts with each atom at precise location at precise time. We admire their work. Perhaps the same way we see the ocean wave and admire its beauty. Its motion is governed by laws of nature.

Washington, DC, USA Natalya Kurochkina

Contents

Chapter 1
Proteins and Protein Structure

Abstract Proteins such as enzymes, channels, signaling molecules and adaptors carry out important functions in living organisms. Fibrous and globular proteins comprise two large groups. Long stretches of coiled coil α-helices in fibers and fibrils, triple helices of collagen, and globular heme binding subunits of hemoglobin give us main representatives and show how diverse these molecules are. Protein polypeptide chain exhibits left-handed and right handed, parallel and antiparallel arrangements of secondary structure elements such as alpha, 3_{10}, polyproline, gamma and pi helices, strands and turns. This chapter describes structural principles of protein molecule, its conformation and relationships between primary, secondary, tertiary, and quaternary structure.

Keywords Fibrous protein · Globular protein · Protein conformation · Amino acid · Protein structure

Fibrous and Globular Proteins

X-ray diffraction of proteins identified distinct patterns of polypeptide chain folding: fibers of α and β types. Diffraction patterns represent different types of molecular conformation. Repeated units of protein molecules, amino acids, fold in regular arrangements. Structure of α-helix was proposed (Pauling et al. 1951). Crystallography brought detailed structure of these arrangements at atomic resolution (Squire and Parry 2017).

For their shape proteins can be subdivided into two large groups: fibrous and globular (Kurochkina 2011; 2012). Fibrous proteins (Fig. 1.1) keratin, myosin, epidermin, and fibrinogen form long rods, whereas globular proteins (Fig. 1.2) myoglobin, immunoglobulin, triose phosphate isomerase, and lysozyme are nearly spherical.

Elongated shape of molecules of fibrous proteins is a consequence of repeated arrangement of amino acids in one of hydrogen bonded conformations: coiled coil structure – a pair of α-helices winded around each other along their entire length;

N. Kurochkina, *Protein Structure and Modeling*,
https://doi.org/10.1007/978-981-13-6601-7_1

Fig. 1.1 Fibrous proteins

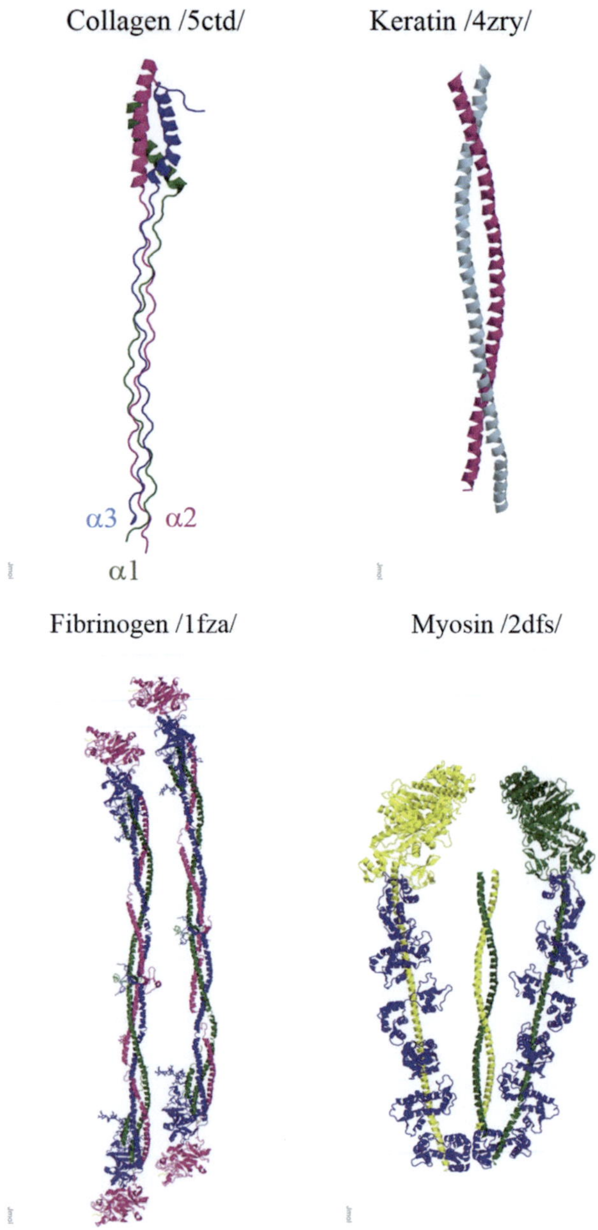

extended chains bound in β-sheets, or collagen three strands in a triple helix. These three major types of hydrogen bonded secondary structures are also present in globular proteins in which length does not deviate significantly from width or height.

Collagen, main component of skin, bone, cartilage, tendon, and ligament, consists of three polypeptide α chains (α1, α2, and α3) in a triple-helical conformation (Goh

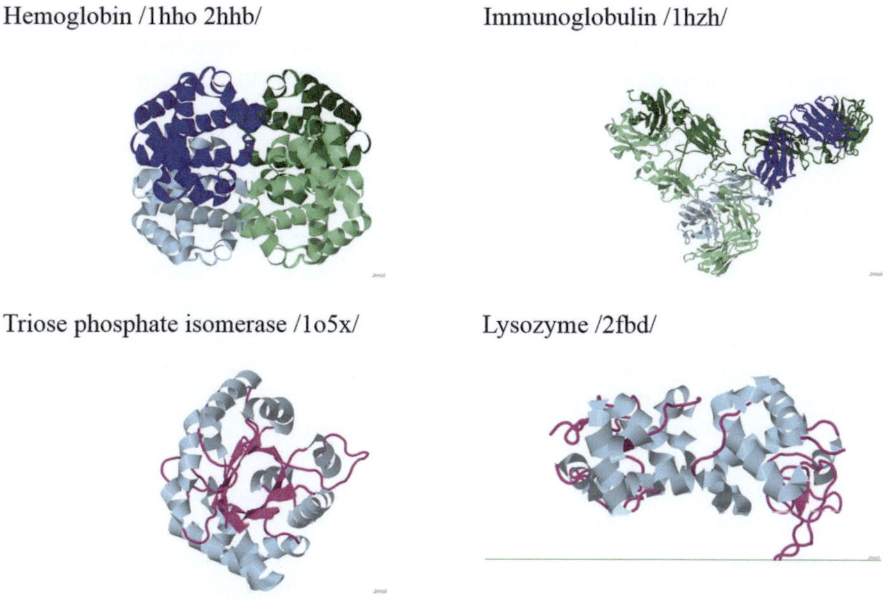

Fig. 1.2 Globular proteins

and Holmes 2017). Homo- and hetero-trimeric types of collagen exist. Totally 28 types of human collagen comprise this most abundant protein of the human organism. Amino acid sequence of collagen triple helix contains repeat of three residues Gly-Pro-Hyp. Hydroxyproline is posttranslational modification at position 4 of prolyl residues, addition of hydroxyl group. Every third residue is glycine. Many types of collagen (I, II, III, V, and VI) form fibrillar structures. Collagen fibrils from a connective tissue of the sea urchin and mammals are very similar and show several levels of hierarchical architecture (Goh and Holmes 2017). Electron micrographs reveal fibrils 37–570 μm long and 25–280 nm in diameter. Collagenous extracellular matrix of connective tissues is a rich source of scaffolds used in a variety of biomedical applications such as wound healing, implants engineering, and drug delivery (Woolfson 2010; Boudko and Bächinger 2016).

Keratin builds soft and hard epithelial tissues of skin, hair, nail and cornea. 54 human keratin genes exist. Combinations of various keratin proteins organize cytoskeleton of epithelial cells, their 10 nm long intermediate filaments. Crystal structures of perinuclear keratin 5 and 14 important for maintenance of the shape of nucleus (Lee et al. 2012) as well as skin keratin 1 (K1) and keratin 10 (K10) (Bunick and Milstone 2017) show parallel association of α-helices of keratin monomers into a heterodimer (Fig. 1.1). Globular domains at both ends of the rod determine specificity of binding. Keratin fibers of human hair are organized in nano-scale grainy structures. The matrix of the keratin fiber cortex is well organized network of keratin associated proteins (Kadir et al. 2017).

Fibrinogen is critical molecule of blood coagulation process in fibrin clot. Targeted to the ER by a signal peptide for glycosylation, cleaved by thrombin to a mature protein, assembled as a dimer $(A\alpha\text{-}B\beta\text{-}\gamma)_2$ of Aα-Bβ-γ half molecules and modified by N-linked oligosaccharides in Golgi, it is released into the circulation (Brennan 2015; Kollman et al. 2009; Pechik et al. 2004; Medved and Nieuwenhuizen 2003). Fibrin/fibrinogen is able to form covering films and protect organism from blood loss and microbial infection after injury.

Myosin α-helical rods are packed in filaments. In the process of muscle contraction, myosin thick filaments move along actin thin filaments (Squire et al. 2017). This process resembles walking (Liu et al. 2006a). Tropomyosin and troponin regulate striated muscle contraction in Ca^{2+} dependent manner. Tropomyosin isoforms specifically bind and stabilize actin filaments regulating assembly of numerous binding partners such as myosin, tropomodulin, formin, Arp2/3 and ADF-cofilin (Gunning et al. 2015). Helices of tropomyosin are 242–284 amino acids long (Barua 2013).

Many proteins including collagen, fibrinogen, myosin and keratin contain elongated and globular domains.

Shape of heme binding globins is typical for globular proteins. Myoglobin and hemoglobin are oxygen carriers in blood. Oxygen reversibly binds to the heme group enclosed by α-helices arranged in a characteristic globin fold, or heme box. Hemoglobin is a heterotetramer comprising two α and two β chains (Kendrew 1959). Brain globins in vertebrates include neuroglobin, myoglobin, hemoglobin, and cytoglobin (Xie and Yang 2016). Mutation of one amino acid residue, glutamine to valine, results in polymerization of hemoglobin and sickle cell disease (SCD). Disruption of interactions between monomers, for example, by hydroxyurea treats patients with SCD (Ferrone 2016). Modern applications of hemoglobin vesicles include blood transfusion, oxygen and carbon monoxide based therapies, photosensitization, and perfusate for transplant organs (Sakai 2017).

Although each group exhibits clearly distinct features (Parry et al. 2008), fibrous and globular proteins have common structural characteristics.

Polypeptide Chain

Amino acids are joined in polypeptide chain by a peptide bond (Fig. 1.3) formed between carbonyl oxygen of amino acid and amide nitrogen of the next amino acid. Repeating atoms, -C-N- C_{i-1}^{a} -C-N- C_{i}^{a} -C-N- C_{i+1}^{a} -C-, form protein backbone. Synthesis of polypeptide chain occurs on ribosome by polymerization reaction.

Rotation around C-N bond is restricted so that dihedral angle (ω) adopts values of 0° (*cis*) or 180° (*trans*). Torsion angles around N-Cα (ϕ) and Cα-C (ψ) bonds show characteristic values for each type of conformation. Side chains conformation is characterized by a set of χ angles. Rotation around each bond results in multiple conformational states. Conformation of polypeptide chain, therefore, can be

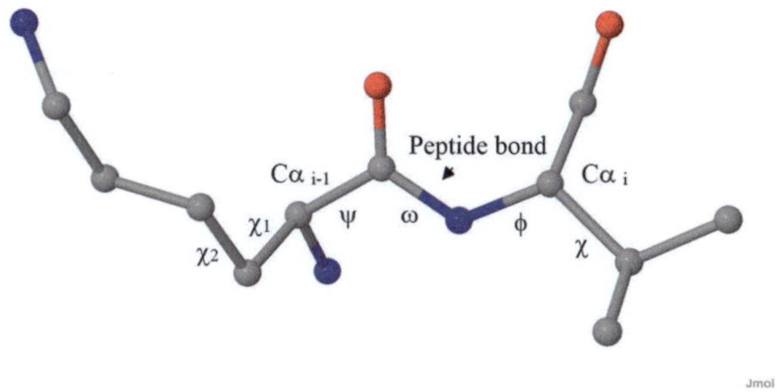

Fig. 1.3 Peptide group

described as a set of dihedral angles. Distribution of main chain dihedral angles (ϕ, ψ) shows that van der Waals interactions restrict allowed angle values (Ramachandran et al. 1963). Ramachandran mapping (plot) summarizes this distribution and represents normally allowed and disallowed (outer limit) configurations of polypeptide chain. Regions in which ϕ, ψ values cluster correspond to left handed and right-handed α-helix, extended β-structure, polyproline type I (PPI) and type II (PPII), 3_{10}, γ-helices, and π-helices.

Each amino acid type exhibits several preferred conformations (or states). Propensities of amino acid to adopt certain conformation are determined by properties of its atomic groups, surrounding amino acids, and solvent. Side chain dihedral angles (χ) also tend to cluster and exhibit several preferred rotameric states (Ponder and Richards 1987). Rotamers that correspond to low energy depend on interactions of proximal atomic groups, secondary and tertiary structure. Relationships between side chain (χ) and backbone (ϕ, ψ) dihedral angles show that rotamer preferences exist to be in particular secondary structure conformation (Schrauber et al. 1993; Dunbrack and Karplus 1993).

Peptides consist of less than and proteins more than hundred amino acids. Polypeptide chain tends to form secondary structure or remain irregular. Amino acids conformation to a large extent is determined by the length of the chain. Conformation of monopeptides, dipeptides, and polypeptides depends on the surrounding amino acids in the chain and adjacent chains.

Twenty Canonical Amino Acids and Their Conformers

Genome of a living organism encodes twenty amino acids. Cell processes: protein and hormone synthesis, energy production, acid/base homeostasis, enzymatic activity, cell signaling, gene expression, and osmosis, rely on amino acids.

Discovered in 1901 in gelatin hydrolysates by Fisher, **proline** (Fig. 1.4) represents unconventional amino acid (Fisher 1901). Proline α-amino group is covalently bonded to Cδ atom of the side chain either in *cis* or *trans* conformation and forms a pyrrolidine ring. The heterocyclic 5-member ring prevents rotation around ϕ angle. The most common ring conformations are envelope and half-chair. In envelope conformation, four atoms are in the same plane whereas the fifth atom is out of the

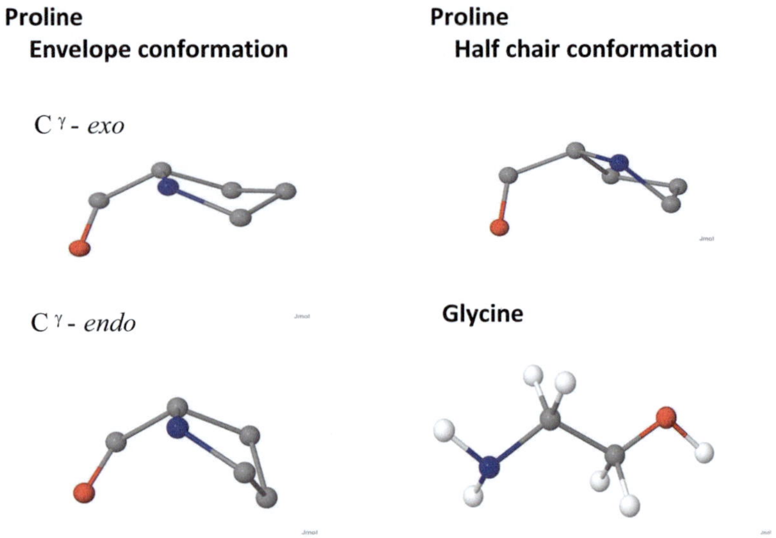

Fig. 1.4 Amino acids: proline and glycine

plane. In half-chair conformation, three atoms are in the same plane whereas two adjacent atoms are on the opposite sides of the plane. Among many possible envelope conformations two are predominant: Cγ-*endo* (C4 or Cγ is on the same side of the plane as C1 carboxylate) and Cγ-*exo* (C4 is on the opposite side of the plane as C1 carboxylate) (Newberry and Raines 2017).

Since proline forms only one backbone hydrogen bond, it is very often located at the N and C termini of secondary structure elements. In the middle of the secondary structure element it bends the structure. Polypeptide fragments with several adjacent prolines exhibit polyproline conformation. In hexapeptides, diproline sequence at the N-terminus generates turns (Kantharaju et al. 2010).

Glycine has the smallest side chain, hydrogen atom. Its participation in secondary structure depends on the surrounding. Small side chain contributes to flexibility of secondary and tertiary structure. Glycine is frequently present in loops and hydrogen bonded motifs of β sheet capping, linking β hairpins to α helices and α helix insertions (Balasco et al. 2018). Together with proline and hydroxyproline, glycine is a component of one of the most abundant proteins in animals, collagen (Li et al. 2017). Glycine and GABA are two inhibitory neurotransmitters of the central nervous system (Gamlin et al. 2018). *N*-substituted glycine oligomers, peptoids, due to their ability to increase membrane permeability and possess flexible secondary structure, are potential peptidomimetic agents for immunotherapies against AD-associated beta amyloid (Young 2018).

For three amino acids: proline, **alanine**, and glycine, only ϕ, ψ, and ω angles describe its backbone conformation. For the remaining amino acids, in addition to ϕ, ψ, and ω angles, χ angles describe side chain conformation.

Aliphatic amino acids: **leucine, valine, alanine, and isoleucine** (Fig. 1.5), mostly comprise protein hydrophobic core. They also can reside on protein surface as a part

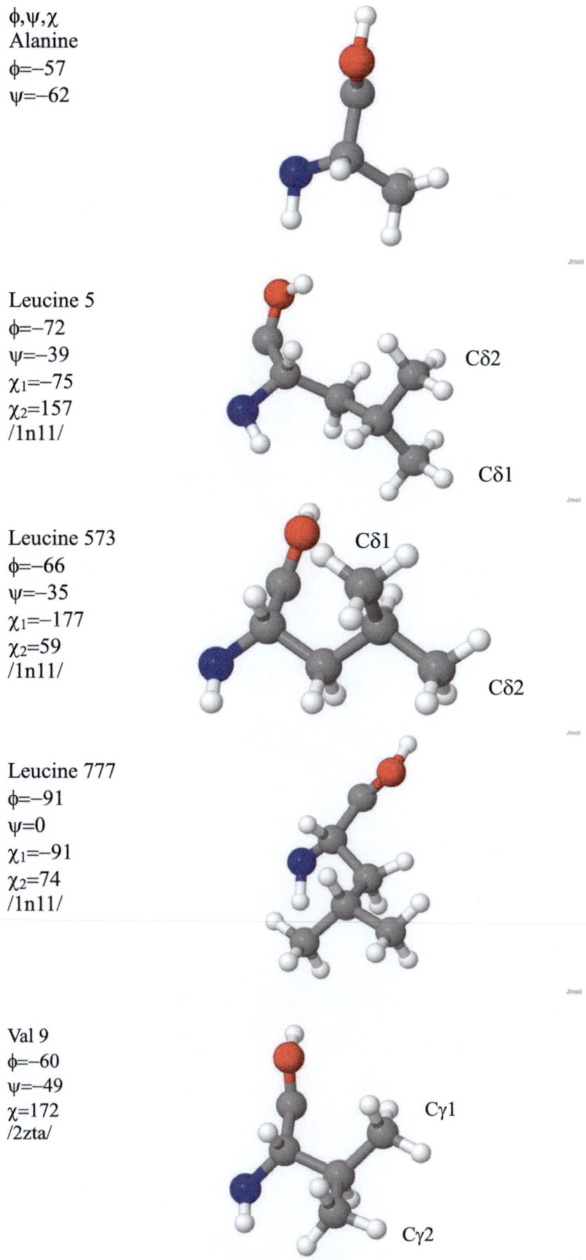

Fig. 1.5 Aliphatic amino acids: alanine, valine, leucine, and isoleucine

Val 789
$\phi=-65$
$\psi=-24$
$\chi=-51$
/1n11/

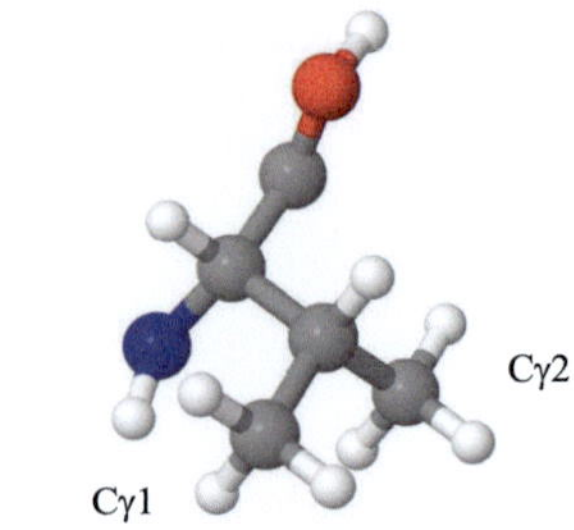

Val 727
$\phi=-73$
$\psi=-8$
$\chi=83$
/1n11/

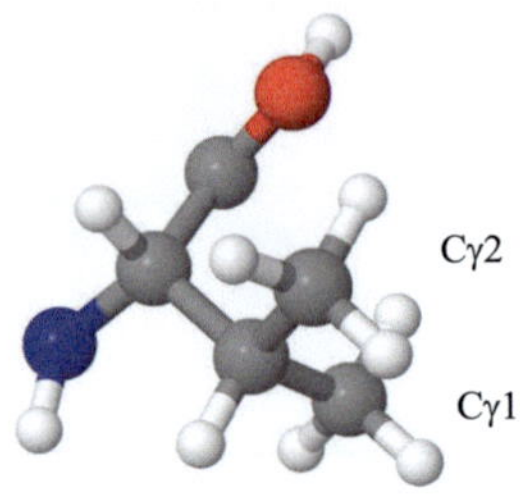

Isoleucine 584
$\phi=-70$
$\psi=-42$
$\chi_1=-61$
$\chi_2=-47$
/1n11/

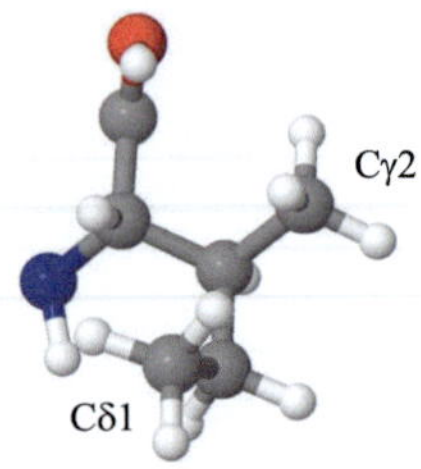

Isoleucine 780
$\phi=-86$
$\psi=-49$
$\chi_1=-150$
$\chi_2=56$
/1n11/

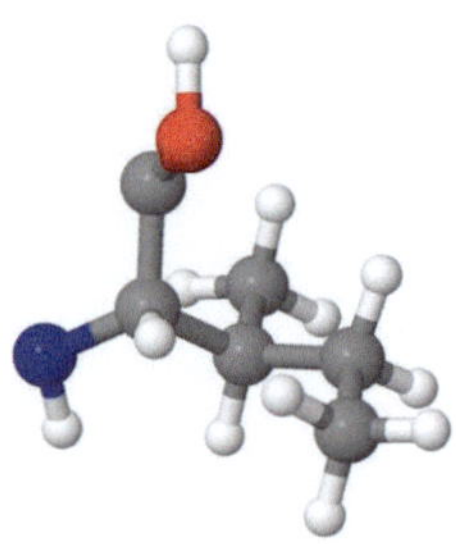

Isoleucine 25
$\phi=-61$
$\psi=-44$
$\chi_1=-64$
$\chi_2=-58$
/5tim/

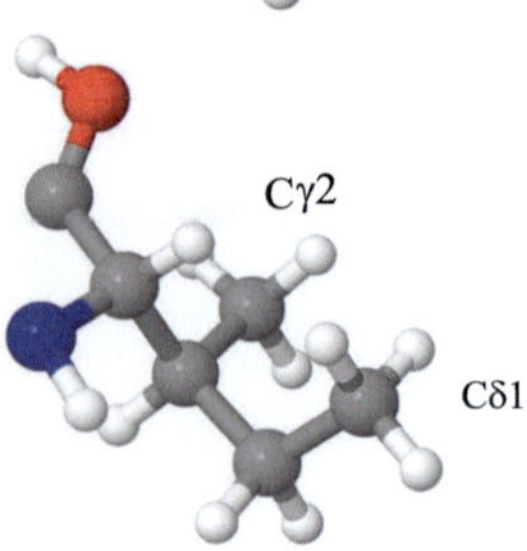

Fig. 1.5 (continued)

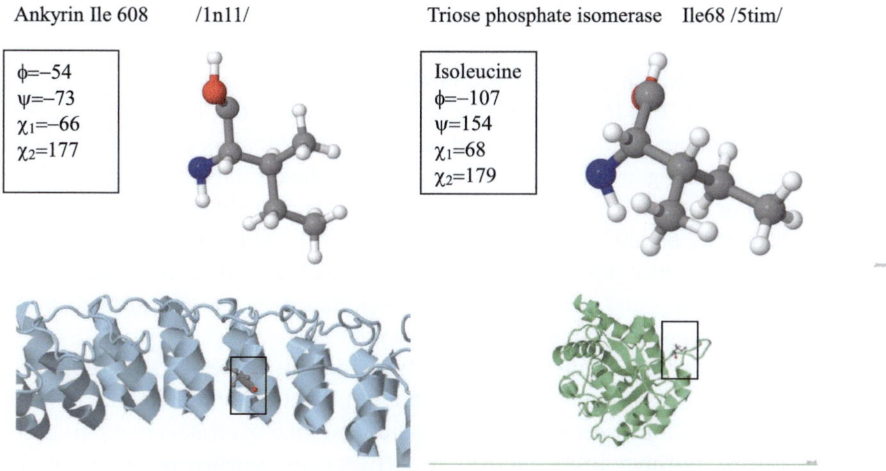

Fig. 1.6 Amino acids: secondary structure dependent side chain conformations of isoleucine

of epitope of protein assembly. Isolated amino acids, monopeptides and dipeptides, can self-assemble into organized arrays and form hydrogels and superstructures such as nanotubes and needles (Subbalakshmi et al. 2017).

Amino acid flexible side chain populates multiple conformational states determined by internal interactions. Side chains exhibit conformational preferences and their conformational states cluster as a set of distinct conformers (Ponder and Richards 1987) (Figs. 1.4 and 1.5) dependent on main chain conformation and tertiary structure. Side chains conformations and backbone conformations depend on each other (Schrauber 1993; Dunbrack and Karplus 1993) (Fig. 1.6).

Side chain conformations which occur most frequently (Ponder and Richards 1987) form a rotamer library. Valine rotamers, for example, cluster around trans, − and + angle values (Fig. 1.5). In the two data sets collected 10 years apart, the most frequent rotamers are almost the same (Ponder and Richards 1987). Secondary structure affects side chain conformation. Isoleucine, for example, exhibits (− trans) χ angles in α helix of ankyrin repeat and (+ trans) χ angles in β structure of triose phosphate isomerase (Fig. 1.6). Similarly, other amino acids exhibit secondary structure dependent rotameric states.

When conformation is considered with respect to the backbone of the helix, we can see that orientation of the side chain end groups favors downward (toward N-terminus of the helix) direction for one of the branched groups and upward (toward C-terminus of the helix) direction for the second group. Both leucine rotamers with χ angles (−64.9, 176.0) and (−176.4, 63.1) point $C_{\delta1}$ group downward (toward the N-terminus of the helix) and $C_{\delta2}$ group upward (toward the C-terminus of the helix). Similarly, valine $C_{\gamma1}$, $C_{\gamma2}$ and isoleucine $C_{\gamma2}$, $C_{\delta1}$ are positioned. If superimposed at C_α atom, each of these rotamers shows distinguished location of its methyl groups (Kurochkina 2008). Therefore, presence of preferred conformations can be explained by ability to form favorable interactions within secondary and tertiary structure.

Methyl groups, which form contacts in intralayer space between two helices are C_β in alanine $C_{\gamma1}$ or $C_{\gamma2}$ in valine, $C_{\delta1}$ or $C_{\gamma2}$ in isoleucine, $C_{\delta1}$ or $C_{\delta2}$ in leucine.

Studies of packing of apolar hydrocarbon chains show that they form layers with distinct types of arrangement. Aliphatic molecules in the crystals of n-paraffins such as n-C29H60 or n-C18H38 are in the shape of plane-zigzag chains of carbon atoms with aligned axes parallel to each other. Chains with odd and even number of carbons pack differently. Increasing the length of the chain results in rearrangement of molecular packing and demonstrates the importance of the end groups in the packing of molecules in crystals (Kitaigorodsky 1973).

Aromatic amino acids, **phenylalanine, tyrosine, histidine**, and **tryptophane** (Fig. 1.7), contain delocalized π electrons in aromatic rings which induce London dispersion forces. Rings interact with each other and their π orbitals overlap. Also, physical forces resulting from interactions between the quadrupole moments of aromatic rings stabilize the conformation (Hou et al. 2018). Phenylalanine is an essential amino acid obtained by an organism through either diet or protein catabolism. Chorismate mutase present in bacteria, fungi, and higher plants but not in humans catalyzes the reaction of the formation of phenylalanine and tyrosine and is an attractive target of design of new drugs against such pathogens as *Mycobacterium tuberculosis* (Khanapur et al. 2017). Phenylalanine/glycine rich nucleoporins are components of selectivity maintenance of nucleocytoplasmic transport via nucleoprotein complex (Ruba and Yang 2016). Phenylalanine with various modifications at C or N terminus (Das et al. 2017) and as dipeptide Phe/Phe (Gan et al. 2017) is base compound of selfassembled gel materials that are used for drug delivery, tissue engineering, and environmental applications. Phenylalanine toxic accumulation in

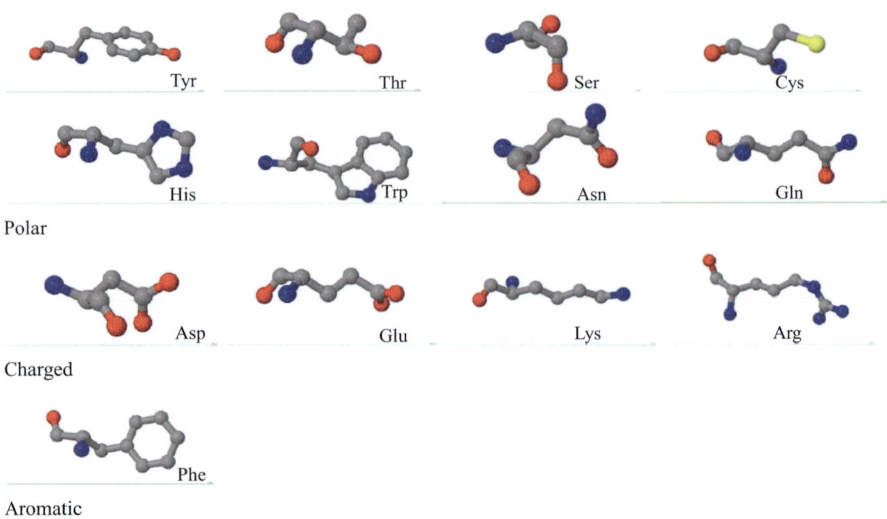

Fig. 1.7 Polar, charged, and aromatic amino acids

the organism occurs because of malfunction of the enzyme phenylalanine hydroxylase (PAH). Phenylketonuria (PKU) is a disease caused by the error of phenylalanine metabolism associated with PAH (Jaffe 2017). Phenylalanine is a precursor of tyrosine: PAH converts Phe to Tyr using molecular oxygen, tetrahydrobiopterin, and a nonheme iron. Tryptophane is an essential amino acid. Tryptophane synthase β subunit catalyses two final steps of tryptophane biosynthesis. Other tryptophane biosynthetic enzymes include anthranilate phosphoribotransferase, anthranilate synthase, phosphoribosyl anthranilate isomerase, and indole 3 glycerol phosphate synthase. Histidine imidazole ring depending on the environment can be neutral (interacting with other aromatic rings via π π clouds) or positively charged (interacting with positively charged Lys/Arg via cation π forces).

Large group of polar and charged amino acids include **cysteine, serine, threonine, aspartic acid, glutamic acid, asparagine, glutamine, lysine,** and **arginine** (Fig. 1.7). Aromatic amino acids **tyrosine, histidine,** and **tryptophane** are also polar. Aspartic acid and glutamic acid carboxylic groups are negatively charged. Lysine ammonium group and arginine guanidinium group are positively charged. Electrostatic interactions in pairs Lys/Asp, Lys/Glu, Arg/Asp, and Arg/Glu between positive and negative charges at a distance of approximately 4 Å are called salt bridges. Posttranslational modifications of polar groups such as lysine ubiquitination and acetylation, arginine monomethylation, tyrosine, serine and threonine phosphorylation and OGlcNAcylation are important for regulation of cellular pathways (Ruba and Yang 2016; Very et al. 2018). Arginine and cysteine are involved in production of gasotransmitters NO and H_2S (Wu 2009). Maintenance of D Ser concentration in brain is important for prevention of neurological disorders (Lee et al. 2017).

Glutamate converts to αketoglutarate that enters tricarboxylic acid (TCA) cycle. Glutamate and aspartate are excitatory neurotransmitters of the central nervous system (Gamlin et al. 2018). Glutamate, but not glutamine, activates excitatory signal and affects neuronal activity. At presynaptic terminals glutamate transporters load glutamate into synaptic vesicles to be released into synaptic clefts as action potential propagates to presynaptic terminals. The glutamate activates glutamate receptors residing on synaptic membranes and preparing for the next signal glutamate transporters remove glutamate. Glutamate/glutamine cycle includes conversion to glutamine by glutamine synthetase in astrocytes and back to glutamate by glutaminase in neurons after being transferred from astrocytes to neurons (Hayashi et al. 2018). Glutamine is a precursor of protein, amino sugars, and nucleotide biosynthesis, involved in the urea cycle/detoxification and considered conditionally essential amino acid since its cell biosynthesis is not always sufficient. Glutamine demand increases in inflammatory and stem cells.

Noncanonical Amino Acids

Although twenty canonical amino acids compose proteins in all organisms, amino acids of other types (Fig. 1.8) can be used in living systems. One of the first discovered noncanonical amino acids, **hydroxyproline** (Hyp), is a component of

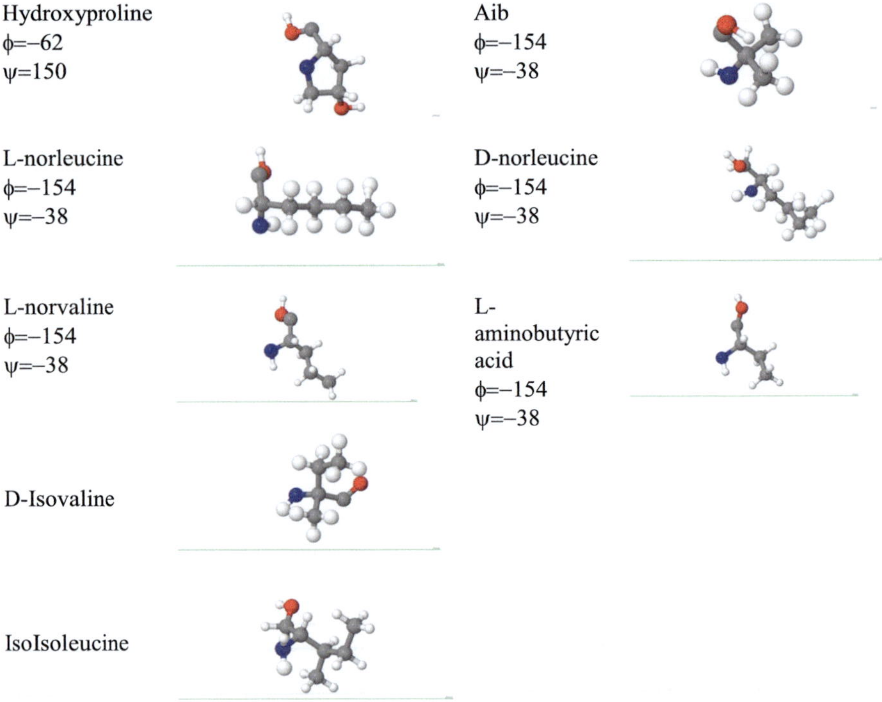

Fig. 1.8 Noncanonical amino acids

collagen (Fisher 1902). Amino acid sequence of collagen triple helix, a polyproline helix (Fig. 1.9), contains repeat of three residues Gly-Pro-Hyp (Fig. 1.10).

Hydroxyproline is posttranslational modification at position 4 of prolyl residues, addition of hydroxyl group (Fisher 1902). The electronic effect of the 4-substituent affects the kinetics of the amide isomerization (Mykhailiuk et al. 2017). Proline hydroxylation stabilizes PPII triple helix (Vitagliano et al. 2001; Berisio and Vitagliano 2012). Replacement of glycine with aza-glycine stabilizes triple helix; melting temperature increases by 8 .6°C (Kasznel et al. 2017).

Modification of hydroxyproline to 4-fluoroproline results in strong inductive effect. This substitution stimulates *cis/trans* peptide bond isomerization and affects conformation of the preceding peptide bond. Also, particular pucker is induced (Newberry and Raines 2017).

Peptidomimetics is one way to extend the set of medicinal compounds by adding more stable and specific analogues. They find applications in the design of new inhibitors, probes, imaging agents, and drugs. **Azapeptides** mimic natural amino acids. In these compounds, α-carbon(s) is substituted by nitrogen. This semicarbazide compound reinforces β-turn conformation (Chingle et al. 2017). Cyclic Cα-tetrasubstituted alpha-amino acids (TAA) incorporation into peptides rigidifies the backbone of the secondary structure and increases stability. Some

A

Dihedral angles of regular structures – secondary structure elements

	$\phi°$	$\psi°$	$\omega°$
Right-handed α-helix	-57	-47	+180
Left-handed α-helix	+57	+47	+180
Parallel-chain pleated sheet	-119	+113	+180
Antiparallel-chain pleated sheet	-139	+135	-178
Polyglycine II	-80	+150	+180
Collagen	-51, -76, -45	+153, +127, +148	+180
Poly(L-proline) I	-83	+15	0
Poly(L-proline) II	-78	+149	+180

B

Secondary structure	Number of residues per turn, n and chirality	Distance per residue, d(Å)	Radius, r(Å)
Pleated parallel β-sheet	±2.0	3.2	1.1
Pleated antiparallel β-sheet	±2.9	3.4	0.9
Twisted parallel/antiparallel b-sheet	-2.3	3.3	1.0
3_{10}-helix	+3.0	2.0	1.9
α-helix, right	+3.6	1.5	2.3
α-helix, left	-3.6	1.5	2.3
π helix	+4.3	1.1	2.8
Collagen helix	-3.3	2.0	1.6

Fig. 1.9 Secondary structure

TAAs alternating with (S)- or (R)-valine show 3_{10} helix conformation, right-handed for (SS) and left-handed for (RR) (Grauer et al. 2009).

Amino acid **aminoisobutyric acid** (Aib) (Fig. 1.8) is one of nonconventional acids branched and achiral at α-carbon (Venanzi et al. 2017). Ideal Aib conformations of left- (ϕ, $\psi = -57^{\circ}, -47^{\circ}$) and righthanded ($\phi$, $\psi = +57^{\circ}, +47^{\circ}$) α-helix are sterically plausible (Burgess and Leach 1973; Marshall and Bosshard 1972). Peptides containing Aib are helical with alpha and 3_{10} mixed conformations. Both left

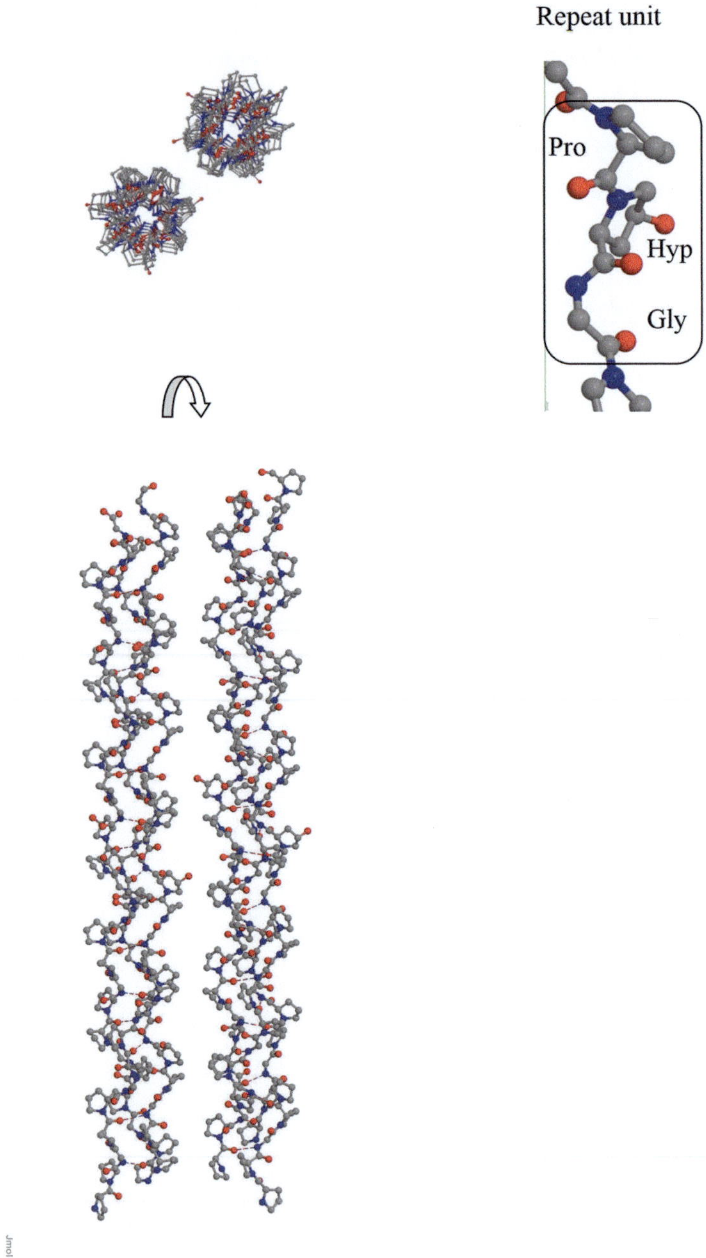

Fig. 1.10 PPII structure of collagen model peptide (Pdb entry 1x1k)

and right handed helices accommodate Aib. Other conformations include PPII and extended (Aravinda et al. 2008). Studies of capped Aib oligomers in gas phase show several preferred conformations. Short chains such as Z-Aib-OH, Z-(Aib)2-OMe, and Z-(Aib)4-OMe (Z is benzyloxycarbonyl) have 3_{10} helical conformation (ϕ, $\psi = -57^\circ, -30^\circ$). Minor states are represented by conformations (ϕ, $\psi = -75^\circ$, -74°) and (ϕ, $\psi = -52^\circ$, 143°). Early stages of polypeptide chain growth show tendency to adopt 3_{10} helix (Gord et al. 2016). Aib oligomers are known to form 3_{10}-helical secondary structures in solution and in crystal. Aib 10-residue long homopeptide forms a 3_{10} helix with three complete turns (Gessmann ét al. 2016). Insertion of α-arylated alanine derivatives such as a-methyl phenylglycine into achiral oligomer of aminoisobutyric acid changes oligomer conformation (Costil et al. 2018).

Alanine to Aib substitution in Fab-binding peptides does not significantly change conformation (Stanfield et al. 1999). Alpha, alpha dimethylation of glycine (Aib), N-methylation of glycine (sarcosine), and various combinations of methylation at positions 1 and 7 in angiotensin II show interdependence and influence on affinity (Cordopatis et al. 1994). Inhibitor of the metalloendopeptidases in which Aib is substituted for an alanine retains potency and possesses higher resistance to degradation (Smith et al. 2000). Peptides of eleven Aib residues with N-terminal acetyl cap (or azide group), C-terminal aminoalcohol and flanked by hydrophobic termini fold into 2.91 nm long 3_{10} helix, approximately membrane bilayer width, and have high antibacterial activity (Adam et al. 2018).

Incorporation of Aib in protein secondary structure: helix, loop, sheet, and turn, even at positions away from the core is destabilizing to the tertiary fold (Reinert et al. 2013 4kgt). Distortion of helix axis in corticosteroid releasing hormone is achieved by a combination of D-Phe12 and Aib15 which results in lack of hormonal function (Spyroulias et al. 2002). Neuropeptide Y analog [Ala^{31}, AIB^{32}]NPY shows deviation from α-helix to more flexible 3_{10} helix/turn structure (Cabrele et al. 2000).

Fungi provide rich source for antimicrobial peptides (Degenkolb and Bruckner 2008). More than 300 peptaibols that contain Aib and are synthesized mainly by non-ribosomal peptide synthetases of *Trichoderma* strains enter a family of antimicrobial peptides (Song et al. 2011). Aib based amphipathic peptides with cyclo (-Arg-Gly-Asp-d-Phe-Cys-) at the C-terminus are used for specific siRNA delivery (Wada et al. 2017). They also are components of Au(III)-peptidodithiocarbamato complexes that improve the tumor selectivity (Boscutti et al. 2018) and lytic peptides designed against triple negative breast cancer (Zhao et al. 2017). Aib-based peptides can form nanotubes filled by water inside hydrophobic channel (Raghavender et al. 2010). **Amino butyric acid** (Aba) linear side chain can substitute valine in hydrophobic core of a dimeric parallel coiled coil (Oshaben and Horne 2014). The neurotransmitter **γ-amino butyritc acid** (GABA), similarly to glycine, is inhibitory neurotransmitter in the CNS. It binds to the receptor molecule GABAR, inhibitory pentameric ligand gated ion channel permeable to chlorine ions (Chen et al. 2018). Inhibition of GABA degradation by GABA aminotransferase is used as treatment of addiction (Juncosa et al. 2018). This pathway is also targeted for the treatment of epilepsy. GABA is a precursor of cyclic derivatives which are used for the development of new drugs to treat neuronal, vascular, and cognitive disorders (Elkin et al. 2018).

Hydrophobic aminoacids S-2-aminobutanoic acid (l-2-aminobutyric acid), S-2-aminopentanoic acid (**l-norvaline**), and S-2-aminohexanoic acid (**l-norleucine**) crystallize as molecular bilayers composed of hydrogen-bonded core and assembled side chains (Görbitz et al. 2016).

Norvaline is an isomer of valine and unlike valine contains linear side chain. Tripeptide N-(benzyl-oxycarbon-yl)glycylglycyl-L-norvaline adopts extended conformation in crystal and forms hydrogen bond with the neighboring molecule (Nicholas et al. 2017). Tripeptides with nonnatural amino acids are good inhibitors as for example was shown for HIV activation response element RNA (Ferner et al. 2009). Tripeptide lysyl-norvalyl-lysine binds in extended β-sheet conformation to oligopeptide-binding protein OppA (Fig. 1.12). It makes hydrogen bonds with OppA and positions its side chains in a large cavity completely shaded away from solvent (Davies et al. 1999).

Norleucine is a hydrophobic amino acid with linear hydrocarbon chain (Fig. 1.8). Solution structure of a D,L-alternating dodecamer of norleucine exhibits double-stranded antiparallel beta-helix (Navarro et al. 2001; Navarro et al. 2002; Navarro et al. 2004) (Fig. 1.12).

Isoisoleucine is a stereoisomer of leucine. It is incorporated in hydrophobic core of designed right handed coiled coil trimer (Harbury et al. 1998). Derivatives, such as 6-diazo-5-oxo-L-norleucine (DON), are used as anticancer agents (Cervantes-Madrid, et al. 2017). DON Inhibition of gamma-glutamyl transpeptidase 1(GGT1) is shown by crystallographic analysis (Terzyan SS et al. 2017).

Noncanonical **gabapentin**, Gpn, is achiral β,β-disubstituted γ-aminoacid with a cyclohexyl ring. Backbone incorporates additional atoms and forms hydrogen bonds of reverse direction (Vasudev et al. 2007, 2009). Side chain preferred conformation about C(α)-C(β) and C(β)-C(γ) bonds are gauche-gauche with NH_i-CO_i and CO_{i-1}-NH_{i+1} backbone hydrogen bonds similar to β-turns of protein polypeptide chain (Balaram et al. 2010).

Isovaline along with Aib is a non-proteinogenic Cα dialkylated amino acid known for its analgesic properties. Isovaline and its derivatives are candidates for new medicinal compounds (Fung et al. 2017). R isovaline homopeptides form the left handed 3_{10} helix (Formaggio et al. 1995). R isovaline is GABA(B) agonist in the thalamus and peripheral tissue and its action is coupled to rectifying K+ channels in cells. Its antiepileptic property is attributed to action on interneurons (Pitman et al. 2015). Antibacterial membrane channel forming peptides zervamicin IIB (Balashova et al. 2000) and other include D-isovaline.

Non-protein amino acid **β-N-methylamino-L-alanine**, a neurotoxin produced by cyanobacteria and by micro-algae, is proposed as an environmental factor in neurodegenerative diseases. Incidence of Amyotrophic Lateral Sclerosis, Parkinson Disease and Dementia indicates this amino acid as causative agent due to its ability to act on the glutamate receptors and contribute to neuronal excitotoxicity (Lance et al. 2018).

Natural and non-natural amino acid derivatives 11C- and 18F-labeled are candidates for imaging in positron emission tomography oncology (Leung 2010).

Secondary Structure

Backbone atoms of polypeptide chain form hydrogen bonds. Each secondary structure element exhibits a regular pattern of hydrogen bonds characterized by a set of parameters: backbone dihedral angles, number of residues per turn, chirality, distance per residue, and radius (Fig. 1.9). Helix is formed if carbonyl oxygen is bonded to amide nitrogen three (3_{10} helix), four (α-helix) or five residues apart (π helix). Hydrogen bonding of carbonyl oxygen to amide nitrogen four (α-helix) residues apart is observed in regular α-helix (3.6 residues per turn) and super coil (3.5 residues per turn). Every third carbonyl oxygen bonded to amide nitrogen in the neighboring strand induces collagen or polyproline helix. If every second carbonyl oxygen is bonded to amide nitrogen in the neighboring strand, β-structure is formed. Contacts between i, i + 3 and i, i + 4 amino acids along the helix or i, i + 2 along the strand stabilize secondary structure (Lim 1974a). Prediction methods of the secondary structure obtain correct results with approximately 70–90% accuracy (Lim 1974b; Garnier et al.1996). However, peptides make transitions between secondary structure conformations; they adopt helical or extended conformation depending on the environment. Stability of secondary structure depends on interactions with other secondary structure elements, membrane, and various ligands.

Polyproline helix Repeated amino acid sequence X-Y-Gly (X = L-proline, Y = 4R-hydroxy-L-proline) forms triple helix. In each chain every third residue is glycine with Pro-Hyp-Gly triplet as main periodically spaced unit. Proline hydroxylation is critical for secondary structure stability. At Y position proline hydroxylation stabilizes whereas at X position inhibits triple helix. Stereo selectivity also contributes to the formation of triple helix: 4S-hydroxy-L-proline diastereoisomer is destabilizing (Vitagliano et al. 2001). Three chains are hydrogen bonded and super coiled (Fig. 1.10).

This right-handed association of three left-handed polyproline (type II) helices comprises collagen. One intrastrand hydrogen bond per Pro-Hyp-Gly triplet stabilizes the structure: carbonyl oxygen of proline is hydrogen bonded to amide nitrogen of the next chain (Shoulders and Raines 2009; Berisio and Vitagliano 2012). Synthetic collagen finds many applications in medicine as for example in design of biomaterials for treatment of wounds.

α-Helix The structure of the α-helix was deciphered from X-ray diffraction of proteins and was proposed by Linus Pauling. Helix shows regular pattern of hydrogen bonds of the protein backbone in which carbonyl oxygen is bonded to amide nitrogen four residues apart (Pauling et al. 1951) (Fig. 1.11). Most naturally occurring helices are left handed, whereas right handed helix is less common. Monomeric helix exhibits values of helical parameters such as bending, number of residues per turn, and height of a turn similar to those of coiled coil dimers which suggests that α-helix possesses intrinsic properties (Strelkov 2002). Super coil structure of leucine zipper gave origin to leucine zipper nomenclature: positions of residues are

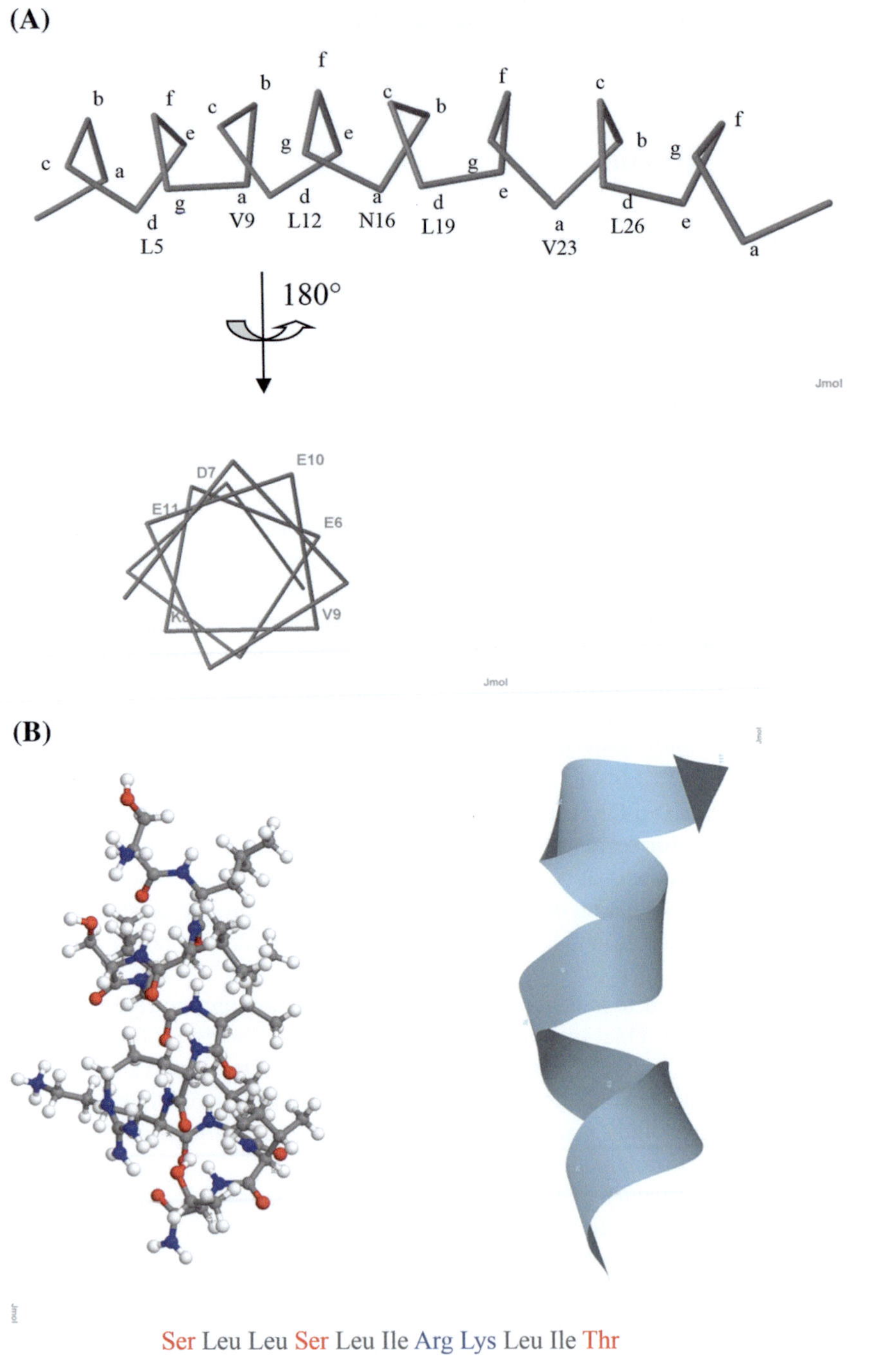

Fig. 1.11 α-Helix (**A**) Backbone and helical wheel; (**B**) Structure of antimicrobial peptide decoralin (Pdb entry 2n9a)

designated by letters **a** through **g** (Fig. 1.11A) for seven consecutive residues of the repeat (Kohn et al. 1977). Spacing of leucine residues is important for super coil core interactions with neighboring helices. Interruptions in sequence pattern such as stutters, four residue insertion at the end of the heptad (Strelkov et al. 2002), and skips (Taylor et al. 2015) result in structure modifications.

Arrangement of hydrophobic and hydrophilic groups on the surface of α-helix is very important structural property. Amphipathic character of the helix and of globular proteins in general is a consequence of their interactions as ionic and polar groups point to the aqueous environment whereas nonpolar groups face the membrane or protein interior (Singer and Nicolson 1972; Branden and Tooze 1991). Helices of leucine zipper dimer show clustering of hydrophobic residues at the interface positions **a** and **d** and hydrophilic residues at the positions interacting with solvent (Zhou 2011). Binding of ligands by α-helix occurs at specific sites of its surface. Hydrophobic patches are narrow in dimeric molecule and expand to the adjacent edges in trimers and tetramers as edges become involved in helix-helix interactions. In membrane helices or helices surrounded by other secondary structure elements rather than polar solvent, all edges are primarily hydrophobic. Hydrophobic interactions determined by specific amino acid combinations are important for structure of these oligomers (Kurochkina 1997; 1998). Contacts on the helical surface determine its binding specificity. Chiral character of contact arrangement is critical for the fold of the entire molecule (Kurochkina and Iadarola 2015a).

Single α-helix is frequently basic element of many drugs. Action of antimicrobial peptides involves amphipathic character of helix surface. Antimicrobial peptides are less than 50 amino acids largely hydrophobic with a positive charge +2–9. They have secondary structure: α-helices, β-sheets, loops and irregular fragments (Duclohier et al. 2010; Tseng et al. 2016). Hydrophobicity and charge are important for the peptide action. Cationic groups interact with anionic microbial surface. Both binding and insertion to membrane induce amphipathic α-helix structure (Tseng et al. 2016). Also, length of the peptide is important for obtaining helical structure and activity. Decoralin (Fig. 1.11B) is an eleven amino acids long cationic α-helical peptide (Guerra et al. 2017). It possesses antimicrobial activity and no hemolytic activity (Konno et al. 2017). In α-helix conformation, hydrophobic residues are on one side whereas hydrophilic on the other side. Designed peptides analogs show variations in cytolytic activity and membrane permeability. DecP-11 has cytolytic and membranolytic activity but no antileishmaanial activity, strong interactions with amino phospholipids of the parasite and aggregation (Martins et al. 2017). Leucine-substituted decoralin analogs $[\mathrm{Leu}]^{10}$-Dec-NH_2 antimicrobial activity similar to that of the parent peptide. $[\mathrm{Leu}]^{8}$-Dec-NH_2 shows two-fold increase in antimicrobial activity and no helical structure. It is also less hemolytic (Torres et al. 2010).

Antifreeze protein is a monomeric α-helix. It binds to ice crystals and possibly acts by lowering freezing point of water in polar fishes and pine needles (Chou 1992).

Oligomers based on achiral imidazole amino acids tend to form helix. Modification of side chain such as attachment of methyl group to the d atom of the ring results in change of secondary structure. In the modified helix, a herringbone, slope of the helix changes sign (Adam et al. 2014; Adam et al. 2017).

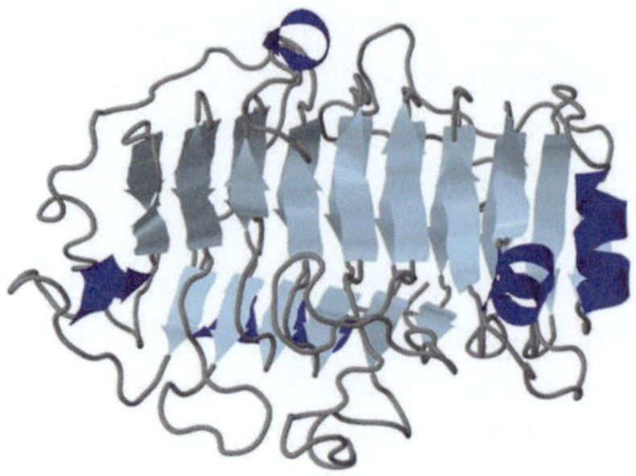

(A) Pectate lyase /1pcl/

(B) Alpha-spectrin SH3 domain /1bk2/

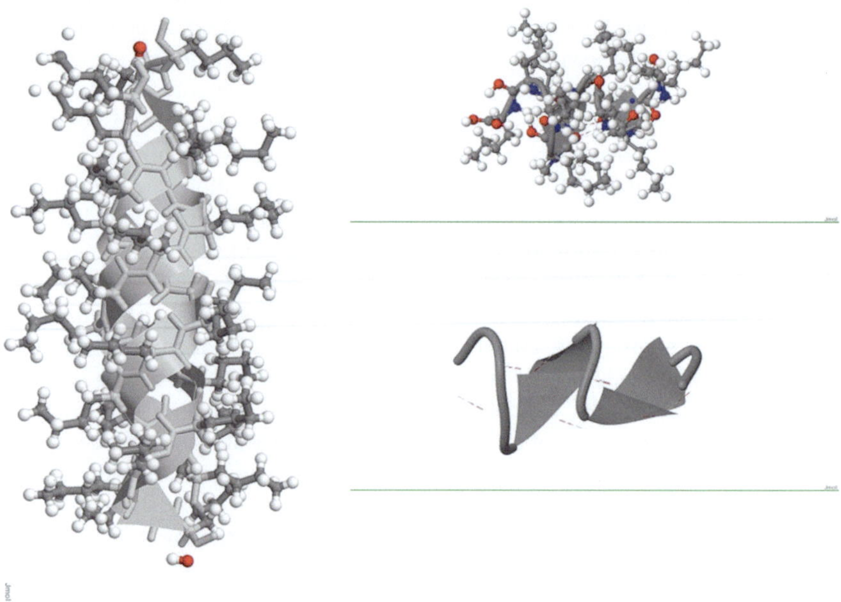

(C) D,L alternating pentadecamer and dodecamer of norleucine: double antiparallel beta-helix /1s1o 1r9v/

(D) Oligo-peptide binding protein complexed with Lysyl-Norvalyl-Lysine in β-sheet conformation /1b6h/

Fig. 1.12 Beta structure (**A**) Parallel beta-helix; (**B**) antiparallel β-structure of SH3 domain; (**C**) double antiparallel beta-helix; (**D**) β-sheet conformation of a peptide bound to the active site

Transmembrane helices form channels and expose hydrophobic faces toward lipids/neighboring helices whereas hydrophilic groups are in protein interior and conduct ions (Brunner et al. 2014 4wis 5oyb). Amphipathic helices in model lipid environments can align parallel to the membrane surface and assemble in nanodiscs (Salnikov et al. 2018).

Short peptides assembled in regular assemblies, helix and sheet, are structural units of biomaterials based scaffolds used for biomedical tissue engineering (Holmes 2002).

Beta-structure Arrangements of parallel/antiparallel strands in which every second amino acid makes hydrogen bonds to the neighboring strand results in β-structure (Fig. 1.12). Twist and bend characterize strand geometry. In globular proteins, most β strands exhibit right handed twist and bend. A group of β strands can form pleated β sheet or a circular arrangement as barrel in solution or membrane (Kikuchi et al. 2018). Novel β-structure is part of host recognition protein of the *Haemophilus influenzae*: His auto transporter (Yeo et al. 2004). Right-handed β-helix domain is essential for bacterial secretion (Yeo et al. 2007).

β–Helix structure forms transmembrane channels. Synthetic model peptides such as D, L-alternating oligonorleucine oligomers provide insight into β–helix organization (Navarro et al. 2001). In nonpolar solvent, twelve residue polynorleucine peptide shows beta(4,4)-helix.

Protein Folds

Proteins can assemble in a variety of shapes. Large oligomeric assemblies as helices, spirals, barrels, and rods can be found in tobacco mosaic protein (Namba et al. 1989), ankyrin repeat proteins (Krzywda et al. 2004), signaling complexes MYDDosome, PIDDosome (Lin et al. 2010), apoptosome (Yuan et al. 2010), inflammasome (De Alba 2009), death inducing signaling complex (DISC) (Scott et al. 2009, Wang et al. 2010), intasome (Yin et al. 2016), Ndc80 complex, site for kinetochore-microtubule attachment (Alushin et al. 2010), and artificiially designed model systems (Aggeli et al. 2001). Protein structural elements are multirepeat assemblies of diverse function and dimensions that involve helical structures of tetratricopeptide (Hinds et al. 2008, Arcy et al. 2010), armadillo/ HEAT (Andrade et al. 2001; Rubinson et al. 2010), leucine rich repeat (Park et al. 2006), and other units.

Most proteins consist of various arrangements of regular (hydrogen bonded secondary structure) and irregular elements. Secondary structure comprises about 80% of residues in a molecule although some proteins contain smaller fraction of hydrogen bonded elements or even are completely disordered. Disordered fragments of polypeptide chain can fold into regular structure making interface with a complimentary regular structure. Secondary structure elements associate parallel or antiparallel to each other exhibiting a variety of tertiary structure motifs. Estimate of the number of different protein folds, or number of different arrangements of secondary structure elements and/or chain topologies, gives 327. Folds are subdivided into all alpha (25%), all beta (20%), alpha/beta (30%), and alpha+beta (25%) categories (Chothia and Janin 1982). Protein complexes assemble via protein core conserved interactions as transient or permanent associations (Sudha et al. 2015).

Proteins contain multiple repeats that arise from duplication of structural unit in the process of evolution (Andrade et al. 2001). Not only short sequence motifs but also larger repeating structures can be found in globular proteins as for instance myohemerithrin two pairs of antiparallel α-helices homologous and related to each other by a two-fold symmetry (Sheriff et al. 1987). Position of apolar and polar residues within repeat as well as type of amino acid residue at each position and complementarity of amino acids from the opposing α-helices are important for the formation of structure. Packing of helices is important determinant of structure (Crick 1953; Chothia et al. 1981). The question how enzyme structure and negative function are related is widely debated. Enzymes of such folds as TIM barrel and Rossmann carry out multiple functions. In contrast, in more than 60% of enzymes, one or two enzymatic functions are observed (Toth-Petroczy and Tawfik 2014).

Parallel, antiparallel, left handed, right handed Parallel and antiparallel arrangements of α-helices can be left handed (positive interhelical angles) and right handed (negative interhelical angles) (Fig. 1.13). The rule to determine sign of an angle

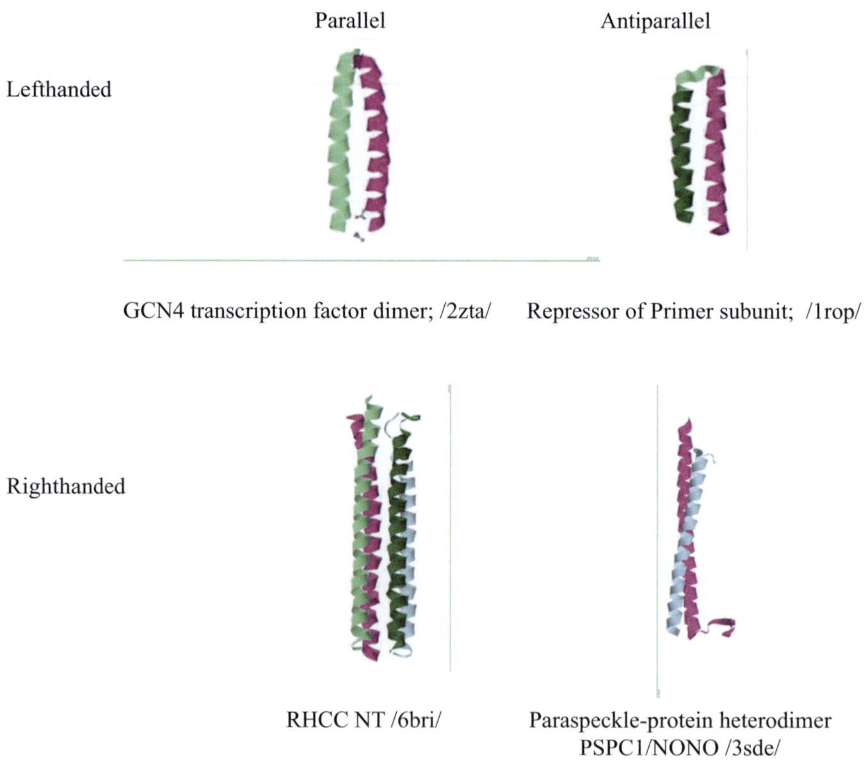

Fig. 1.13 Parallel/antiparallel left/right handed helix arrangement

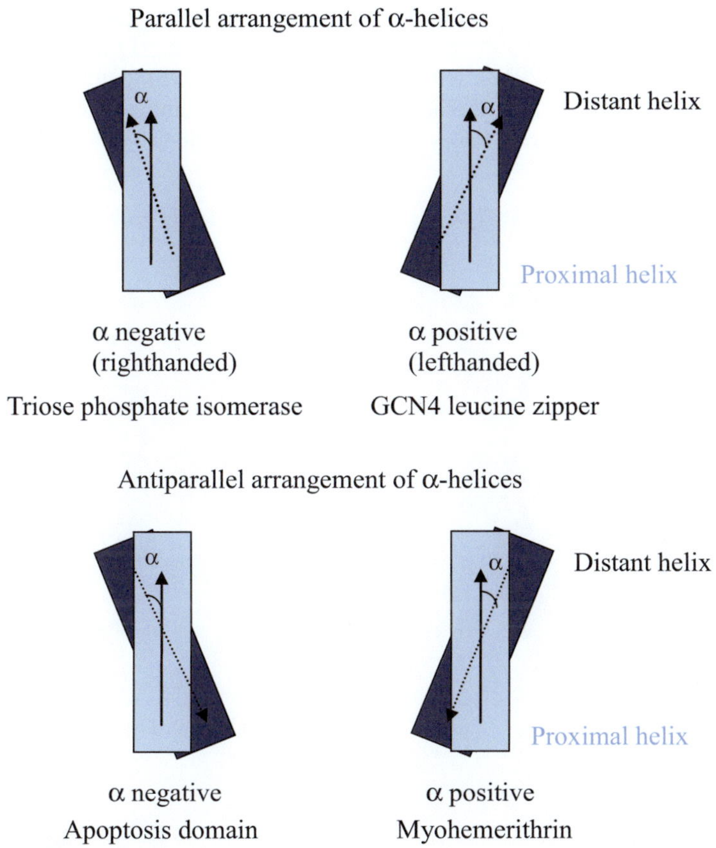

Fig. 1.14 Positive and negative interhelical angles in parallel and antiparallel arrangements of α-helices

assumes that angle is positive if helix distant from the viewer is rotated clockwise relative to proximal helix and negative if it is rotated counterclockwise (Fig. 1.14). Parallel interfaces with positive angles are observed in GCN4 leucine zipper, subtilisin, carboxypeptidase, dihydrolipoamide dehydrogenase, glutathione-s transferase and other proteins. Beta-lactamase three parallel interfaces in the helical portion of the protein have positive helix-helix angle. Parallel interfaces with negative angles occur in such proteins as TIM-barrel, RHCC NT, and ankyrin repeats.

Parallel coiled coil structure is arrangement of two or more parallel helices. Parallel coiled-coil can be recognized by a heptad repeat of leucine residues, 11-residue hendecad, or 15-residue pentadecad repeats. Coiled-coil has slightly tighter helix resulting in 3.5 rather than 3.6 residues per turn. It is present in many proteins involved in transcription regulation, muscle contraction, blood coagulation, neuronal development, cell membrane adhesion and other processes. Coiled coil domains of globular and fibrous proteins very often function as dimerization vehicles and associate in various oligomers such as dimers (O'Shea 1991), trimers (Spraggon 1997; Le Rumeur et al. 2012), tetramers (Sutton et al. 1998), pentamers (Oxenoid 2005), hexamers, or dimers of trimers both parallel and antiparallel to each other (Fig. 1.15). They can be found in cytoplasmic and membrane compartments carrying

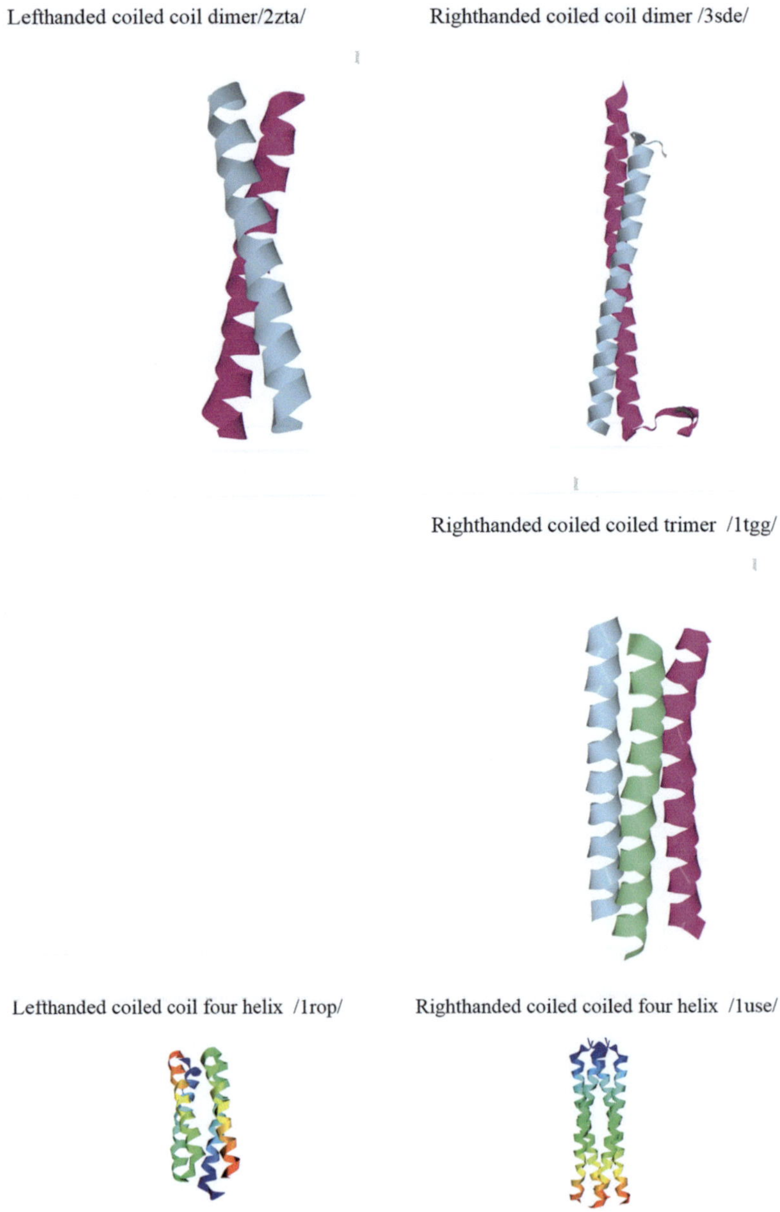

Fig. 1.15 Coiled coil

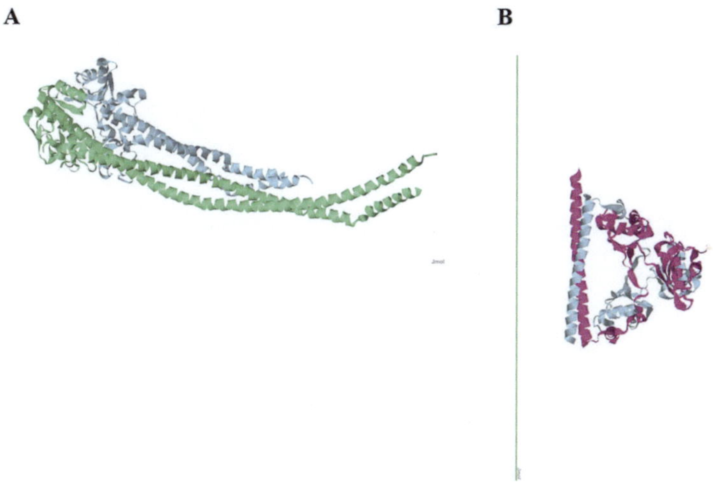

Fig. 1.16 Coiled coil in proteins. (**A**) Yeast SMC2-SMC4 hinge domain of chromosomes protein 2 /4rsi/; (**B**) DNA-binding P52/P100 complex /3sde/

out multiple functions (Fig. 1.16). Oligomers of GCN4 transcription factor, one of the early identified coiled coil molecules, were extensively studied by x-ray crystallography (O'Shea 1991, Harbury 1993) and provided scaffold for numerous design efforts. Coiled coil homodimeric domain of the cGMP-dependent protein kinase Iα functions in the NO-mediated relaxation of the vascular myosin muscle (Schnell 2005). Cardiac phospholamban homopentamer is in a shape of a flower (Fig. 1.17) held together by leucine/isoleucine zipper motif along membrane-spanning helices (Oxenoid 2005). Heterotrimeric parallel four-helix bundle of the synaptic fusion complex (Sutton et al. 1998) and homo-oligomeric parallel arrangement of α-helical antiparallel bundles of the HIV/SIV gp41 (Caffrey et al. 1998) and influenza virus haemagglutinin (Stevens et al. 2004) show core organization similar to leucine zipper. The assembly of a heterotrimer itself leads to the membrane fusion of a synaptic complex while HIV/SIV gp41 and influenza virus haemagglutinin undergo conformational changes in the fusion protein that leads to viral fusion. Toroids and rings contain rows of parallel helices of various lengths. Sodium translocating ATPase contains a rotor ring composed of parallel protomers varying in length between 8 and 24 (Chap. 5) (Vollmar et al. 2009).

GCN4 homodimer leucine zipper structure have shown that leucines separated by seven amino acid residues (positions **d**) and valines/asparagines (positions **a**) form a 3–4 hydrophobic repeat. Being main contributors to secondary structure interfaces, hydrophobic amino acids pack the core of the helix/helix interface (O'Shea 1991). Substitutions at the interface between helices can rearrange interface packing and oligomerization state (Harbury 1993). Mutations of valine and asparagine to leucine at position **a** or leucine to isoleucine at position **d** rearrange dimer to all-parallel tetramer. If both core positions, **a** and **d,** are occupied by isoleucine, structure rearranges to a parallel trimer.

A Parallel arrangement of helices

GCN4 transcription factor dimer; coiled coil /2zta/

(B)

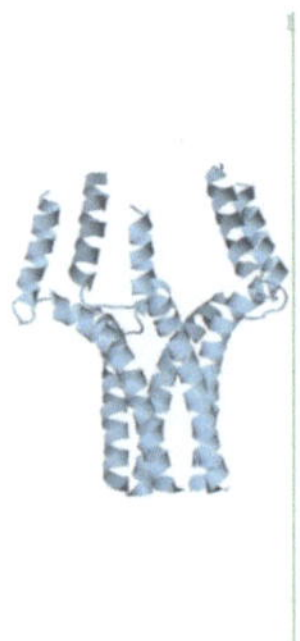

Fig. 1.17 Folds with parallel, antiparallel, and parallel/antiparallel helix arrangement
Parallel helix-helix interfaces of (**A**) GCN4 leucine zipper, (**B**) Cardiac phospholamban pentamer/1zll/

Amino acids at core positions **a** and **d** from each helix of the dimer comprise complementary pairs and assemble in layers. In the structure of coiled-coil domain of cGMP-dependent protein kinase Iα, phospholamban pentamer (Oxenoid 2005), transcription factor MAX (Ferre-D'Amare 1993), and fibrinogen (Spraggon 1997, Madrazo 2001) the same amino acid combination of leucine and isoleucine occupies **a** and **d** positions. Aromatic residues can be accommodated at helix interface core positions as in vimentin.

Helix in a coiled coil structure (3.5 residues per turn) results in a tighter arrangement of amino acids compared to α-helix (3.6 residues per turn). This arrangement is very often found in parallel/antiparallel dimers, trimers, dimers of dimers, tetramers, pentamers, hexamers, and other oligomerization states.

Spacing of leucine residues is important for super coil core interactions. Interruptions in sequence pattern introduce new structural features. Stutter is a four residue insertion at the end of the heptad found in intermediate filament protein vimentin. Amino acid alanine at position **a** of the heptad repeat interacts with phenylalanine of the insertion, and this conserved sequence is an essential component of the stutter in vimentins and desmins of vertebrates. Insertion results in local unwinding of the coiled coil with increase of coiled coil pitch and radius: 5. 22 Å and 160.5 Å compared to typical 5.0 Å and 128.5 Å αHelix pitch and radius remain the same. Whereas insertion does not perturb coiled coil arrangement, it produces nearly parallel helices, shift in coiled coil phase, and affects oligomerization states

(Strelkov et al. 2002). Skips in the structure of coiled coil rod domain of the myosin thick filaments result in local unfolding and hinges which are necessary for the myosin assembly (Taylor et al. 2015).

Left-handed super coiling occurs most frequently. Naturally occurring right-handed super coiled structures are very rare. Parallel right-handed coiled coil tetramer of surface layer protein tetrabrachion from *Staphylothermus marinus* is characterized by 11-residue repeats and in contrast to left-handed tightly packed structures contains large water-filled cavities (Stetefeld et al. 2000). Stability of this structure at extreme pH, salt, temperature, and pressure conditions, increased permeability in tumor blood vessels, and ability to accommodate heavy metals in cavities was used to make it a carrier of platinum containing drug cisplatin (Eriksson et al. 2009). Right-handed α-helical coiled coil is found in tetramerization domain of the vasodilator-stimulated phosphoprotein (VASP) with 15-residue repeat exhibiting a characteristic pattern of hydrophobic residues combined with salt bridges (Kuhnel et al. 2004). Transmembrane and cytoplasmic right-handed coiled coil heterodimeric domains of integrins, cell adhesion receptors transmitting extracellular signals across the membrane, contain a hendecad (11) repeat (Yang et al. 2009). Designed right-handed super coiled trimers (Plecs et al. 2004) and tetramers (Harbury et al. 1998) show that hydrophobic core is highly organized.

Antiparallel coiled coil structure is arrangement of two or more antiparallel helices. Similarly to domains of parallel coiled coil, domains of antiparallel coiled coil in globular and fibrous proteins provide dimerization interfaces and associate as oligomers: dimers (O'Shea 1991), trimers (Le Rumeur et al. 2012), tetramers, dimers of trimers, etc. Coiled coils may contain combinations of parallel and antiparallel arrangements.

Two antiparallel helices of Cox17 in mitochondrial energy transfer chain, stabilized by two disulfide bonds, bind copper ion (Banci et al. 2011). Many viral fusion proteins in HIV (Liu et al. 2009), Ebola virus (Malashkevich et al. 1999) and infectious salmon anemia virus (ISAV) (Cook et al. 2015) have antiparallel three helix proteins that mediate viral host endosomal membrane fusion. Long antiparallel helices occur as coiled coil of SIN Nombre virus nucleocapsid protein, serine t-RNA synthetase, DNA binding transcription factor STAT-4, activator of transcription STAT-1, two-stranded homodimer PROP, DNA-binding protein SSO 10, colicin e3 and other proteins.

Very often antiparallel coiled coil exhibits specific sequence patterns such as heptad repeat of alanines or other small residues named alacoil (Gernert et al. 1997). Alacoil, which carries alanines or other small residues in every seventh position (Gernert et al. 1995), is found in several proteins including repressor of primer and ferritin. Heterodimeric right-handed coiled coil of *Thermus thermophilus* A-type ATPase peripheral stalk is a part of assembly of proton translocating ATPases, rotary enzymes that couple ion translocation and ATP synthesis/hydrolysis. It contains both hendecad (11) and quindecad (15) repeats (Lee et al. 2010). Right-handed coiled coil heterodimeric domain of human paraspeckle-protein PSPC1/NONO contains not a heptad but a hendecad (11) repeat. It assembles RNA recognition domains along the 20 Angstrom long channel and is important for nuclear organization and gene regulation (Passon et al. 2012) (Fig. 1.16B).

The bacterial flagellar filament switches between left- and right-handed super coiled forms when bacteria switch their swimming mode between running and tumbling. Packing interactions of a protein flagellin are responsible for super coiling of a filament. Pairs of R-type straight protofilaments studied by x-ray crystallography exhibit long stretches of antiparallel coiled coil (Samatey et al. 2001). Left- and right-handed super coiled dimers coexist in human lamin A (Kapinos et al. 2011). Transitions between left- and right-handed super coiled forms are also observed in trimeric auto transporter adhesins. This molecule carries heptad repeats of 7/2, 15/4, and 19/5 periodicity which provides basis for these transitions (Alvarez et al. 2010).

Coiled coil structure found numerous applications in design of biomaterials, protein engineering, and synthetic biology (Woolfson et al. 2017).

Parallel helices occur in many cytosolic (leucine zipper) and membrane (phospolamban) proteins (Fig. 1.17). Large group of enzymes contains 8α/8β fold named TIM barrel by the name of triose phosphate isomerase. Eight parallel α-helices surround eight-stranded β-barrel. All parallel α-helices form short interfaces along 1–2 turns of the helix. Each helix contacts two other helices of the barrel, one helix preceding and one helix following in amino acid sequence. Crystallographic structures of triosephoshate isomerase at high resolution for *T. brucei* triosephosphate isomerase in complex with glycerol-3-phosphate and sulfate (Wierenga 1991), *L. mexicana* triosephosphate isomerase in complex with 2-phosphoglycolate (Kursula 2003), *S. cereviciae* triosephosphate isomerase (Lolis 1990) in complex with phosphoglycolohydroxamate (Davenport 1991) and dihydroxyacetone phosphate (Jogl 2003), *P. falciparum* triosephosphate isomerase in complex with 2-phosphoglycerate (Parthasarathy 2003), *G. gallus* triosephosphate isomerase in complex with phosphoglycolohydroxamate, pyruvate kinase from *Orictolagus cuniculus* and *Homo sapiens* (Smith 2003), malate synthase from *E. coli* (Anstrom 2003) and *M. tuberculosis* (Smith 2003), methylmalonyl CoA mutase (Mancia 1996), fructose-1,6-biphospate aldolase from *Thermus aquaticus* (Izard 2004), *Drosophila melanogaster* (Hester 1988), *Homo sapiens* (Gamblin 1990), *Orictolagus cuniculus* (St Jean 2007), KDPG aldolase from *E. coli*, Spider's venom phospholipase D (Masood et al. 2018), β-mannanases (Sharma et al. 2018) and other proteins are solved. In spite of low sequence similarity, all triose phosphate isomerase enzymes have similar structure with 8α/8β domain. The 8α/8β fold also includes antiparallel arrangement so that one helix and one strand reverse direction at the junction of N terminus and C-terminus of the barrel in enolase from *S. cereviciae* (Lebioda 1991) *H. sapiens* (Chai 2004), *T. Brucei* (Da Silva Giotto 2003), and epimerase from *B. subtilis* (Klenchin 2004). Cellulases, enzymes of 8α/8β fold that hydrolyze the cellulose producing short polysaccharide units as a result of cleavage of β-1,4-bond, are important in manufacturing of wide variety of products from food and textile to biofuels (Dadheech et al. 2018). B-D galactosidase, enzyme catalyzing hydrolysis of terminal β-linked galactose units and implicated in rare lysosomal storage disease, function as a homodimer (Shimizu et al. 2013). Rossmann fold gives another example of parallel α-helices around β-structure but organized as sheet.

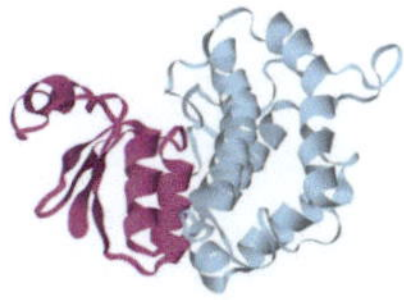

Thioredoxin /2yoi/ GST /1gne/

Fig. 1.18 Thioredoxin and GST N-terminal thioredoxin like (magenta) and C-terminal (blue) domains /1gne/

Whereas helices of TIM barrel fold form short interfaces, helices of Rossmann fold have weak interactions. In a group of enzymes radical-based chemistry is coordinated by a combined 8α/8β and Rossmann fold domains (Dowling et al. 2012).

A pair of parallel helices on one side of a β-sheet is an important structural unit of Glutathione S-transferase (GST), a detoxifying enzyme, which catalyses conjugation of glutathione to xenobiotic hydrophobic compounds (toxins, pharmaceuticals, pesticides) in order to make them more soluble and easily excreted (Fig. 1.18). This enzyme is present virtually in all organisms. Its cellular localization divides GSTs into cytosolic, mitochondrial, or microsomal. A classification of GSTs based on substrate/inhibitor specificity, amino acid sequence homology, structure similarity, and immunological identity divides all members of this family into alpha, mu, pi, theta, and omega classes (Sheehan et al. 2001). Crystallographic structures of GSTs exist already for many members of each class.

Cytosolic GSTs are folded as molecules with two domains, the N-terminal thioredoxin-like domain of $\alpha_1\beta\alpha_2\beta\beta\alpha_3$ fold, which is also a part of glutaredoxin, glutathione peroxidase, bacterial Dsba enzyme, and glutathione reductase, and the C-terminal α-helical domain. A GST-like domain is also reported in a part of Elongation factor 1B from *Saccharomyces cerevisiae* (Wells et al. 1993; Jeppesen et al. 2003). The N-terminal (α_1) and C-terminal (α_3) helices of the N-terminal GST domain are located on the same side of the β-sheet and interact with each other and with C-terminal α-helical domain. The middle helix (α_2) of the N-terminal GST domain is on the opposite side of the β-sheet. GST contains a conserved glutathione binding G-site within domain 1 and a xenobiotic substrate binding H-site in the domain 2 that is not conserved and accommodates various ligands (Sheehan et al. 2001). The interface between N-terminal and C-terminal helices of the N-terminal GST domain exhibits both positive and negative interhelical angles in various species. Glutathione S-transferase from *Rattus norvegicus,* pi and theta class *Mus musculus,* pi and mu class *Homo sapiens, Gallus gallus, Caenorhabditis elegans,* and *Drosophila melanogaster* exhibited positive interhelical angles similar to annexin. Interfaces with negative interhelical angle were found to be present in glutathione S-transferase from *Ochrobactrum anthropi*, *Anopheles dirus*, *Anopheles gambiae, Aegilops tauschii, Oryza sativa, Burkholderia xenovorans, Arabidopsis thaliana,* theta and omega class *Homo sapiens,* and zeta class *Mus musculus.*

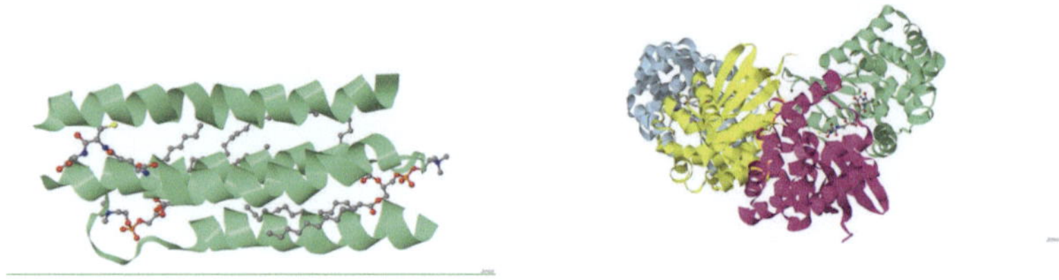

Fig. 1.19 Mitochondrial cytosolic and microsomal GSTs in complex with GSH

Mitochondrial GSTs (class Kappa) contain one domain of a conserved thioredoxin like fold (Figs. 1.18 and 1.19), are more similar to protein disulfide isomerase, dsbA, and function as homodimers (Ladner et al. 2004 1r4w; Poncalli et al. 2015).

Microsomal GSTs are membrane-bound; their tertiary structure differs from a thioredoxin-like fold of cytosolic GSTs (Fig. 1.19). Microsomal glutathione S-transferases I, II, and III are members of a common gene family together with 5-Lipoxygenase-activating protein (FLAP) and leukotriene-C4 (LTC4) synthase (Iacobsson et al. 1997). However, microsomal GSTs and cytosolic GSTs possess a common four-helical structure, which is present in a transmembrane region of LTC4 (Ago et al. 2007) and FLAP (Ferguson et al. 2007). They also have GSH peroxidase activity. Active site of MGST1 is at the interface of trimer of tetramers (Kuang et al. 2017 5i9k). Its binding of GSH shows that ligand conformation differs from that of bound to the soluble form (Holm et al. 2006).

GST is a biomarker due to its polymorphic variations associated with exposure to environmentally hazardous substances: air pollutants, carcinogens, heavy metal, etc. Gstp1 epigenetic modifications are early diagnostic cancer biomarkers. Five mutations are particularly important as potential sources of structural changes that influence detoxification process possibly causing carcinogenesis. Pi class GST P1 is over expressed in many cancers (Hollman et al. 2016).

GST interface of two parallel helices demonstrates rearrangement of mutual orientation of helices depending on combinations of amino acids at helix/helix interface or upon ligand binding (Chap. 3 Ligand binding). Comparison of these combinations in GSTs from various organisms shows correlation between amino acids and tertiary structure changes. Since these two parallel helices bind glutathione and other ligands, there is a link between GST conformation and ligand conformation (Kurochkina and Choekyi 2011). The site of ligand binding in DHAR shows that the two parallel helices have different orientation in ligand free and ligand bound state (Fig. 1.20).

GST structural family includes soluble chloride channels *Homo sapiens* CLIC1 (Dulhunty et al. 2001), CLIC4 (Li et al. 2006), *Caenorhabditis elegans* excretory canal abnormal protein 4, *Mus musculus* CLIC6 (Ferofontov et al. 2018) and *Drosophila melanogaster* CLIC (Littler et al. 2008), which are similar to omega class GSTs. Similarly to other intracellular ion channels such as annexin, BCL-XL, and some bacterial toxins, these chloride channels function as soluble and membrane-bound proteins; their conductance increases in membrane and depends

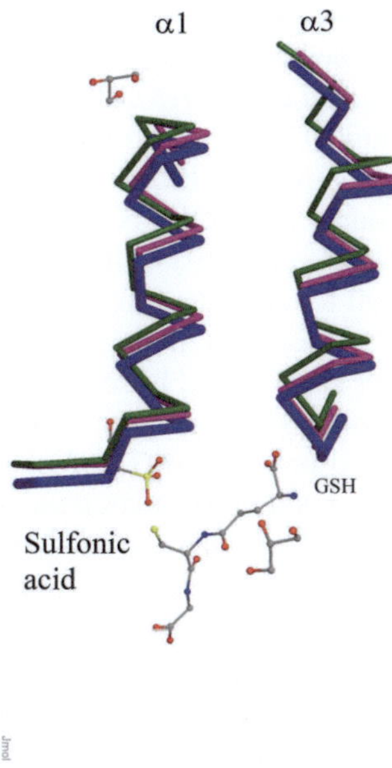

Fig. 1.20 *Arabidopsis thaliana* DHAR. Comparison of helix helix (α1α3) orientation in ligand free (green 5el8/ and ligand bound (magenta /5ela/ blue / 5elg/) forms

on lipid composition as well as monomeric/reduced or dimeric/oxidized oxidation state (Harrop et al. 2001; Al Khamici et al. 2016).

Unlike GST, CLIC1 or 6 do not bind GSH in solution. One of recently determined crystal structures of mouse CLIC6 shows amino terminus of helix α1 with the canonical 379Cys-Pro-Phe-Ser382 motif conserved in all CLICs and involved in ligand binding and oligomerization (Ferofontov et al. 2018).

CLICs are involved in atrial fibrillation, pulmonary hypertension, cell cycle, bone resorption, deafness, familial goiter, and kidney and hair cells stereo cilia function (Jiang et al. 2018).

Pairs of helices as substructures occur in completely unrelated proteins. GST and annexin, for example, contain subdomain that exhibits similar structure. Annexins consist of four homologous domains (domain I, II, III and IV), each comprising five α-helices as an assembly of a four-helix bundle with an extra helix C that connects helix-turn-helix substructures AB and DE (Huber et al. 1990). The interior residues of AB substructures of domain II and domain IV form the annexin ion channel (Hoffman et al. 2000). The annexin domains differ in their ligand binding properties such as binding of phospholipids is Ca^{2+}-dependent manner, binding of sulfatide by annexin V to regulate coagulability in the blood stream (Ida et al. 2004), binding of heparin by annexin A2 (Shao et al. 2006) or specific binding of glycosaminoglycans by annexins IV, V, and VI (Ishitsuka et al. 1998).

Four helix motif This motif represents an arrangement of four α-helices with parallel or antiparallel interfaces (Review: Harris et al. 1994; Kohn et al. 1997). One of the most frequently observed arrangements of helices is all antiparallel named four-α-helix bundle. The four-α-helical motif is found in hemerythrins (Sheriff et al. 1987), ferritins (Andrews et al. 1989), tobacco mosaic virus coat protein (Namba and Stubbs 1986), cytochrome b_{562}, cytochrome c' (Mathews et al. 1985), transcription factors (Banner et al. 1987), membrane M2 proton channel of influenza A virus (Schnell et al. 2008; Pielak et al. 2009), Ca + dependent membrane fusion proteins (Harsini et al. 2018), and other proteins. The four-α-helical structure also represents a unit of higher assemblies (Chou et al. 1988). The subunit of a tobacco mosaic virus coat protein, which is a four-α-helix bundle, assembles in a helical rod wrapped around viral RNA (Chap. 6). Three turns of the helical rod contain 49 subunits (Namba et al. 1989, Bhyravbhatla et al. 1998). Bacterioferritin assembly is spherical shell for storage of iron (Frolow et al. 1994).

Some of four helix structures are super coiled. SNARE complex is an assembly of four super coiled helices, one from syntaxin, one from VAMP2 (vesicle associated membrane protein 2 or synaptobrevin), and two from SNAP25 (synaptosome associated protein 25). The four helix domain is flanked by three helix Habc and two coiled coil helix linker domains. At synaptic and neuromuscular junctions SNARE complex mediate exocytosis of neurotransmitter carrying vesicles. It brings together two lipid bilayers, one from target membrane, another from vesicle membrane, in nucleotide independent manner by generating force to overcome electrostatic repulsion between membranes and dehydrate and bend the bilayers (Zhang 2017). Bacteriorhodopsin and annexin, for instance, combine four-α-helix bundle substructures as a part of 5-, 6-, 7-, 8-helical assemblies. The subunit of a tobacco mosaic virus coat protein, which is a four-α-helix bundle, assembles in a helical rod wrapped around viral RNA. Three turns of the helical rod contain 49 subunits (Namba et al. 1989, Bhyravbhatla et al. 1998). Angles between helices of the four-α-helix bundle in these assemblies can be negative, positive, or a combination (Fig. 1.15). Arrangement of α-helices, which exhibits negative interhelical angles also exists in the six-helix motif (Chap. 6) of the proteins that are components of oligomeric signaling complexes MYDDosome, PIDDosome (Lin et al. 2010), apoptosome (Yuan et al. 2010), inflammosome (de Alba 2009), and death inducing signaling complex (DISC) (Scott et al. 2009; Wang et al. 2010). Domain carrying six-helix bundle fold with a Greek key topology (Lasker et al. 2005) is the major unit involved in assembly and belongs to apoptosis (DD) superfamily (Weber and Vincenz 2001), which includes caspase recruitment domain (CARD), pyrin and death effector domain (DED). The helical assembly of subunits MYDDosome forms a tower shaped structure 110 Å in height and 70 Å in diameter comprising **6** myeloid differentiation primary response proteins (MYD88), 4 interleukin-1 receptor-associated kinases 4 (IRAK4), and 4 IRAK2, each interacting within the complex via their apoptosis domains (DD) (Lin et al. 2010). Similar domain-domain interfaces are observed in PIDDosome PIDD-RAIDD complex (Park et al. 2007) although mode of assembly resembles double-stranded helical oligomer (Lin et al. 2010).

Parallel Antiparallel Alcohol dehydrogenase, β-lactamase, annexin and other proteins contain alternating interfaces of positive and negative angles and parallel/antiparallel orientation. Alcohol dehydrogenase from *Equus caballus* α/β fold includes several parallel and antiparallel interfaces. Parallel helices A, B, and C of alcohol dehydrogenase from *Equus caballus* (Li 1994) and parallel and antiparallel interfaces of beta-lactamase from *Bacillus licheniformis* (Knox 1991), *E. coli* (Thomas 2005), and *Mycobacterium fortuitum* are arranged as if they form an arc of TIM-barrel. Both contain α/β fold. Similarly beta-lactamase from *Bacillus licheniformis* (Knox 1991), *E. coli* (Thomas 2005), and *Mycobacterium fortuitum* which has α/β fold including both parallel and antiparallel interfaces contains an arc of TIM-barrel with one parallel and one antiparallel interface.

Toroids Modular assembly of repeat units result in structures shaped as toroids: ankyrin, tetratricopeptide repeat (TPR) ARM/HEAT, PUM, LRR, TAL, PPR, SLR, and other types (Chap. 5). Secondary structure elements in these repeats stack parallel to each and form toroids, helices of helices (Blatch and Lassle 1999).

Helix interactions and packing Parallel and antiparallel arrangements of α-helices in proteins can have either positive or negative interhelical angles. Positive sign of an angle is assigned for packing of the distant helix clockwise relative to the proximal helix and negative sign of an angle if distant helix is directed counterclockwise from the proximal helix as in Fig. 1.14. Very important question is to identify the determinants of packing associated with handedness.

Small structural units located at the interfaces of the two interacting α-helices are shown to determine to great extent the type of the interface. These units are composed mainly of core positions **a** and **d** in the leucine zipper nomenclature (Hodges et al. 1972). When two α-helices form an interface, amino acids at these positions come into close contact, form hydrophobic core and contribute significantly to specificity of interaction. Heptad repeat of leucine residues in a leucine zipper homodimer results in leucine being at position **d** of the helix/helix interface. Within the same helix, leucine makes hydrophobic interactions with valines/asparagines at positions **a**. This **ad** unit is complementary to **a'd'** unit of a second helix. In hendecad (11) repeat, leucine appears in position **a**, whereas in quindecad (15) repeat in position **e**. The core formed by positions **ad** of one helix and positions **a'd'** unit of the second helix differs in the two arrangements.

Although leucine zipper nomenclature was first introduced for coiled coil structures, it was successfully applied for all interactions of α-helices in globular proteins. Identical and/or similar sequence combinations at positions **a** and **d** that are responsible for the same type of fold exist in fibrous and globular proteins.

Each type of helix is stabilized by specific interactions of atomic groups. Periodicity of contacts distinguishes one type of helix from another. Interactions along helix axis between every second amino acid in polypeptide chain, (i, i + 2), results in β-structure, whereas interactions between every third (i, i + 3) amino acid in polyproline helix. In both types of secondary structure, amino acids of one strand are hydrogen bonded to the amino acids from a neighboring strand. Interactions between every forth amino acid in polypeptide chain (i, i + 4), every fifth amino acid (i, i + 5), or every sixth amino acid in polypeptide chain (i, i + 6) characterize 3_{10}, α and π helices correspondingly. Hydrogen bond is to the amino acid 3, 4, or 5 residues apart. As a result, surface of a secondary structure element with hydrogen bonded backbone and protruding side chains is formed.

Pattern of hydrogen bonds between (i, i + 4) amino acid residues along the polypeptide backbone and van der Waals interactions between (i, i + 3) and (i, i + 4) residues are characteristic for α-helix. Grooves of the helix host polar hydrogen bonded atoms of amide nitrogen and carbonyl oxygen enclosed by the ridges of α-carbons with extending side chains (Fig. 1.21A). Surface of the helix backbone

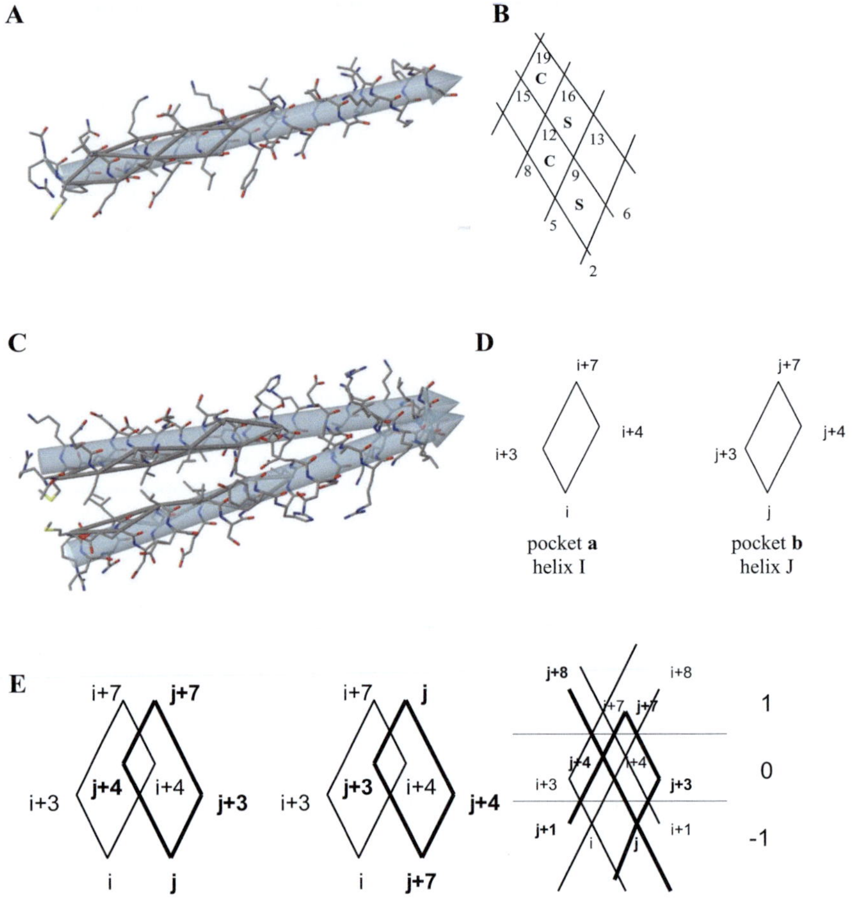

Fig. 1.21 Cylinder approximation of the leucine zipper helix backbone in 3D (**A**) and on the plane (**B**). Pockets on the helix surface: two central (C) and two side (S) pockets. (**C**) Cylinders/helices of the leucine zipper dimer. (**D**) Designation of the pockets on the two interacting helices, I and J. (**E**) Interactions of the pockets on the helices I and J

can be approximated by a cylinder. If we unfold helix cylinder to a plane by cutting it along the helix axis, the picture of lines connecting α carbon atoms will look like rows of parallel lines, or sides of adjacent parallelograms, in which each vertex represents an amino acid (Fig. 1.21B). This schematic view of rows of ridges on the surface of the helix formed by amino acids (i, i $\pm$ 1), (i, i $\pm$ 3), and (i, i $\pm$ 4) help to identify key features of its surface important for interactions. Amino acids (i, i + 3, i + 4, i + 7) are vertices of the edges that surround a "knob/hole", interaction unit of the helix. Let us call this unit, i. e. amino acids (i, i + 3, i + 4, i + 7), a pocket (Fig. 1.21 CD; Appendix B). When pocket **a** of helix I contacts pocket **b** of helix J, one side chain, a "knob", from pocket **a** packs into the "hole" of pocket **b**, and vice versa (Fig. 1.21 E).

The questions how secondary structure elements interact and what driving forces underline favorable interactions arise. Non-polar and hydrogen-bonded polar atomic groups form hydrophobic bonds (Kauzman 1959). Protein folding and unfolding experiments, data obtained for aqueous dissolution of hydrophobic model compounds suggest principles of hydrophobic interactions (Privalov an 1988). The important concept of accessible surface area (Richards 1977, Lee and Richards 1971) introduces a link between interior/exterior atomic groups and solvent. Energy of hydrophobic interactions derived from the data on free energy of transfer of amino acid side chains from organic solvents to water was found to be proportional to accessible surface area for each amino acid side chain (Chothia 1975). Contribution of multiple factors such as solvent, intermolecular/ intramolecular bonds, and entropy effects is integrated in order to correlate estimated and measured quantities (Chou 1988).

Packing of the secondary structure elements as they interact with each other to form a compact globule or elongated chains according to "knobs into holes" model (Crick 1953) suggests that one secondary structure element interacts with another element so that side chains (knobs) fit into spaces of hydrogen bonded backbone (holes). On the helix surface, protruding side chains and backbone form ridges alternating with groves. Proposed "ridges into grooves" packing gives basis for many observed preferred orientations of interacting secondary structure elements (Chothia 1981). Helix-helix preferred orientations which occur at angles of 20° and − 70° are consistent with "knobs into holes" packing (Crick 1953). "Ridges into grooves" packing explains preference for −50° interhelical angle (Chothia 1981) and orthogonal packing of beta pleated sheets (Chothia et al. 1982). Side chain rotamers form layers in globin hydrophobic cores (Efimov 1979). Certain combinations of rotational isomers of hydrophobic and some hydrophilic side chains pack in regular patterns at helix interfaces. Both α-helical and β-structural hydrophobic layers stack to form hydrophobic core. The two types of helix-helix packing, polar and apolar, add to previously identified discrete values of interhelical angles. Apolar packed α-helices exhibit three discreet values of angles between the axes of helices. Polar packing explains continuous changes in angle between the axes of interacting helices in the interval between −30 and 30 degrees depending on the shape and dimensions of the hydrophobic cluster. Polypeptides as helical structures are tightly packed, form folded fragments of native conformation, and existence of several frequently observed patterns (packing arrangements of α-helices, β-sheets, triple helices) can be explained in terms of local interaction energies, nonbonded energy of close packed atoms in the complementary surfaces and electrostatic energy of helix dipoles (Scheraga 1982). Many approaches are based on development of energy functions that target the origin of the right- and left-handed twist (Chou et al. 1990) reproducing structures observed in native proteins (Chothia 1973). Although interaction energy of the loops and loop-helix contacts in the four-helix structure plays a role in the stability of the structure (Chou et al. 1992) and loop length contributes to the stable fold, importance of the core residues of the interacting secondary structure elements for the protein three-dimensional structure is shown for association of α-helices and β-sheets in native and designed molecules: a pair of α-helices of GCN4 transcription factor (O'Shea et al. 1991), a dimer of two-helical fragments resulting in a four-helix motif of ROP protein (Paliakasis and Kokkinidis 1991), an active recombinant Fv fragment of permutated (Brinkmann et al. 1997). Deletion of the loop regions connecting the secondary structure elements does not impair the folding of the molecule as shown by the structure of regular four-helix bundle protein geometry (Paliakakis and Kokkinidis 1991) and design experiments. Rearrangement of loop connections does not affect activity of recombinant Fv fragment of antibody molecule which supports the hypothesis that structure of the core formed by interacting secondary structure elements depends on the core interactions and remains intact in permutated molecules (Brinkmann 1997). Role of the core residues in the folding of the protein is demonstrated by systematic analysis of ROP protein and its mutants which are not influenced by changes of the loop region but sensitive to changes of the hydrophobic core side-chains.

To embrace large number of packing arrangements and interhelical angles, we consider a model in which "knobs into holes" packing arranges same types of "ridges and grooves" from the two interacting helices with involvement of side chains that form specific tightly packed complementary units in which interhelical angle depends on amino acid composition and rotameric states. Contacts on the surface of the helix interacting with another helix follow a regular pattern (Kurochkina 2007, 2008) and involve a row of central pockets and a row of side pockets from each helix (Fig. 1.21 AB).

There are many ways holes can accommodate knobs; some amino acid combinations at the knob/hole site are preferred at parallel whereas other at antiparallel left/right handed helix/helix arrangements. Repeating sequence combinations occur at positions **a** and **d** of parallel/antiparallel helix-helix interfaces. They correlate with values of interhelical angles (or handedness) not only in homologous proteins but also within the same protein and in nonhomologous unrelated proteins. For groups of observed repeating combinations the correlation between the size of amino acid and magnitude of the interhelical angle exists (Kurochkina 2007). Distinct sequence combinations in leucine zipper and TIM-barrel proteins correspond to distinct arrangement of parallel helices which makes possible to identify parallel interfaces with positive (as in leucine zipper) and negative (as in TIM-barrel) interhelical angles (Kurochkina 2008). More evidence is provided by the prediction of interhelical angles for glutathione S-transferase, DHAR, intracellular chloride channel, ankyrin repeats, annexin, and other molecules from various sources (Kurochkina and Choekyi 2011).

In parallel and antiparallel types of packing, correlation between specific position of methyl groups and interhelical angle exists. Position of the side chain methyl groups gives the basis for the preference of a particular helix-helix arrangement. Side chains from the two interacting helices in the layer associate in accord with specific arrangement of their methyl groups. A good correlation was observed between the position of side chain methyl groups and the interhelical angle initially in a set of 200 proteins (Kurochkina 2008) and later in more and more proteins. Hydrophobic interactions are as specific as polar and charge interactions. Mutations at core positions of helix-helix interfaces can significantly modify interhelical angles and fold of the protein binding (Kurochkina and Choekyi 2011). They can lead to misfolded structure and impaired function. Helix-helix interfaces are key determinants of protein structure and protein assemblies (Kurochkina and Iadarola 2015a).

Hydrophobic interactions are specific by the unique shape and positioning of methyl groups. Polar interactions are specific due to geometry of the hydrogen bonded atoms. Charge interactions are specific with geometry of the salt bridge forming oppositely charged or repelling similarly charged residues. Combined to form a hydrophobic core, all types of interactions balance each other.

Mutations in the hydrophobic core can drastically change fold. In GCN4 LZ, mutations of valine/asparagine to leucine at position **a** or leucine to isoleucine at position **d** rearrange molecule to all parallel tetramer. Also, parallel trimer structure is observed if both **a** and **d** positions are mutated to isoleucine (Harbury et al. 1993) whereas antiparallel trimer if asparagines at position a are exchanged for alanine (Holton and Alber 2004). Coil ser peptide, a trimer with up–up–down topology (Lovejoy et al. 1992, 1993), converts to parallel trimer as a result of mutation of leucine 9 to cysteine (Touw et al. 2007). Hydrophobic interactions dominate at the helix interfaces and are critically important for the stability of protein structure (Moll et al. 2001; Moitra et al. 1997; Wagschal et al. 1999). Residues at positions **a** and **d** of the dimeric parallel coiled coils pack in layers, and their methyl groups interact similarly to layers of n-paraffins (Kitaigorodsky 1973). When mutated, hydrophobic residues repack to new arrangements inside the core and specific (**a**, **d**)–(**a**', **d**') interactions change with new combinations of residues at positions (**a**, **d**)–(**a**', **d**') replacing previous arrangements. All GC mutants undergo core arrangements and very often exhibit lower packing density of interior with cavities and channels (Yadav et al. 2005; Liu et al. 2006b).

Amino acid composition of helix-helix interfaces and conformations of side chains in the contact regions are very important for the arrangement of helices.

Nomenclature for designated positions (Fig 1.11A) of the two interacting helices originates in coiled coil structure and is applied to any pair of the interacting helices. Letters **a**-**g** designate the first helix and **a**′-**g**′ the second helix. In leucine zipper molecule, positions **d**-**d**′ at the interface between helices host leucines whereas positions **a**-**a**′ contain valines, asparagines, or other residues. Residues at positions **a** and **d** are inside the protein and to a large extent removed from contacts with solvent. They form a complementary interface and have to fit together in the space between helices being consistent with interactions of entire interface.

Side chain groups from the two interacting helices form layers. GCN4 leucine zipper layers are examples of ideally fitted groups formed so that leucine repeats from each helix sandwich valines (or asparagines) between them. In these layers, valine 9 (position **a**) and 9′ (**a**′) in the middle layer direct their $C_{\gamma 1}$ groups upward toward $C_{\delta 2}$ groups of leucine 12 (**d**) and 12′ (**d**′) but their $C_{\gamma 2}$ groups downward toward $C_{\delta 1}$ groups of leucine 5 (**d**) and 5′ (**d**′) (Fig. 1.22).

If each amino acid combination at the secondary structure interface is represented by its name (L-leucine, V-valine, and so on) or by interacting atomic groups (for example, methyl groups, which form contacts in intralayer space), interacting groups C_β in alanine (designated as β), $C_{\gamma 1}$ or $C_{\gamma 2}$ in valine (γ), $C_{\delta 1}$ or $C_{\gamma 2}$ in isoleucine (δ, γ), $C_{\delta 1}$ or $C_{\delta 2}$ in leucine (δ), a set of characteristic combinations represents this interaction. These interlayer and intralayer methyl group contacts occur at average

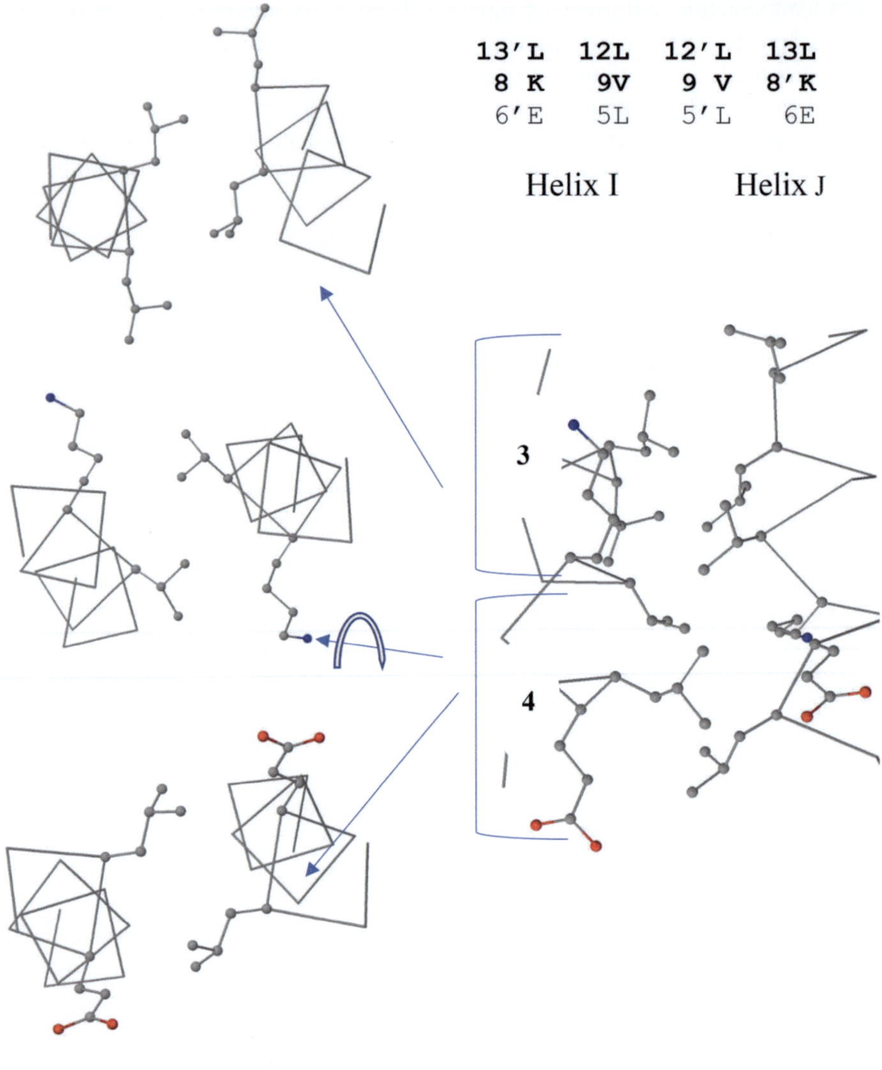

Fig. 1.22 Helix helix interactions in the leucine zipper molecule. The two interacting helices, I and J. Two types of pockets, "3" and "4". Three layers (two types of pockets) comprise one repeat unit

carbon/carbon distance of 3.8 Å. Combination of atomic groups where two δ groups at positions **d, d'** and two γ groups at positions **a, a'** contact in intralayer space corresponds to parallel leucine zipper amino acid combination of two leucines and two valines. Therefore, (**d**, **a**)-(**d'**, **a'**) layer for leucine zipper heptad is represented by $\binom{VV}{LL}$ amino acid combination or $\binom{\gamma\gamma}{\delta\delta}$ combination of interacting atomic groups while (**a**, **d**)-(**a'**, **d'**) layer by $\binom{LL}{VV}$ and $\binom{\delta\delta}{\gamma\gamma}$ correspondingly (Kurochkina 2008).

Interactions of the methyl groups occur as $C_{\delta 2}{}^{5}$ -$C_{\delta 2}{}^{5'}$ facing upward with $C_{\gamma 1}{}^{9}$ -$C_{\gamma 1}{}^{9'}$ facing downward at the boundary of (**d**,**d'**)-(**a**,**a'**) or as $C_{\delta 1}{}^{12}$ -$C_{\delta 1}{}^{12'}$ facing downward with $C_{\gamma 2}{}^{9}$ -$C_{\gamma 2}{}^{9'}$ facing upward at the boundary of (**a**,**a'**)-(**d**,**d'**). GCN4 leucine zipper interface has symmetric positioning of the end methyl groups relative to helix axes, results in 22° angle between helices and can be taken as reference point to compare methyl group interactions at other helix-helix interfaces.

"3" part of the 3–4 repeat has residues at positions **a** and **d** separated by three other residues. Positions **a** and **a'** are in the lower layer while **d** and **d'** in the upper layer as Val 9, Val 9′, Leu 12, Leu 12′ of leucine zipper. In amino acid sequence position **d** residue is followed by position **a** residue. "4" part of the 3–4 repeat has residues at positions **a** and **d** separated by four other residues. Positions **d** and **d'** are in the lower layer while **a** and **a'** in the upper layer as Leu 5, Leu 5′, Val 9, Val 9′ of leucine zipper and position **a** residue is followed by position **d** residue in amino acid sequence (Figs. 1.22 and 1.23).

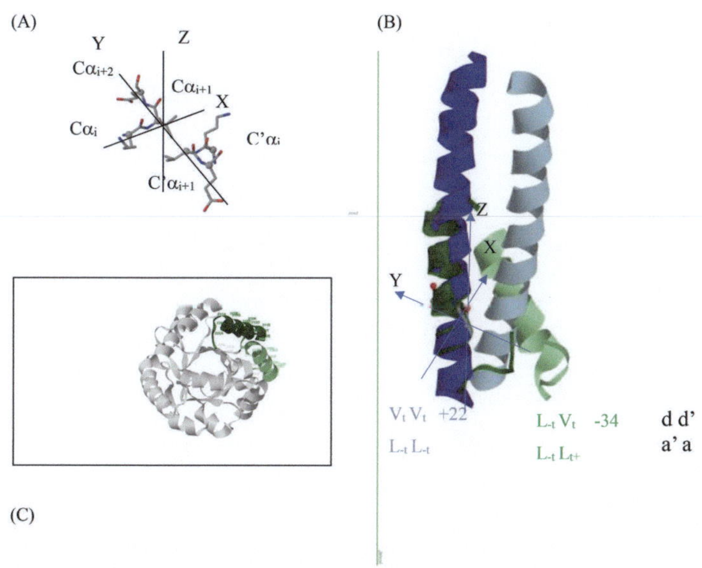

Fig. 1.23 Structure comparison in coordinate system of the layer. (**A**) Coordinate system; (**B**) Parallel interfaces of GCN4 Leucine zipper (blue) and fructose 1,6-biphosphate aldolase (green), amino acid combinations at the core positions a and d and interhelical angles. "4" pockets of the two protein superimpose so that Val 9, Leu 12, Val 9′, Leu 12′ of GCN4 correspond to Leu 214, Val 217, Leu 247, Leu 250 of fructose 1,6-biphosphate aldolase. (**C**) fructose 1,6-biphosphate aldolase (grey); the two helices (green)

Change of amino acids at positions **a, a′** and **d, d′** rearranges methyl group and their ability to form complementary interfaces. They exhibit the same, larger positive as well as negative interhelical angle. Comparison of the amino acids in positions **a** and **d** for different types of interfaces shows that that they can be subdivided into groups according to their amino acid content and interhelical angle (Fig. 1.23).

Within a layer, interactions are preserved if “3”/“4” interfaces have Val-like residues in positions **a** and Leu-like residues in position **d.** Conservation at the level of atomic groups results in equivalent pockets (Fig. 1.24). Sequence combination of two leucines in **d** layer and two isoleucines in **a** layer is equivalent to the combination of two leucines in **d** layer and two valines in **a** layer. This combination is repeated with positive interhelical angles in CGMP-dependent protein kinase Iα, Aα-Bβ interface of bovine fibrinogen, Aα-γ interface of human fibrinogen and leucine zipper interface of MAX transcription factor. Combination $\binom{LL}{II}$ is equivalent to $\binom{LL}{VV}$ because they both contribute $\binom{\gamma\gamma}{\delta\delta}$ interaction to the “3” part of the repeat.

Interaction at parallel interface of the “3/4” repeat results in the two types of alternating pockets “3” and “4” with layers $\binom{d\ d'}{a'a}$ and $\binom{a'a}{d\ d'}$ correspondingly. Interaction at antiparallel interface of the “3/4” repeat results in the two types of alternating pockets “3” and “4” with layers $\binom{d\ a'}{d'a}$ and $\binom{d'a}{d\ a'}$ (Fig. 1.25).

“4” part $\binom{\gamma\gamma}{\delta\delta}$ $\binom{VV}{LL}$ $\begin{matrix}a'\ a\\ d\ d'\end{matrix}$ “3” part $\binom{\delta\delta}{\gamma\gamma}$ $\binom{LL}{VV}$ $\begin{matrix}d\ d'\\ a'\ a\end{matrix}$

Equivalent $\binom{LL}{II}$ $\binom{LL}{VI}$ $\binom{LI}{II}$ $\binom{LI}{VI}$

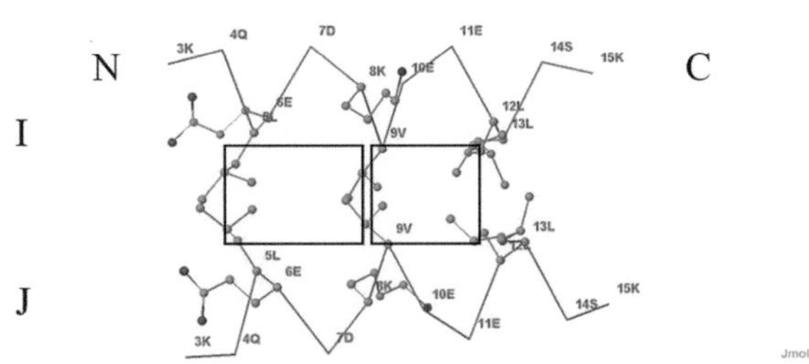

Fig. 1.24 Equivalent pockets. GCN4 LZ Pocket is a **reference** point. Residues from the two helices I and J interact so that side chains at positions **a** and **d** form layers: in the “4” part, Leu 5, Leu 5′, Val 9, Val 9′; in the “3” part, Val 9, Val 9′, Leu 12, Leu 12′. Interactions of the methyl groups occur as $C_{\delta2}{}^{5}$ -$C_{\delta2}{}^{5'}$ facing C terminus and $C_{\gamma1}{}^{9}$ -$C_{\gamma1}{}^{9'}$ facing N terminus at the boundary of (d,d′)-(a,a′) or as $C_{\delta1}{}^{12}$ -$C_{\delta1}{}^{12'}$ facing N terminus with $C_{\gamma2}{}^{9}$ -$C_{\gamma2}{}^{9'}$ facing C terminus at the boundary of (a, a′)-(d,d′). All other pockets with the same arrangement of interacting methyl groups are **equivalent** pockets

Residues from the two helices I and J interact so that side chains at positions **a** and **d** form layers: in the “4”

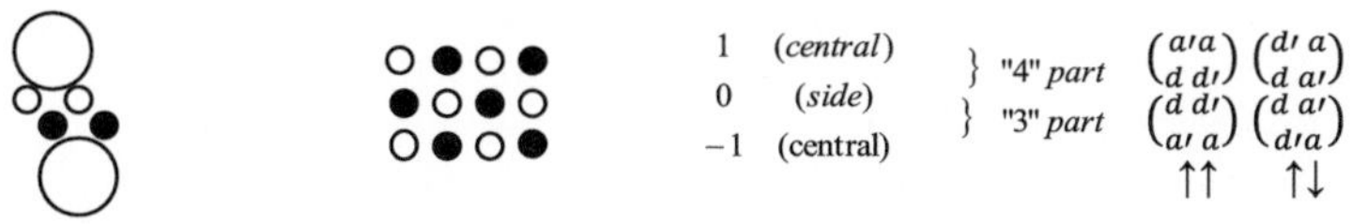

Fig. 1.25 Side chain layers at parallel (↑↑) and antiparallel (↑↓) interfaces. Repeating combinations and equivalent pockets. GCN4 LZ pocket is a **reference** point.!LZs Tim Ank Residues from the two helices I and J interact so that side chains at positions **a** and **d** form layers: in the “4”

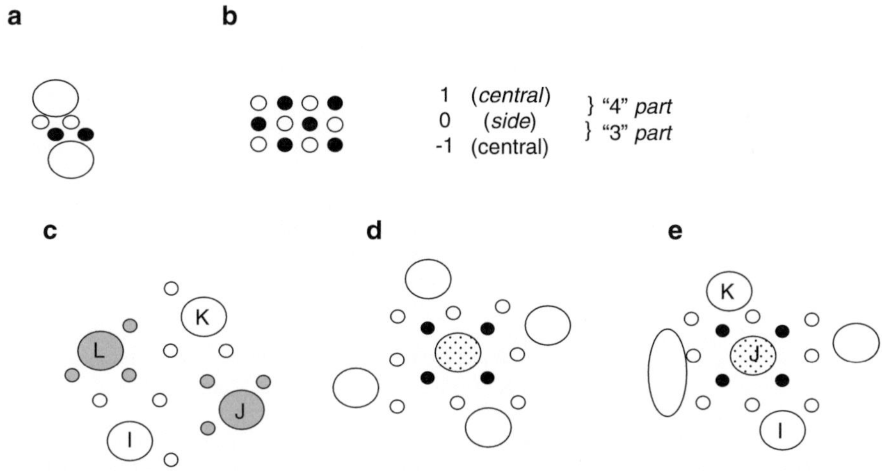

Fig. 1.26 Arrangements of the interacting helices

Antiparallel helix/helix interface also consists of stacked alternating (**a**, **d**)-(**a’**, **d’**) layer and (**d**, **a**)–(**d’**, **a’**) layers. In VPR protein, proline/betaine transporter, and seryl t-RNA synthetase repeated amino acid combinations are (**a**, **d**)-(**a’**, **d’**) layer $\binom{LI}{LI}$ $\binom{\delta\gamma}{\delta\gamma}$ followed by (**d**, **a**)–(**d’**, **a’**) layer with equivalent pockets with $\binom{IR}{LI}$ and $\binom{VL}{LI}$ $\binom{\gamma\delta}{\delta\delta}$. Within (**a**, **d**)-(**a’**, **d’**) layer, $\binom{\delta\gamma}{\delta\gamma}$ pockets $\binom{IV}{LI}$, $\binom{LV}{LI}$, $\binom{LV}{LV}$, and $\binom{LI}{LV}$ are equivalent, whereas within (**d**, **a**)–(**d’**, **a’**) layer, $\binom{\gamma\delta}{\delta\delta}$ pockets $\binom{VL}{LI}$, $\binom{VI}{LI}$, and $\binom{VL}{LL}$ are equivalent (Kurochkina 2008).

Pairs of helices assemble in various arrangements (Fig. 1.26). Large variety of folds is generated by multiple combinations of these arrangements.

References

Adam A, Haberhauer G, Wölper C (2017) Bio-inspired herringbone foldamers: strategy for changing the structure of helices. J Org Chem 82(8):4203–4215. https://doi.org/10.1021/acs.joc.7b00185. Epub 2017 Apr 12

Adam C, Peters AD, Lizio MG, Whitehead GFS, Diemer V, Cooper JA, Cockroft SL, Clayden J, Webb SJ (2018) The Role of terminal functionality in the membrane and antibacterial activity of peptaibol-mimetic aib foldamers. Chemistry 24(9):2249–2256. https://doi.org/10.1002/chem.201705299. Epub 2018 Jan 17. PubMed PMID: 29210477

Aggeli A, Nyrkova IA, Bell M, Harding R, Carrick L, McLeish TCB, Semenov AN, Boden N (2001) Hierarchical self-assembly of chiral rod-like molecules as a model for peptide beta -sheet tapes, ribbons, fibrils, and fibers. Proc Natl Acad Sci USA 98:11857–11862

Ago H, Kanaoka AH, Irikura D et al (2007) Crystal structure of a human membrane protein involved in cysteinyl leukotriene biosynthesis. Nature 448:609–612

De Alba E (2009) Structure and interdomain dynamics of apoptosis-associated speck-like protein containing a CARD (ASC). J Biol Chem 284:32932–32941

Al Khamici H, Hossain KR, Cornell BA, Valenzuela SM (2016) Investigating sterol and redox regulation of the ion channel activity of CLIC1 using tethered bilayer membranes. Membranes (Basel). 6(4). pii: E51. PubMed PMID: 27941637; PubMed Central PMCID: PMC5192407

Alushin GM, Ramey VH, Pasqualato S, Ball DA, Grigorieff N, Musacchio A, Nogales E (2010) The Ndc80 kinetochore complex forms oligomeric arrays along microtubules. Nature 467:805–810

Alvarez BH, Gruber M, Ursinus A, Dunin-Horkawicz S, Lupas AN, Zeth K (2010) A transition from strong right-handed to canonical left-handed supercoiling in a conserved coiled-coil segment of trimeric autotransporter adhesins. J Struct Biol 170(2):236–245. https://doi.org/10.1016/j.jsb.2010.02.009. Epub 2010 Feb 21. PubMed PMID: 20178846

Andrade MA, Perez-Iratxeta C, Ponting CP (2001) Protein repeats: structures, functions, and evolutions. J Struct Biol 134:117–131

Andrews SC, Smith JM, Guest JR, Harrison PM (1989) Amino acid sequence of the bacterioferritin (cytochrome b1) of Escherichia coli-K12. Biochem Biophys Res Commun 158(2):489–496. PubMed PMID: 2644932

Aravinda S, Shamala N, Balaram P (2008) Aib residues in peptaibiotics and synthetic sequences: analysis of nonhelical conformations. Chem Biodivers 5(7):1238–1262. https://doi.org/10.1002/cbdv.200890112

Arcy SD, Davies OR, Blundell TL, Bolanos-Garcia VM (2010) Defining the molecular basis of BubR1 kinetochore interactions and APC/C-CDC20 inhibition. J Biol Chem 285:14764–14776

Balasco N, Smaldone G, Ruggiero A, De Simone A, Vitagliano L (2018) Local structural motifs in proteins: Detection and characterization of fragments inserted in helices. Int J Biol Macromol. pii: S0141-8130(18)32822–32828. https://doi.org/10.1016/j.ijbiomac.2018.07.047. [Epub ahead of print] PubMed PMID: 30017977

Balashova TA, Shenkarev ZO, Tagaev AA, Ovchinnikova TV, Raap J, Arseniev AS (2000) NMR structure of the channel-former zervamicin IIB in isotropic solvents. FEBS Lett 466 (2-3):333–336. PubMed PMID: 10682854

Banci L, Bertini I, Cefaro C, Ciofi-Baffoni S, Gallo A (2011) Functional role of two interhelical disulfide bonds in human Cox17 protein from a structural perspective. J Biol Chem 286 (39):34382–34390. https://doi.org/10.1074/jbc.M111.246223. Epub 2011 Aug 4. PubMed PMID: 21816817; PubMed Central PMCID: PMC3190761

Banner DW, Kokkinidis M, Tsernoglou D (1987) Structure of the ColE1 rop protein at 1.7 A resolution. J Mol Biol 196(3):657–675. PubMed PMID: 3681971

Barua B (2013) Periodicities designed in the tropomyosin sequence and structure define its functions. Bioarchitecture 3(3):51–56. https://doi.org/10.4161/bioa.2561

Berisio R, Vitagliano L (2012) Polyproline and triple helix motifs in host-pathogen recognition. Curr Protein Pept Sci 13(8):855–865. https://doi.org/10.2174/138920312804871157

Bhyravbhatla B, Watowich SJ, Caspar DL (1998) Refined atomic model of the four-layer aggregate of the tobacco mosaic virus coat protein at 2.4-Å resolution. Biophys J 74:604–615

Blatch GL, Lassle M (1999) The tetratricopeptide repeat: a structural motif mediating protein-protein interactions. Bioassays 21:932–939

Boscutti G, Nardon C, Marchiò L, Crisma M, Biondi B, Dalzoppo D, Dalla Via L, Formaggio F, Casini A, Fregona D (2018) Anticancer gold(III) peptidomimetics: from synthesis to in vitro and ex vivo biological evaluation. ChemMedChem. https://doi.org/10.1002/cmdc.201800098. [Epub ahead of print] PubMed PMID: 29570944

Boudko SP, Bächinger HP (2016) Structural insight for chain selection and stagger control in collagen. Sci Rep 6:37831. https://doi.org/10.1038/srep37831

Branden, C., Tooze, J (1991) Introduction to protein structure.

Brennan SO (2015) Variation of fibrinogen oligosaccharide structure in the acute phase response: possible haemorrhagic implications. BBA Clinical 3:221–226. https://doi.org/10.1016/j.bbacli.2015.02.007

Brinkmann U, Di Carlo A, Vasmatzis G, Kurochkina N, Beers R, Lee B, Pastan I (1997) Stabilization of a recombinant Fv fragment by base loop interconnection and VH-VL permutation. J Mol Biol 268:107–117

Brunner JD, Lim NK, Schenck S, Duerst A, Dutzler R (2014) X-ray structure of a calcium-activated TMEM16 lipid scramblase. Nature 516(7530):207–212. https://doi.org/10.1038/nature13984. Epub 2014 Nov 12. PubMed PMID: 25383531

Bunick CG, Milstone LM (2017) The X-ray crystal structure of the keratin 1-keratin 10 helix 2B heterodimer reveals molecular surface properties and biochemical insights into human skin disease. J Invest Dermatol;137(1):142–150.: https://doi.org/10.1016/j.jid.2016.08.018. Epub 2016 Sep 3.

Burgess AW, Leach SJ (1973) An obligatory alpha-helical amino acid residue. Biopolymers 12 (11):2599–2605. PubMed PMID: 4780721

Cabrele C, Langer M, Bader R, Wieland HA, Doods HN, Zerbe O, Beck-Sickinger AG (2000) The first selective agonist for the neuropeptide YY5 receptor increases food intake in rats. J Biol Chem 275(46):36043–36048

Caffrey M, Cai M, Kaufman J, Stahl SJ, Wingfield PT, Covell DG, Gronenborn AM, Clore GM (1998) Three-dimensional solution structure of the 44 kDa ectodomain of SIV gp41. Embo J 17:4572–4584

Cervantes-Madrid D, Dominguez-Gomez G, Gonzalez-Fierro A, Perez-Cardenas E, Taja-Chayeb L, Trejo-Becerril C, Duenas-Gonzalez A (2017) Feasibility and antitumor efficacy in vivo, of simultaneously targeting glycolysis, glutaminolysis and fatty acid synthesis using lonidamine, 6-diazo-5-oxo-L-norleucine and orlistat in colon cancer. Oncology Letters 13(3):1905–1910. https://doi.org/10.3892/ol.2017.5615

Chen Q, Wells MM, Arjunan P, Tillman TS, Cohen AE, Xu Y, Tang P (2018) Structural basis of neurosteroid anesthetic action on GABA(A) receptors. Nat Commun 9(1):3972. https://doi.org/10.1038/s41467-018-06361-4. PubMed PMID: 30266951; PubMed Central PMCID: PMC6162318

Chingle R, Proulx C, Lubell WD (2017) Azapeptide synthesis methods for expanding side-chain diversity for biomedical applications. Acc Chem Res 50(7):1541–1556. https://doi.org/10.1021/acs.accounts.7b00114. Epub 2017 Jun 9

Chothia C (1973) Conformation of twisted beta-pleated sheets in proteins. J Mol Biol 75:295–302

Chothia C (1975) Structural invariants in protein folding. Nature 254:304–308

Chothia C, Levitt M, Richardson D (1981) Helix to helix packing in proteins. J Mol Biol 145:215–250

Chothia C, Janin J (1982) Orthogonal packing of beta-pleated sheets in proteins. Biochemistry 21 (17):3955–3965. PubMed PMID: 6751382

Chou KC (1988) Review: Low-frequency collective motion in biomacromolecules and its biological functions. Biophysical Chemistry 30:3–48

Chou KC (1992) Energy-optimized structure of antifreeze protein and its binding mechanism. J Mol Biol 223:509–517

Chou KC, Nemethy G, Scheraga HA (1990) Review: Energetics of interactions of regular structural elements in proteins. Acc Chem Res 23:134–141

Chou KC, Maggiora GM, Nemethy G, Scheraga HA (1988) Energetics of the structure of the four-alpha-helix bundle in proteins. Proc Natl Acad Sci USA 85:4295–4299

Chou KC, Maggiora GM, Scheraga HA (1992) The role of loop-helix interactions in stabilizing four-helix bundle proteins. Proc Natl Acad Sci USA 89:7315–7319

Cook JD, Soto-Montoya H, Korpela MK, Lee JE (2015) Electrostatic architecture of the Infectious Salmon Anemia Virus (ISAV) Core fusion protein illustrates a Carboxyl-Carboxylate pH Sensor. J Biol Chem 290(30):18495–18504. https://doi.org/10.1074/jbc.M115.644781. Epub 2015 Jun 16. PubMed PMID: 26082488; PubMed Central PMCID: PMC4513110

Cordopatis P, Manessi-Zoupa E, Theodoropoulos D, Bossé R, Bouley R, Gagnon S, Escher E (1994) Methylation in positions 1 and 7 of angiotensin II. A structure-activity relationship study. Int J Pept Protein Res 44(4):320–324. PubMed PMID: 7875933

Costil R, Fernández-Nieto F, Atkinson RC, Clayden J (2018) α-Methyl phenylglycines by asymmetric α-arylation of alanine and their effect on the conformational preference of helical Aib foldamers. Org Biomol Chem doi: 10.1039/c8ob00551f. [Epub ahead of print] PubMed PMID: 29595846

Crick F (1953) Acta Crystallogr 6:689

Dadheech T, Shah R, Pandit R, Hinsu A, Chauhan PS, Jakhesara S, Kunjadiya A, Rank D, Joshi C (2018) Cloning, molecular modeling and characterization of acidic cellulase from buffalo rumen and its applicability in saccharification of lignocellulosic biomass. Int J Biol Macromol 113:73–81. https://doi.org/10.1016/j.ijbiomac.2018.02.100. Epub 2018 Feb 15. PubMed PMID: 29454942

Das T, Häring M, Haldar D, Díaz DD (2017) Phenylalanine and derivatives as versatile low-molecular-weight gelators: design, structure and tailored function. Biomater Sci 6(1):38–59. https://doi.org/10.1039/c7bm00882a. Review. PubMed PMID: 29164186

Davies TG, Hubbard RE, Tame JR (1999) Relating structure to thermodynamics: the crystal structures and binding affinity of eight OppA-peptide complexes. Protein Sci 8(7):1432–1444

Dowling DP, Croft AK, Drennan CL (2012) Radical use of Rossmann and TIM barrel architectures for controlling coenzyme B12 chemistry. Annu Rev Biophys 41:403–427. https://doi.org/10.1146/annurev-biophys-050511-102225. Review. PubMed PMID: 22577824

Dulhunty A, Gage P, Curtis S et al (2001) The glutathione transferase structural family includes a nuclear chloride channel and a ryanodine receptor calcium release channel modulator. J Biol Chem 276:3319–3323

Efimov AV (1979) Packing of alpha-helices in globular proteins. Layer-structure of globin hydrophobic cores. J Mol Biol 134(1):23–40

Elkin I, Maris T, Melkoumov A, Hildgen P, Banquy X, Leclair G, Barrett C (2018) Crystal structure of 2-oxopyrrolidin-3-yl 4-(2-phenyl-diazen-1-yl)benzoate. Acta Crystallogr E Crystallogr Commun 6;74(Pt 4):458–460. doi: 10.1107/S205698901800333X. eCollection 2018 Apr 1. PubMed PMID: 29765745; PubMed Central PMCID: PMC5946967

Eriksson M, Hassan S, Larsson R, Linder S, Ramqvist T, Lövborg H, Vikinge T, Figgemeier E, Müller J, Stetefeld J, Dalianis T, Ozbek S (2009) Utilization of a right-handed coiled-coil protein from archaebacterium Staphylothermus marinus as a carrier for cisplatin. Anticancer Res 29(1):11–18. PubMed PMID: 19331128

Ferguson AD, Mckeever BM, Wisniewski D et al (2007) Crystal structure of inhibitor-bound human 5-lipoxygenase-activating protein. Science 317:510–512

Ferner J, Suhartono M, Breitung S, Jonker HR, Hennig M, Wöhnert J, Göbel M, Schwalbe H (2009) Structures of HIV TAR RNA-ligand complexes reveal higher binding stoichiometries. Chembiochem 10(9):1490–1494. https://doi.org/10.1002/cbic.200900220

Ferrone FA (2016) Sickle cell disease: Its molecular mechanism and the one drug that treats it. Int J Biol Macromol 93(Pt A):1168–1173. https://doi.org/10.1016/j.ijbiomac.2016.09.073. Epub 2016 Sep 22

Ferofontov A, Strulovich R, Marom M, Giladi M, Haitin Y (2018) Inherent flexibility of CLIC6 revealed by crystallographic and solution studies. Sci Rep 8(1):6882. https://doi.org/10.1038/s41598-018-25231-z. PubMed PMID: 29720717; PubMed Central PMCID: PMC5931990

Fisher E (1901) Z Physiol Chem 33:151

Fisher E (1902) Chem Ber 35:2660

Formaggio F, Crisma M, Bonora GM, Pantano M, Valle G, Toniolo C, Aubry A, Bayeul D, Kamphuis J (1995) (R)-isovaline homo-peptides adopt the left-handed 3(10)-helical structure. Pept Res 8(1):6–15. PubMed PMID: 7756755

Fung T, Asiri YI, Wall R, Schwarz SKW, Puil E, MacLeod BA (2017) Variations of isovaline structure related to activity in the formalin foot assay in mice. Amino Acids 49(7):1203–1213. https://doi.org/10.1007/s00726-017-2421-6. Epub 2017 Apr 21. PubMed PMID: 28432424

Gamlin CR, Yu WQ, Wong ROL, Hoon M (2018) Assembly and maintenance of GABAergic and Glycinergic circuits in the mammalian nervous system. Neural Dev 13(1):12. https://doi.org/10.1186/s13064-018-0109-6. Review. PubMed PMID: 29875009; PubMed Central PMCID: PMC5991458

Gernert KM, Surles MC, Labean TH, Richardson JS, Richardson DC (1995) Alacoil: a very tight, antiparallel coiled-coil of helices. Protein Science 4:2252–2260

Goh KL, Holmes DF (2017) Collagenous extracellular matrix biomaterials for tissue engineering: lessons from the common sea urchin tissue. Int J Mol Sci 18(5):901. https://doi.org/10.3390/ijms18050901

Görbitz CH, Karen P, Dušek M, Petrícek V (2016) An exceptional series of phase transitions in hydrophobic amino acids with linear side chains. IUCrJ 3(Pt 5):341–353. https://doi.org/10.1107/S2052252516010472

Gord JR, Hewett DM, Hernandez-Castillo AO, Blodgett KN, Rotondaro MC, Varuolo A, Kubasik MA, Zwier TS (2016) Conformation-specific spectroscopy of capped, gas-phase Aib oligomers: tests of the Aib residue as a 310-helix former. Phys Chem Chem Phys 18(36):25512–25527

Grauer AA, Cabrele C, Zabel M, König B (2009) Stable right- and left-handed peptide helices containing C(alpha)-tetrasubstituted alpha-amino acids. J Org Chem 74(10):3718–3726. doi: https://doi.org/10.1021/jo900222g. PubMed PMID: 19354242

Guerra ME, Fadel V, Maltarollo VG, Baldissera G, Honorio KM, Ruggiero JR, Dos Santos Cabrera MP (2017) MD simulations and multivariate studies for modeling the antileishmanial activity of peptides. Chem Biol Drug Des 2017 Mar 7. https://doi.org/10.1111/cbdd.12970. [Epub ahead of print]

Gunning PW, Hardeman EC, Lappalainen P, Mulvihill DP (2015) Tropomyosin - master regulator of actin filament function in the cytoskeleton. J Cell Sci 128(16):2965–2974. https://doi.org/10.1242/jcs.172502. Epub 2015 Aug 3

Harbury PB, Zhang T, Kim PS, Alber T (1993) A switch between two-, three-. and four-stranded coiled coils in GCN4 leucine zipper mutants. Science 262(5138):1401–1407

Harbury PB, Plecs JJ, Tidor B, Alber T, Kim PS (1998) High-resolution protein design with backbone freedom. Science 282(5393):1462–1467. PubMed PMID: 9822371

Harris NL, Presnell SR, Cohen FE (1994) Four helix bundle diversity in proteins. J Mol Biol 236:1356–1368

Harrop SJ, DeMaere MZ, Fairlie WD et al (2001) Crystal structure of a soluble form of the intracellular chloride ion channel CLIC1 (NCC27) at 1.4-A resolution. J Biol Chem 276:44993–45000

Harsini FM, Chebrolu S, Fuson KL, White MA, Rice AM, Sutton RB (2018) FerA is a membrane-associating four-helix bundle domain in the Ferlin family of membrane-fusion proteins. Sci Rep;8(1):10949. https://doi.org/10.1038/s41598-018-29184-1. PubMed PMID: 30026467; PubMed Central PMCID: PMC6053371.

Hodges RS, Sodek J, Smillie LB et al (1972) Amino-acid sequence of rabbit skeletal tropomyosin and its coiled-coil structure. Proc Natl Acad Sci 69:3800–3804

Hollman AL, Tchounwou PB, Huang HC (2016) The Association between gene-environment interactions and diseases involving the human GST superfamily with SNP variants. Int J Environ Res Public Health 13(4):379. https://doi.org/10.3390/ijerph13040379. Review. PubMed PMID: 27043589; PubMed Central PMCID: PMC4847041

Holm PJ, Bhakat P, Jegerschold C, Gyobu N, Mitsuoka K, Fujiyoshi Y, Morgenstern R, Hebert H (2006) Structural basis for detoxification and oxidative stress protection in membranes. J Mol Biol *360*:934–945

Holmes TC (2002) Novel peptide-based biomaterial scaffolds for tissue engineering. Trends Biotechnol 20(1):16–21. Review. PubMed PMID: 11742673

Holton A, Alber J (2004) Automated protein crystal structure determination using ELVES. Proc Natl Acad Sci 101:1537–1542

Hou Q, Bourgeas R, Pucci F, Rooman M (2018) Computational analysis of the amino acid interactions that promote or decrease protein solubility. Sci Rep 8(1):14661. https://doi.org/10.1038/s41598-018-32988-w. PubMed PMID: 30279585; PubMed Central PMCID: PMC6168528

Huber R, Romish J, Paques EP (1990) The crystal and molecular structure of human annexin V, an anticoagulant protein that binds to calcium and membranes. Embo J 9:3867–3874

Ida M, Sato A, Matsumoto I et al (2004) Human annexin V binds to sulfatide: contribution to regulation of blood coagulation. J Mol Biol 135:583–588

Ishitsuka R, Kojima K, Utsumi H et al (1998) Glycosaminoglycan Binding Properties of Annexin IV, V, and VI. J Biol Chem 273:9935–9941

Jaffe EK (2017) New protein structures provide an updated understanding of phenylketonuria. Mol Genet Metab 121(4):289–296. https://doi.org/10.1016/j.ymgme.2017.06.005. Epub 2017 Jun 15. Review. PubMed PMID: 28645531; PubMed Central PMCID: PMC5549558

Jeppesen MG, Ortiz P, Shepard W et al (2003) The Crystal Structure of the Glutathione *S*-Transferase-like Domain of Elongation Factor 1B from *Saccharomyces cerevisiae*. J Biol Chem 278:47190–47198

Jiang Q, Li K, Lu WJ, Li S, Chen X, Liu XJ, Yuan J, Ding Q, Lan F, Cai SQ (2018) Identification of small-molecule ion channel modulators in C. elegans channelopathy models. Nat Commun 9 (1):3941. https://doi.org/10.1038/s41467-018-06514-5. PubMed PMID: 30258187; PubMed Central PMCID: PMC6158242

Juncosa JI, Takaya K, Le HV, Moschitto MJ, Weerawarna PM, Mascarenhas R, Liu D, Dewey SL, Silverman RB (2018) Design and Mechanism of (S)-3-Amino-4-(difluoromethylenyl) cyclopent-1-ene-1-carboxylic Acid, a Highly Potent ?-Aminobutyric Acid Aminotransferase Inactivator for the Treatment of Addiction. J Am Chem Soc 140(6):2151–2164. https://doi.org/10.1021/jacs.7b10965. Epub 2018 Jan 30. PubMed PMID: 29381352; PubMed Central PMCID: PMC5812813

Kadir M, Wang X, Zhu B, Liu J, Harland D, Popescu C (2017) The structure of the "amorphous" matrix of keratins. J Struct Biol 198(2):116–123. doi: 10.1016/j.jsb.2017.04.001. Epub 2017 Apr 5

Kantharaju RS, Aravinda S, Shamala N, Balaram P (2010) Helical conformations of hexapeptides containing N-terminus diproline segments. Biopolymers 94(3):360–370. https://doi.org/10.1002/bip.21395

Kapinos LE, Burkhard P, Herrmann H, Aebi U, Strelkov SV (2011) Simultaneous formation of right- and left-handed anti-parallel coiled-coil interfaces by a coil2 fragment of human lamin A. J Mol Biol 408(1):135–146. https://doi.org/10.1016/j.jmb.2011.02.037. Epub 2011 Feb 24. PubMed PMID: 21354179

Kasznel AJ, Zhang Y, Hai Y, Chenoweth DM (2017) Structural basis for aza-glycine stabilization of collagen. J Am Chem Soc 2017 Jul 19;139(28):9427–9430. https://doi.org/10.1021/jacs.7b03398. Epub 2017 Jul 6.

Kendrew JC (1959) Structure and function in myoglobin and other proteins. Fed Proc 18(2, Part 1):740–751. PubMed PMID: 136722

Khanapur M, Alvala M, Prabhakar M, Shiva Kumar K, Edwin RK, Sri Saranya PS, Patel RK, Bulusu G, Misra P, Pal M (2017) Mycobacterium tuberculosis chorismate mutase: A potential target for TB. Bioorg Med Chem 25(6):1725–1736. https://doi.org/10.1016/j.bmc.2017.02.001. Epub 2017 Feb 4. Review. PubMed PMID: 28202315

Kikuchi N, Fujiwara K, Ikeguchi M (2018) ß-strand twisting/bending in soluble and transmembrane ß-barrel structures. Proteins. https://doi.org/10.1002/prot.25576. [Epub ahead of print] PubMed PMID: 30019770

Kitaigorodsky AI (1973) Molecular crystals and molecules. Academic Press, New York

Kohn WD, Mant CT, Hodges RS (1977) Helical protein assembly motifs. J Biol Chem 272:2583–2586

Kollman JM, Pandi L, Sawaya MR, Riley M, Doolittle RF (2009) Crystal structure of human fibrinogen. Biochemistry 48(18):3877–3886. https://doi.org/10.1021/bi802205g

Konno K, Rangel M, Oliveira JS, Dos Santos Cabrera MP, Fontana R, Hirata IY, Hide I, Nakata Y, Mori K, Kawano M, Fuchino H, Sekita S, Neto JR (2017) Decoralin, a novel linear cationic alpha-helical peptide from the venom of the solitary eumenine wasp Oreumenes decoratus. Peptides 28(12):2320–2327. Epub 2007 Sep 29

Krzywda S, Brzozowski AM, Higashitsuji H, Fujita J, Welchman R, Dawson S, Mayer RJ, Wilkinson AJ (2004) The Crystal Structure of Gankyrin, an Oncoprotein Found in Complexes with Cyclin-dependent Kinase 4, a 19 S Proteasomal ATPase Regulator, and the Tumor Suppressors Rb and p53. J Biol Chem *279*:1541–1545

Kuang Q, Purhonen P, Ålander J, Svensson R, Hoogland V, Winerdal J, Spahiu L, Ottosson-Wadlund A, Jegerschöld C, Morgenstern R, Hebert H (2017) Dead-end complex, lipid interactions and catalytic mechanism of microsomal glutathione transferase 1, an electron crystallography and mutagenesis investigation. Sci Rep 7(1):7897. https://doi.org/10.1038/s41598-017-07912-3. PubMed PMID: 28801553; PubMed Central PMCID: PMC5554250

Kurochkina N (2007) J Theor Biol 247:110–121

Kurochkina N (2008) J Theor Biol 255:188–198

Kurochkina N (2011) Common structural characteristics of fibrous and globular proteins. In: Haggerty LM (ed) Protein Structure. Nova Science Publishers, Inc, Hauppauge. https://www.springer.com/us/book/9783319200972

Kurochkina N, Choekyi T (2011) Helix-helix interfaces and ligand binding. J Theor Biol 283:92–102

Ladner JE, Parsons JF, Rife CL, Gilliland GL, Armstrong RN (2004) Parallel evolutionary pathways for glutathione transferases: structure and mechanism of the mitochondrial class kappa enzyme rGSTK1-1. Biochemistry 43(2):352–361. PubMed PMID: 14717589

Lance E, Arnich N, Maignien T, Biré R (2018) Occurrence of β N-methylamino-l-alanine (BMAA) and Isomers in Aquatic Environments and Aquatic Food Sources for Humans. Toxins (Basel) 10 (2):pii: E83. https://doi.org/10.3390/toxins10020083. Review.PubMed PMID: 29443939; PubMed Central PMCID: PMC5848184

Lasker MV, Gajjar MM, Nair SK (2005) Molecular Structure of the IL-1R-Associated Kinase-4 Death Domain and Its Implications for TLR signaling. J Immun 175:4175–4179

Lee C-H, Kim M-S, Chung BM, Leahy DJ, Coulombe PA (2012) Structural basis for heteromeric assembly and perinuclear organization of keratin filaments. Nat Struct Mol Biol 19(7). https://doi.org/10.1038/nsmb.2330

Lee B, Richards FM (1971) The interpretation of protein structures: estimation of static accessibility. J Mol Biol 55:379–400

Lee H, Song C, Hong YS, Kim MS, Cho HR, Kang T, Shin K, Choi SH, Hyeon T, Kim DH (2017) Wearable/disposable sweat-based glucose monitoring device with multistage transdermal drug delivery module. Sci Adv 3(3):e1601314. doi:10.1126/sciadv.1601314. eCollection 2017 Mar. PubMed PMID: 28345030; PubMed Central PMCID: PMC5342654

Lee LK, Stewart AG, Donohoe M, Bernal RA, Stock D (2010) The structure of the peripheral stalk of Thermus thermophilus H+-ATPase/synthase. Nat Struct Mol Biol. 17(3):373–378. https://doi.org/10.1038/nsmb.1761. Epub 2010 Feb 21. PubMed PMID: 20173764; PubMed Central PMCID: PMC2912985

Leung K. (2010) (R)-3-[(18)F]Fluoro-2-methyl-2-N-(methylamino)propanoic acid. [updated 2010 Jul 29]. Molecular Imaging and Contrast Agent Database (MICAD) [Internet]. Bethesda (MD): National Center for Biotechnology Information (US); 2004-2013. Available from http://www.ncbi.nlm.nih.gov/books/NBK45198/ PubMed PMID: 20662134.

Li Y, Zhou M, Hu Q, Bai XC, Huang W, Scheres SH, Shi Y (2017) Mechanistic insights into caspase-9 activation by the structure of the apoptosome holoenzyme. Proc Natl Acad Sci U S A 114(7):1542–1547. https://doi.org/10.1073/pnas.1620626114. Epub 2017 Jan 31. PubMed PMID: 28143931; PubMed Central PMCID: PMC5320974

Lin S-C, Lo Y-C, Wu H (2010) Helical assembly in the MyD88–IRAK4–IRAK2 complex in TLR/IL-1R signaling. Nature 465:885–890

Littler DR, Harrop SJ, Brown LJ et al (2008) Comparison of vertebrate and invertebrate CLIC proteins: the crystal structures of Caenorhabditis elegans EXC-4 and Drosophila melanogaster DmCLIC. Proteins 71:364–378

Liu J, Deng Y, Dey AK, Moore JP, Lu M (2009) Structure of the HIV-1 gp41membrane-proximal ectodomain region in a putative prefusion conformation. Biochemistry 48(13):2915–2923. https://doi.org/10.1021/bi802303b. PubMed PMID: 19226163; PubMed Central PMCID: PMC2765501

Liu J, Taylor DW, Krementsova EB, Trybus KM, Taylor KA (2006a) Three-dimensional structure of the myosin V inhibited state by cryoelectron tomography. Nature 442(7099):208–211. Epub 2006 Apr 16.

Liu J, Deng Y, Zheng Q, Cheng CS, Kallenbach NR, Lu M (2006b) A parallel coiled-coil tetramer with offset helices. Biochemistry 45(51):15224–15231. Epub 2006b Nov 29. PubMed PMID: 17176044

Lovejoy B, Choe S, Cascio D, McRorie DK, DeGrado WF, Eisenberg D (1993) Crystal structure of a synthetic triple-stranded alpha-helical bundle. Science 259:1288–1293

Lovejoy B, Le TC, Luthy R, Cascio D, O'Neil KT, DeGrado WF, Eisenberg D (1992) X-ray grade crystals of a designed a-helical coiled coil. Protein Sci 1:956–957

Malashkevich VN, Schneider BJ, McNally ML, Milhollen MA, Pang JX, Kim PS (1999) Core structure of the envelope glycoprotein GP2 from Ebola virus at 1.9-A resolution. Proc Natl Acad Sci U S A. 96(6):2662–2667. PubMed PMID: 10077567; PubMed Central PMCID: PMC15825

Martins DB, Vieira MR, Fadel V, Santana VAC, Guerra MER, Lima ML, Tempone AG, Dos Santos Cabrera MP (2017) Membrane targeting peptides toward antileishmanial activity: design, structural determination and mechanism of interaction. Biochim Biophys Acta pii: S0304-4165(17):30248–30249. https://doi.org/10.1016/j.bbagen.2017.08.003. [Epub ahead of print]

Masood R, Ullah K, Ali H, Ali I, Betzel C, Ullah A (2018) Spider's venom phospholipases D: A structural review. Int J Biol Macromol 107(Pt A):1054–1065. https://doi.org/10.1016/j.ijbiomac.2017.09.081. Epub 2017 Sep 23. Review. PubMed PMID: 28951301

Medved L, Nieuwenhuizen W (2003) Molecular mechanisms of initiation of fibrinolysis by fibrin. Thromb Haemost. 2003 Mar;89(3):409–419.

Moitra J, Szilak L, Krylov D, Vinson C (1997) Leucine is the most stabilizing aliphatic amino acid in the d position of a dimeric leucine zipper coiled coil. Biochemistry 36:12567–12573

Moll JR, Ruvinov SB, Pastan I, Vinson C (2001) Designed heterodimerizing leucine zippers with a ranger of pI's and stabilities up to 10–15 M. Protein Sci 10:649–655

Mykhailiuk PK, Kubyshkin V, Bach T, Budisa N (2017) Peptidyl-prolyl model study: how does the electronic effect influence the amide bond Conformation? J Org Chem. https://doi.org/10.1021/acs.joc.7b00803. [Epub ahead of print]

Namba K, Stubbs G (1986) Structure of tobacco mosaic virus at 3.6 Angstroms resolution. Implications for assembly. Science 231:1401–1406

Namba K, Pattanayek R, Stubbs G (1989) Visualization of protein-nucleic acid interactions in a virus. Structure of intact tobacco mosaic virus at 2.9 Angstrom resolution by X-ray fiber diffraction. J Mol Biol 208:307–325

Navarro E, Fenude E, Celda B (2002) Solution structure of a D, L-alternating oligonorleucine as a model of double-stranded antiparallel beta-helix. Biopolymers 64(4):198–209

Navarro E, Fenude E, Celda (2004) Conformational and structural analysis of the equilibrium between single- and double-strand beta-helix of a D,L-alternating oligonorleucine. Biopolymers 73(2):229–241

Navarro E, Tejero R, Fenude E, Celda B (2001) Solution NMR structure of a D,L-alternating oligonorleucine as a model of beta-helix. Biopolymers 59(2):110–119

Newberry RW, Raines RT (2017) 4-Fluoroprolines: conformational analysis and effects on the stability and folding of peptides and proteins. Top Heterocycl Chem 48:1–25. https://doi.org/10.1007/7081_2015_196

Oshaben KM, Horne WS (2014) Tuning assembly size in Peptide-based supramolecular polymers by modulation of subunit association affinity. Biomacromolecules 15(4):1436–1442. https://doi.org/10.1021/bm5000423. Epub 2014 Mar 17

O'Shea EK, Klemm JD, Kim PS, Alber T (1991) Crystal structure of GCN4 leucine zipper, a two-stranded parallel coiled coil. Science 254:539

Paliakasis CD, Kokkinidis M (1991) The stability of the four-α-helix bundle motif in proteins. Protein Eng 4:849

Park HH, Logette E, Raunser S, Cuenin S, Walz T, Tschopp J, Wu H (2007) Death domain assembly mechanism revealed by crystal structure of the oligomeric PIDDosome core complex. Cell 128:533–546

Parry DAD, Fraser RD, Squire JM (2008) Fifty years of coiled-coils and α-helical bundles: a close relationship between sequence and structure. J Struct Biol 163:258–269

Passon DM, Lee M, Rackham O, Stanley WA, Sadowska A, Filipovska A, Fox AH, Bond CS (2012) Structure of the heterodimer of human NONO and paraspeckle protein component 1 and analysis of its role in subnuclear body formation. Proc Natl Acad Sci USA 109(13):4846–4850. https://doi.org/10.1073/pnas.1120792109. Epub 2012 Mar 13. PubMed PMID: 22416126; PubMed Central PMCID: PMC3324020

Pauling L, Corey RB, Branson HR (1951) The structure of proteins: two hydrogen-bonded helical configurations of the polypeptide chain. Proc Natl Acad Sci USA 37:205

Pechik I, Madrazo J, Mosesson MW, Hernandez I, Gilliland GL, Medved L (2004) Crystal structure of the complex between thrombin and the central "E" region of fibrin. Proc Natl Acad Sci USA 101(9):2718–2723. https://doi.org/10.1073/pnas.0303440101

Pielak RM, Schnell JR, Chou JJ (2009) Mechanism of drug inhibition and drug resistance of influenza A M2 channel. Proc Natl Acad Sci USA 106:7379–7384

Pitman KA, Borgland SL, MacLeod B, Puil E (2015) Isovaline does not activate GABA (B) receptor-coupled potassium currents in GABA(B) expressing AtT-20 cells and cultured rat hippocampal neurons. PLoS One 10(2):e0118497. https://doi.org/10.1371/journal.pone.0118497. eCollection 2015. PubMed PMID: 25706125; PubMed Central PMCID: PMC4337901

Plecs JJ, Harbury PB, Kim PS, Alber T (2004) Structural test of the parameterized-backbone method for protein design. J Mol Biol 342(1):289–297. PubMed PMID: 15313624.

Ponder JW, Richards FM (1987) Tertiary templates for proteins. J Mol Biol 193:775

Raghavender US, Kantharaju Aravinda S, Shamala N, Balaram P (2010) Hydrophobic peptide channels and encapsulated water wires. J Am Chem Soc 132(3):1075–1086. https://doi.org/10.1021/ja9083978

Ramachandran GN, Ramakrishnan C, Sasisekharan V (1963) Stereochemistry of polypeptide chain configurations. J Mol Biol 7:95–99

Reinert ZE, Lengyel GA, Horne WS (2013) Protein-like tertiary folding behavior from heterogeneous backbones.J Am Chem Soc 135(34):12528–12531. https://doi.org/10.1021/ja405422v. Epub 2013 Aug 15

Richards FM (1977) Areas, volumes, packing, and protein structure. Ann Rev Biophys Bioeng 6:151–176

Ruba A, Yang W (2016) O-GlcNAc-ylation in the nuclear pore complex. Cell Mol Bioeng 9 (2):227–233. https://doi.org/10.1007/s12195-016-0440-0. Epub 2016 Apr 26. PubMed PMID: 28638491; PubMed Central PMCID: PMC5475274

Rubinson EH, Gowda AS, Spratt TE, Gold B, Eichman BF (2010) An unprecedented nucleic acid capture mechanism for excision of DNA damage. Nature 468:406–411

Le Rumeur E, Hubert JF, Winder SJ (2012) A new twist to coiled coil. FEBS Lett 586 (17):2717–2722. https://doi.org/10.1016/j.febslet.2012.05.004. Epub 2012 May 11. Review. PubMed PMID: 22584055

Sakai H (2017) Overview of potential clinical applications of Hemoglobin Vesicles (HbV) as artificial red cells, evidenced by preclinical studies of the academic research consortium. J Funct Biomater 8(1):10. https://doi.org/10.3390/jfb8010010

Salnikov ES, Anantharamaiah GM, Bechinger B (2018) Supramolecular organization of apolipo-protein-A-I-derived peptides within disc-like arrangements. Biophys J 115(3):467–477. https://doi.org/10.1016/j.bpj.2018.06.026. Epub 2018 Jul 11. PubMed PMID: 30054032; PubMed Central PMCID: PMC6085177

Samatey FA, Imada K, Nagashima S, Vonderviszt F, Kumasaka T, Yamamoto M, Namba K (2001) Structure of the bacterial flagellar protofilament and implications for a switch for supercoiling. Nature 410:331–337

Schrauber H, Eisenhaber F, Argos P (1993) Rotamers: to be or not to be? An analysis of amino acid side-chain conformations in globular proteins. J Mol Biol 230(2):592–612

Scott FL, Stec B, Pop C, Dobaczewska MK, Lee JJ, Monosov E, Robinson H, Salvesen GS, Schwarzenbacher R, Riedl SJ (2009) The Fas/FADD death domain complex structure unravels signaling by receptor clustering. Nature 457:1019–1022

Shao C, Zhang F, Kemp MM, Linhardt RJ, Waisman DM, Head JF, Seaton BA (2006) Crystallo-graphic analysis of calcium-dependent heparin binding to Annexin A2. J Biol Chem 281:31689–31695

Sharma K, Dhillon A, Goyal A (2018) Insights into structure and reaction mechanism of β-mannanases. Curr Protein Pept Sci 19(1):34–47. https://doi.org/10.2174/1389203717666161013115724. Review. PubMed PMID: 27739373

Sheehan D, Meade G, Foley VM, Dowd CA (2001) Structure, function and evolution of glutathione transferases: implications for classification of non-mammalian members of an ancient enzyme superfamily. Biochem J 360:1–16

Sheriff S, Hendrickson WA, Smith JL (1987) Structure of myohemerythrin in the azidomet state at 1.7/1.3 angstroms resolution. J Mol Biol 197:273–296

Shoulders MD, Raines RT (2009) Collagen structure and stability. Annu Rev Biochem 78:929–958. https://doi.org/10.1146/annurev.biochem.77.032207.120833

Smith AI, Lew RA, Shrimpton CN, Evans RG, Abbenante G (2000) A novel stable inhibitor of endopeptidases EC 3.4.24.15 and 3.4.24.16 potentiates bradykinin-induced hypotension. Hypertension 35(2):626–630. PubMed PMID: 10679508

Song X, Zhang Y, Wang Y (2011) Antimicrobial peptides peptaibols from trichoderma—a review. Wei Sheng Wu Xue Bao 51(4):438–444. Review. Chinese. PubMed PMID: 21796977

Spyroulias GA, Papazacharias S, Pairas G, Cordopatis P (2002) Monitoring the structural consequences of Phe12—>D-Phe and Leu15—>Aib substitution in human/rat corticotropin releasing hormone. Implications for design of CRH antagonists. Eur J Biochem 269(24):6009–6019. PubMed PMID: 12473096

Squire JM, Paul DM, Morris EP (2017) Myosin and actin filaments in muscle: structures and interactions. Subcell Biochem 82:319–371. https://doi.org/10.1007/978-3-319-49674-0_11

Stanfield R, Cabezas E, Satterthwait A, Stura E, Profy A, Wilson I (1999) Dual conformations for the HIV-1 gp120 V3 loop in complexes with different neutralizing fabs. Structure 7(2):131–142. PubMed PMID: 10368281

Stetefeld J, Jenny M, Schulthess T, Landwehr R, Engel J, Kammerer RA (2000) Crystal structure of a naturally occurring parallel right-handed coiled coil tetramer. Nat Str Biol 7:772–776

Stevens J, Corper AL, Basler CF, Taubenberger JK, Palese P, Wilson IA (2004) Structure of the uncleaved human H1 hemagglutinin from the extinct 1918 influenza virus. Science 303:1866–1870

Strelkov SV, Herrmann H, Geisler N, Wedig T, Zimbelmann R, Aebi U, Burkhard P (2002) Conserved segments 1A and 2B of the intermediate filament dimer: their atomic structures and role in filament assembly. EMBO J 21(6):1255–1266. PubMed PMID: 11889032; PubMed Central PMCID: PMC125921

Subbalakshmi C, Basak P, Nagaraj R (2017) Self-assembly of t-Butyloxycarbonyl protected dipeptide methyl esters composed of leucine, isoleucine and valine into highly organized structures from alcohol and aqueous alcohol mixtures. Biopolymers https://doi.org/10.1002/bip.23033. [Epub ahead of print]

Sudha G, Singh P, Swapna LS, Srinivasan N (2015) Weak conservation of structural features in the interfaces of homologous transient protein–protein complexes. Protein Sci A Publ Protein Soc 24(11):1856–1873. https://doi.org/10.1002/pro.2792

Sutton RB, Fasshauer F, Jahn R, Brunger AT (1998) Crystal structure of a SNARE complex involved in synaptic exocytosis at 2.4 Å resolution. Nature 395:347–353

Taylor KC, Buvoli M, Korkmaz EN, Buvoli A, Zheng Y, Heinze NT, Cui Q, Leinwand LA, Rayment I (2015) Skip residues modulate the structural properties of the myosin rod and guide thick filament assembly. Proc Natl Acad Sci USA 112(29):E3806–E3815. https://doi.org/10.1073/pnas.1505813112. Epub 2015 Jul 6. PubMed PMID: 26150528; PubMed Central PMCID: PMC4517226

Touw DS, Nordman CE, Stuckey JA, Pecorano VL (2007) Identifying important structural characteristics of arsenic resistance proteins by using designed three-stranded coiled-coils. Proc Natl Acad Sci 104:11969–11974

Tseng T-S, Wang S-H, Chang T-W, Wei H-M, Wang Y-J, Tsai K-C et al (2016) Sarkosyl-induced helical structure of an antimicrobial peptide GW-Q6 plays an essential role in the binding of surface receptor OprI in Pseudomonas aeruginosa. PLoS ONE 11(10):e0164597. https://doi.org/10.1371/journal.pone.0164597

Vasudev PG, Ananda K, Chatterjee S, Aravinda S, Shamala N, Balaram P (2007) Hybrid peptide design. Hydrogen bonded conformations in peptides containing the stereochemically constrained gamma-amino acid residue, gabapentin. J Am Chem Soc 129(13):4039–4048. Epub 2007 Mar 10

Vasudev PG, Chatterjee S, Shamala N, Balaram P (2009) Gabapentin: a stereochemically constrained gamma amino acid residue in hybrid peptide design. Acc Chem Res 42 (10):1628–1639. https://doi.org/10.1021/ar9001153. Review. PubMed PMID: 19572698

Venanzi M, Gatto E, Formaggio F, Toniolo C (2017) The importance of being Aib. Aggregation and self-assembly studies on conformationally constrained oligopeptides. J Pept Sci 23(2):104-116. https://doi.org/10.1002/psc.2956. Epub 2017 Jan 5. Review. PubMed PMID: 28054413

Very N, Vercoutter-Edouart AS, Lefebvre T, Hardivillé S, El Yazidi-Belkoura I (2018 Oct 9) Cross-Dysregulation of O-GlcNAcylation and PI3K/AKT/mTOR Axis in Human Chronic Diseases. Front Endocrinol (Lausanne). 9:602. doi: 10.3389/fendo.2018.00602. eCollection 2018. Review. PubMed PMID: 30356686; PubMed Central PMCID: PMC6189293

Vitagliano L, Berisio R, Mazzarella L, Zagari A (2001) Structural bases of collagen stabilization induced by proline hydroxylation. Biopolymers 58(5):459–464

Vollmar M, Schlieper D, Winn M, Büchner C, Groth G (2009) Structure of the c14 rotor ring of the proton translocating chloroplast ATP synthase. J Biol Chem 284(27):18228–18235. https://doi.org/10.1074/jbc.M109.006916. Epub 2009 May 7. PubMed PMID: 19423706; PubMed Central PMCID: PMC2709358

Yadav MK, Redman JE, Leman LJ, Alvarez-Gutiérrez JM, Zhang Y, Stout CD, Ghadiri MR (2005) Structure-based engineering of internal cavities in coiled-coil peptides. Biochemistry 44 (28):9723–9732. PubMed PMID: 16008357; PubMed Central PMCID: PMC1779508

Yang J, Ma YQ, Page RC, Misra S, Plow EF, Qin J (2009) Structure of an integrin alphaIIb beta3 transmembrane-cytoplasmic heterocomplex provides insight into integrin activation. Proc Natl Acad Sci USA 106(42):17729–17734. https://doi.org/10.1073/pnas.0909589106. Epub 2009 Oct 1. PubMed PMID: 19805198; PubMed Central PMCID: PMC2764936

Young SC (2018) A systematic review of antiamyloidogenic and metal-chelating peptoids: two structural motifs for the treatment of Alzheimer's disease. Molecules 23(2). pii: E296. https://doi.org/10.3390/molecules23020296. Review. PubMed PMID: 29385058

Wada SI, Takesada A, Nagamura Y, Sogabe E, Ohki R, Hayashi J, Urata H (2017) Structure-activity relationship study of Aib-containing amphipathic helical peptide-cyclic RGD conjugates as carriers for siRNA delivery. Bioorg Med Chem Lett;27(24):5378–5381. https://doi.org/10.1016/j.bmcl.2017.11.018. Epub 2017 Nov 10. PubMed PMID: 29157863

Wagschal K, Lavigna P, Mant C, Hodges RS (1999) The role of position a in determining the stability and oligomerization state of a-helical coiled coils: 20 amino acid stability coefficients in the hydrophobic core of proteins. Protein Sci 8:2312–2329

Weber CH, Vincenz C (2001) The death domain superfamily: a tale of two interfaces? TIBS 26:475–481

Wells WW, Yang Y, Deits TL, Gan ZR (1993) Thioltransferases. Adv Enzymol Relat Areas Mol Biol;66:149–201. Review. PubMed PMID: 8430514

Woolfson DN (2010) Building fibrous biomaterials from -helical and collagen-like coiled-coil peptides. Biopolymers 94:118–127

Xie L, Yang S (2016) Brain globins in physiology and pathology. Med Gas Res 6(3):154–163. https://doi.org/10.4103/2045-9912.191361

Yeo HJ, Yokoyama T, Walkiewicz K, Kim Y, Grass S, Geme JW (2007) The structure of the Haemophilus influenzae HMW1 pro-piece reveals a structural domain essential for bacterial two-partner secretion. J Biol Chem 282(42):31076–31084. Epub 2007 Aug 14. PubMed PMID: 17699157

Yin Z, Shi K, Banerjee S, Pandey KK, Bera S, Grandgenett DP, Aihara H (2016) Crystal structure of the Rous sarcoma virus intasome. Nature 530(7590):362–366. https://doi.org/10.1038/nature16950. PubMed PMID: 26887497; PubMed Central PMCID: PMC4881392

Zhang Y (2017) Energetics, kinetics, and pathway of SNARE folding and assembly revealed by optical tweezers. Protein Sci;26(7):1252–1265. https://doi.org/10.1002/pro.3116. Epub 2017 Mar 8. Review. PubMed PMID: 28097727; PubMed Central PMCID: PMC5477538.

Zhao H, Qin X, Yang D, Jiang Y, Zheng W, Wang D, Tian Y, Liu Q, Xu N, Li Z (2017) The development of activatable lytic peptides for targeting triple negative breast cancer. Cell Death Discov 3:17037. https://doi.org/10.1038/cddiscovery.2017.37. eCollection 2017. PubMed PMID: 29263848; PubMed Central PMCID: PMC5629628

Zhou GP (2011) The structural determinations of the leucine zipper coiled-coil domains of the cGMP-dependent protein kinase I alpha and its interaction with the myosin binding subunit of the myosin light chains phosphase. Proteins Pept Lett 18:966–978

Chapter 2
Conformational Transitions

Abstract Protein molecules are dynamic bodies. They form and disassemble complexes, channel substrates, and rotate, changing conformational states to perform their functions. Small motions in the enzyme active site, large rearrangements of subunits and sliding along the filaments involve conformations changes.

Conformational transitions occur in many forms: change in orientation of hemoglobin subunits in oxygenated/deoxygenated states, proteolytic cleavage and conversion of precursors to mature molecules of fibrin/fibrinogen, helix coil transitions or changes between left handed and right handed orientation of helices.

The transitions of the molecule or molecular assembly largely depend on conformational changes of the constituent parts, amino acids. Rotameric states of the backbone and side chains change and these changes add up and propagate to provide large scale motions of the molecular parts. Micro motions and macro motions provide basis for conformational transitions.

Keywords Protein conformation · Amino acid · Dynamics · Switch · Fluctuation

To perform their functions proteins act as flexible molecules that constantly make transitions and switch between various conformational states. Multiprotein assemblies in biological processes work in joint action of many components and execute consorted motions performing their function.

Secondary structure elements can change conformation from regular hydrogen bonded to disordered or from one secondary structure type to another. Microtubule-stabilizing agents (MSAs) zampanolide and epothilone A used for cancer treatment bind to the taxane pocket of ß-tubulin and induce conformational change of the M-loop into a short helix interacting by their side chains (Prota et al. 2013). Ends of secondary structure elements in solution constantly undergo transitions between regular hydrogen bonded and disordered conformation. We can see these transitions in ensembles of fluctuating NMR structures of proteins which exhibit stable immobile core and unwinding ends. In crystallographic structures, helix ends also order upon ligand binding as in GST (Kurochkina and Choekyi 2011) or upon

N. Kurochkina, *Protein Structure and Modeling*,
https://doi.org/10.1007/978-981-13-6601-7_2

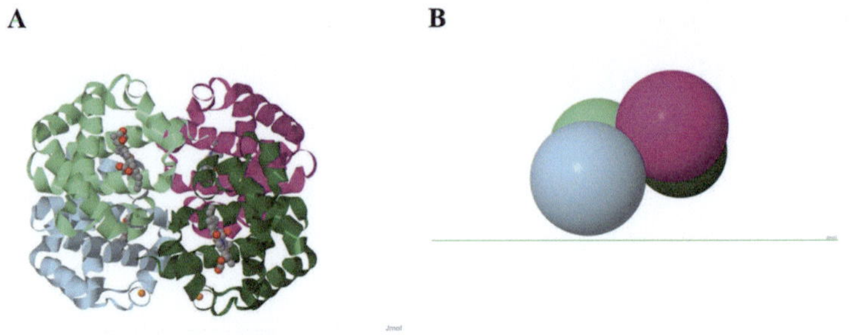

Fig. 2.1 Hemoglobin tetramer in complex with heme /4hhb /. (A) Tertiary structure and (B) subunit arrangement

posttranslational modifications such as phosphorylation/dephosphorylation of TRPA1 (Hall et al. 2018).

Hemoglobin Hemoglobin (Fig. 2.1) carries oxygen in blood. It consists of four subunits, each subunit binding heme group. Low oxygen concentration transforms hemoglobin to tense (T) state whereas high oxygen concentration to relaxed (R) state (Fermi et al. 1984).

Mutation on the surface of hemoglobin asparagine (polar) to valine (nonpolar) causes polymerization and sickle cell disease. Stabilization of state transition is very important for the treatment of sickle cell anemia. Mutated sickle hemoglobin in T (deoxygenated) state polymerizes. Sickle erythrocytes are rigid and exhibit increased adhesion and impaired blood flow in vessels which causes pain and organ damage. Agents that stabilize the hemoglobin in higher oxygen affinity R-state and/or destabilize the lower oxygen affinity T-state are used to treat patients with sickle cell disease (Oder et al. 2016).

Trimeric autotransporter adhesins (TAAs) Transitions between left- and right-handed super coiled forms are observed in trimeric autotransporter adhesins (Alvarez et al. 2010). TAAs, pathogenicity factors in proteobacteria, contain a conserved membrane anchor through the outer membrane as 12-stranded beta-barrel which is a left-handed parallel β-roll. The proteins pass through the pore of this barrel. After the transport is complete, the pore is occluded by a trimeric coiled coil. This domain exhibits canonical heptad (7/2) sequence periodicity followed by another coiled coil which has pentadecad (15/4) periodicity. Crystallographic structure shows that the transition between these two periodicities in the coiled-coil stalk of the *Yersinia* adhesin YadA results in transition between left handed and right handed coiled coil conformations (Alvarez et al. 2010).

The *Escherichia coli* TAA, Ig-binding protein (Fig. 2.2) contains both a YadA-like head domain and the coiled-coil domain (stalk). In addition, it contains an insert as minidomain which introduces 120 degrees rotation eliminating the twist between the stalk N terminal right-handed superhelix, and C terminal left handed one. Folding of the domain is important for making zipper-like contacts between cells. Protein forms a fibrillar layer covering the cell surface and necessary for biofilm

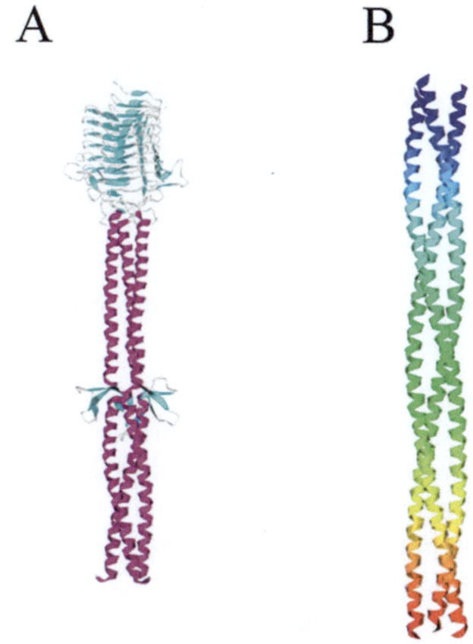

Fig. 2.2 Transitions between (A) LH and (B) RH coiled coil in trimeric autotransporter adhesin / 2xqh 2xzr /

formation (Leo et al. 2011). Autoagglutination and biofilm formation are key factors that determine bacterial virulence.

TAAs bind ligands in its hydrophilic core. Chloride ions are coordinated by three pointing inward glutamine or asparagine residues in position **d** of the coiled-coil. Asn in position **d** also binds anions in SadA and UspA1. Unique to Ig-binding protein is chlorine binding by threonine in position **a** of the RTD sequence. RxD motif is recognized for organizing a polar interior network in left-handed coiled coil. Two residues in position **d**, H421 and Q428, coordinate the water molecule and form a three-way hydrogen bond. Also, two sequences YxD contain Thr: in the position **a** and in the position **h**.

Molecular machines such as kinesin and myosin *biomotors* support complex mechanical operations in cellular processes of cell division and intracellular transport. Potential of these devices is explored for applications in artificial nanosystems (van den Heuvel and Dekker 2007). Molecular motors perform the movement of intracellular vesicles, chromosomes, and messenger RNA–protein complexes on tracks of actin filaments and microtubules. Allostery or communications in these enzymes occurs at large distances of several nanometer. Amazing coordination of the chemical events in the active site with events of binding and release at the track along which substrate moves occurs. Small movements in the active site are amplified to generate large scale movements (Spudich et al. 2011).

Dynein and *kinesin* move cargo along microtubules in opposite directions (Chap. 6). Dynein major components include a motor with six AAA+ domains (AAA1 to AAA6) and four extensions and stalk interacting with microtubules via microtubule binding domain (MTBD). Four of the AAA+ domains are catalytic and two are accessory. ATP hydrolysis induces stalk helix sliding. Stalk helix slides to release the dynein motor from the microtubule. The two helices of the stalk (CC1 and

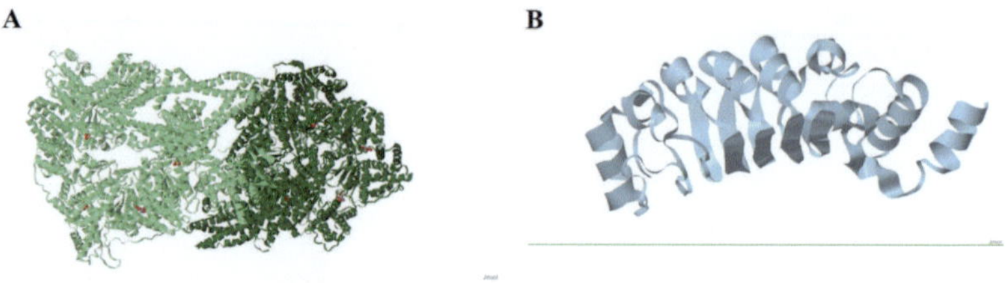

Fig. 2.3 Dynein motor domain in complex with ATP /4akg / (A) and outer arm dynein light chain 1 /1ds9/ (B)

CC2) provide sliding motion. A sliding movement of CC2 relative to CC1 is induced by pulling of the extension domain. It causes bulging of CC2 leading to the low microtubule affinity state of the microtubule binding domain (MTBD). These two helices in a straight conformation associate with strong affinity state of MTBD toward extension domain. When ATP binds to AAA1, AAA1 adopts the closed ring conformation, AAA6L/AAA5S rotate and transfer the motion to the extension domain, the MTBD dissociate from microtubule and the linker bending results in the contact of its N-terminus with AAA2/AAA3. After ATP-hydrolysis, the dynein motor rebinds to the microtubule and the linker performs the power stroke. The linker transition from a bent to a straight conformation generates the force for cargo movement (Schmidt and Carter 2016). Alternative model of this process studied by electron micrographs suggests that heads and stalk move relative to the tail (Ueno et al. 2008).

The dynein motor (Fig. 2.3A), a ring of six AAA+ subunits, transfers motion to all parts of the molecule. Dynein multiprotein assembly contains one to three heavy chains and numerous accessory chains. The N-terminal tail part of the heavy chain is the site of cargo binding and assembly platform for the accessory chains (Schmidt and Carter 2016).

Chlamydomonas flagellar outer arm dynein, light chain 1 (LC1) associates with the nucleotide binding region within the gamma heavy chain motor domain (Fig. 2.3B). The C-terminal domain of LC1, which consists of six leucine-rich repeats that folds as a cylindrical right handed spiral is suggested to interact with the assembly site of the heavy chain (Wu et al. 2000). *Chlamydomonas* ciliary and flagella motions require phosphorylation of inner arm dynein fI1 and ankyrin repeats protein (Ikeda et al. 2009).

Kinesin-5 (K5) motors move cargoes along microtubule during cell division in ATP dependent manner. Binding to microtubules initiates conformational transition in the regions surrounding nucleotide binding (Bodey et al. 2009).

Rotary motors of *Thermotoga maritima* consist of approximately 25 types of proteins located in the cell membrane and use proton gradient as energy source. Protein complex comprising FliG, FliM and FliN is a component of rotation and clockwise/counter-clockwise switching in the bacterial flagellum. It is central for the bacterium flagellar assembly. The middle and C-terminal domains of FliG comprising two globular domains with helix as a linker and extended segment containing

Gly-Gly motif are coordinated by FliM that regulates the relative orientation of the domains suggesting structural components involved in switching motion of the flagellar. FliG structure contains three helix modules that resemble ARM repeats of β catenin. Hydrophobic residues in the core of the assembly are particularly important for function of the motor (Brown et al. 2002).

Proteins containing helix repeats ankyrin, ARM/HEAT, PUM, TAL, and other are toroids in shape. Stacked repeat units assemble as helix of helices. These regular arrangements can exhibit shape of a cylinder, disc, ring, or spiral (Chap. 5). Conformations of the individual amino acids add up to determine shape of the entire assembly.

One of the functions that require repeats flexibility is binding ligands. Armadillo repeats of β catenin, transcription coactivator, provide groove binding surface for such ligands as cadherins, Tcf/LEF1, and APC. Tcfs transcription factors induce DNA bends. How LEF1 bends DNA also depends on the complex with catenin.

Catenin/cadherin complex is the link to actin cytoskeleton, tyrosine phosphorylation, growth factor signaling, regulation of cell adhesion, mobility, and polarity (Huber et al. 1997).

In helical assemblies, change in shape is achieved by rearrangement of contacts on the surface of the two interacting helices. Also, main chain and side chains switch rotameric states in molecules that undergo conformational transition. Therefore, change of rotamers combined with rearrangement of contacting core residues occurs to provide conformations change on the molecular scale. Micromotions give rise to macromotions.

Helix repeats proteins β-karyopherin (Imp13) imports and exports cargoes in and out of the nucleus. Comparison of the Imp13–Ubc9 and Imp13–Mago-Y14 complexes shows the karyopherin possess remarkable plasticity. Its conformation shows a closed ring in one complex and an open superhelix in the second complex when bound to the two different cargoes (Grünwald et al. 2011 2xwu). Conformational changes occur in PP2A HEAT repeats on binding catalytic subunits to regulatory B subunits (Groves et al. 1999), ribonuclease inhibitor leucine rich repeat (Kobe and Deisenhoffer 1996) and other helical repeats.

TAL repeats containing protein BurrH wrapps around the DNA molecule which is in Bform. The protein changes conformation by compression 23 A° along the longitudinal axis upon DNA binding. The protein bends as a spiral around the DNA (Stella et al. 2014). TAL repeats proteins are very important for DNA editing. Accuracy of DNA editing and integrity of genome are very important for applications of this method in synthetic biology and medicine. Specific gene modifications and precise editing are the goals that can be achieved via specific protein–DNA interactions.

Prion proteins Formation of pathogenic β-sheet aggregates from α-helix prion proteins demonstrates polypeptide chain conformational transitions under various conditions (Zhou 2011; Zhou and Huang 2013).

TAU (Fig. 2.4) interactions with microtubules show short peptide fragments on the surface of microtubule. Tau is regulated during development. In axonal compartments it binds and stabilizes microtubules. Hyperphosphorylation stimulates

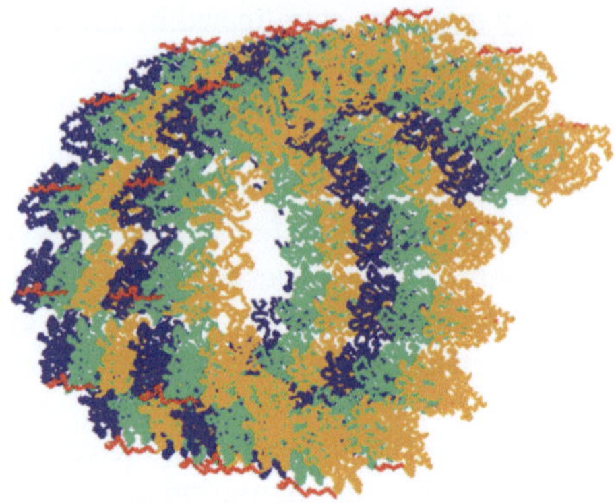

Fig. 2.4 Tau (red) complex with microtubules; tubulin subunits α and β alternate blue, green, and brown /6cvn/

TAU detachment from microtubules and aggregation. Formation of fibrils is implicated in Alzheimer's disease. Cryo-EM structure of tau/tubulin complex shows the tau tubulin-binding repeats bound to the surface of the protofilament in extended conformation, which stabilize the polymerization interfaces between tubulin dimers (Kellogg et al. 2018). Tubulin consists of two subunits, α and β. C terminus of subunit β subunit is important site of assembly (Joe et al. 2009).

Tau is a dynamic molecule that regulates microtubules polymerization. It obtains regular structure when bound to other molecules. Disordered fragments of tau are candidates for binding other ligands. Hyperphosphorylation of tau (Binukumar et al. 2013), conformational mobility associated with axonal transport inhibition (Cox et al. 2016), phosphorylation dependent cellular localization, assembly and abnormal aggregation of neurofilaments (Boumil et al. 2018) are processes associated with neurodegenerative diseases. Phosphorylation protects neurofilaments from proteolysis and stimulates calcium dependent assembly of neurofilaments and cytoskeleton in axons, whereas dephosphorylation leads to more dynamic chains which are able to repair stationary chains (Rudrabhatla et al. 2009). Cytoskeletal components regulate specificity, shape, rigidity, dynamics of association, and function of the multiprotein assemblies (Kurochkina et al. 2018).

Amino acid transporters (Chap. 4), transmembrane proteins, have outward facing and inward facing conformational states. For many transporters such as ASCT1, ASCT2, SLC1A4, AND SLC1A5 which transport alanine, serine, glutamate/glutamine, cysteine and other amino acids tertiary structure and conformational changes were extensively studied (Kaplan et al. 2018). The transporters of the glutamate/glutamine cycle are transmembrane proteins that use electrochemical gradient of sodium, potassium, and hydrogen ions between the membranes to transfer glutamate (Hayashi 2018). The transporter trimerization domain and transport domain move separately as rigid bodies during amino acid transport across the membrane. Glutamate transporter GluTPh transport domain moves 18 Angstrom toward the cytoplasm when switching between outward facing and inward facing conformational states. Similarly, aspartate transporter transport domain moves 15 Angstrom (Verdon et al. 2014; 4p19). In inward facing conformation, substrate binding site faces the cytoplasmic side of the membrane bilayer. When tips of substrate binding hairpins interfere with substrate binding site, the transporter is in inward occluded conformation. Hairpins movement opens the gate allowing the substrate to dissociate from the transporter into the cytoplasm. Each subunit of the trimeric transporter can adopt different conformations being on different stages of the transport process. The substrate binding is sodium dependent and sodium binding stabilizes helical hairpin at the extracellular gate in closed conformation. Outward facing conformation is favourable in the sodium/substrate bound state. Three dimensional structures of the glutamate (Verdon et al. 2014) and aspartate (Boudker et al. 2007) transporters at each step of the transport process show coupling of ion binding to substrate processing and conformational transitions of opening/closing the gate.

CLIC Cellular functions of CLICs involve conformational transitions and dimerization. CLIC exhibits conformational transitions between globular-soluble and membrane-bound forms. In CLIC1, local unfolding of the amino terminal domain in oxidized state and transition from monomer to dimer creates a channel. Also, the amino terminus can translocate through membrane (Littler et al. 2004). Environmental changes induce frequency of molecular fluctuations and tertiary structure rearrangements. Membrane induced oligomerization stabilizes the structure (Ferofontov et al. 2018). CLIC4 contains putative nuclear localization sequence that binds importin α and exhibit helix conformation in the native form but extended conformation in importin α bound form (Mynott et al. 2011).

Fibrinogen/fibrin Conformational transitions occur upon proteolysis. Exposure of specific assembly sites and fibrinogen to fibrin transition after thrombin cleavage of fibrinopeptides from six paired disulfide bonded chains $(A\alpha\ B\beta\ \gamma)_2$, polymerization of $(\alpha\ \beta\ \gamma)_2$ fibrin monomers into protofibrils and fiber formation are key processes that suggest important knob:hole and topology driven interactions (Zhmurov et al. 2016; Huang et al. 2014). Conformational changes in fibrin/fibrinogen were extensively studied with fragments derived from fibrinogen: fragment D, and fibrin D-D. E(1). Fibrin specific epitopes are not accessible in D or D-D fragments as tested by binding of monoclonal specific antibody, tPA, or plasminogen. The Alpha epitope becomes accessible after the beta- and gamma-modules are proteolytically removed from D. Fibrin specific epitopes become accessible in the D-D.E(1) complex; therefore, the DD.E interaction leads to their exposure. Fibrin assembly is driven by the interaction between complementary sites of the D and E fragments. After the D region changes conformation, plasminogen- and tPA-binding sites become accessible. These changes regulate fibrin assembly and fibrinolysis (Yakovlev et al. 2000).

References

Alvarez BH, Gruber M, Ursinus A, Dunin-Horkawicz S, Lupas AN, Zeth K (2010) A transition from strong right-handed to canonical left-handed supercoiling in a conserved coiled-coil segment of trimeric autotransporter adhesins. J Struct Biol 170(2):236–245. https://doi.org/10.1016/j.jsb.2010.02.009. Epub 2010 Feb 21. PubMed PMID: 20178846

Binukumar BK, Shukla V, Amin ND, Reddy P, Skuntz S, Grant P, Pant HC (2013) Topographic regulation of neuronal intermediate filaments by phosphorylation, role of peptidyl-prolyl isomerase 1: significance in neurodegeneration. Histochem Cell Biol 140(1):23–32. https://doi.org/10.1007/s00418-013-1108-7

Bodey AJ, Kikkawa M, Moores CA (2009) 9-Angström structure of a microtubule-bound mitotic motor. J Mol Biol 388(2):218–224. https://doi.org/10.1016/j.jmb.2009.03.008. Epub 2009 Mar 10. PubMed PMID: 19285086

Boudker O, Ryan RM, Yernool D, Shimamoto K, Gouaux E (2007) Coupling substrate and ion binding to extracellular gate of a sodium-dependent aspartate transporter. Nature 445 (7126):387–393. Epub 2007 Jan 17. PubMed PMID: 17230192

Boumil EF, Vohnoutka R, Lee S, Pant H, Shea TB (2018) Assembly and turnover of neurofilaments in growing axonal neurites. Biol Open 7(1). pii: bio028795). https://doi.org/10.1242/bio.028795

Brown PN, Hill CP, Blair DF (2002) Crystal structure of the middle and C-terminal domains of the flagellar rotor protein FliG. EMBO J 21(13):3225–3234. PubMed PMID: 12093724; PubMed Central PMCID: PMC126082

Cox K, Combs B, Abdelmesih B, Morfini G, Brady ST, Kanaan NM (2016) Analysis of isoform-specific tau aggregates suggests a common toxic mechanism involving similar pathological conformations and axonal transport inhibition. Neurobiol Aging 47:113–126. https://doi.org/10.1016/j.neurobiolaging.2016.07.015

Fermi G, Perutz MF, Shaanan B, Fourme R (1984) The crystal structure of human deoxyhaemoglobin at 1.74 A resolution. J Mol Biol 175(2):159–174. PubMed PMID: 6726807

Ferofontov A, Strulovich R, Marom M, Giladi M, Haitin Y (2018) Inherent flexibility of CLIC6 revealed by crystallographic and solution studies. Sci Rep 8(1):6882. https://doi.org/10.1038/s41598-018-25231-z. PubMed PMID: 29720717; PubMed Central PMCID: PMC5931990

Groves MR, Hanlon N, Turowski P, Hemmings BA, Barford D (1999) The structure of the PROTEIN Phosphatase 2A PR65/A Subunit Reveals the Conformation of Its 15 Tandemly Repeated HEAT Motifs. Cell *96*:99–110

Hall BE, Prochazkova M, Sapio MR, Minetos P, Kurochkina N, Binukumar BK, Amin ND, Terse A, Joseph J, Raithel SJ, Mannes AJ, Pant HC, Chung MK, Iadarola MJ, Kulkarni AB (2018) Phosphorylation of the Transient Receptor Potential Ankyrin 1 by Cyclin-dependent Kinase 5 affects Chemo-nociception. Sci Rep 8(1):1177. https://doi.org/10.1038/s41598-018-19532-6. PubMed PMID: 29352128; PubMed Central PMCID: PMC5775258

Hayashi MK (2018) Structure-Function Relationship of Transporters in the Glutamate-Glutamine Cycle of the Central Nervous System. Int J Mol Sci 19(4):pii: E1177. https://doi.org/10.3390/ijms19041177. Review. PubMed PMID: 29649168; PubMed Central PMCID: PMC5979278

Huang L, Hsiao JP, Powierza C, Taylor RM 2nd, Lord ST (2014) Does topology drive fiber polymerization? Biochemistry 53(49):7824–7834. https://doi.org/10.1021/bi500986z. Epub 2014 Dec 5. PubMed PMID: 25419972; PubMed Central PMCID: PMC4270379

Huber AH, Nelson WJ, Weis WI (1997) Three-Dimensional Structure of the Armadillo Repeat Region of *b*-Catenin. Cell *90*:871–882

Ikeda K, Yamamoto R, Wirschell M, Yagi T, Bower R, Porter ME, Sale WS, Kamiya R (2009) A novel ankyrin-repeat protein interacts with the regulatory proteins of inner arm dynein f (I1) of Chlamydomonas reinhardtii. Cell Motil Cytoskeleton 66(8):448–456. https://doi.org/10.1002/cm.20324. PubMed PMID: 19021242; PubMed Central PMCID: PMC3102495

Joe PA, Banerjee A, Ludueña RF (2009) Roles of beta-tubulin residues Ala428 and Thr429 in microtubule formation in vivo. J Biol Chem 284(7):4283–4291. https://doi.org/10.1074/jbc.M807491200. Epub 2008 Dec 13. PubMed PMID: 19074767; PubMed Central PMCID: PMC3837401

Kaplan E, Zubedat S, Radzishevsky I, Valenta AC, Rechnitz O, Sason H, Sajrawi C, Bodner O, Konno K, Esaki K, Derdikman D, Yoshikawa T, Watanabe M, Kennedy RT, Billard JM, Avital A, Wolosker H (2018) ASCT1 (Slc1a4) transporter is a physiologic regulator of brain d-serine and neurodevelopment. Proc Natl Acad Sci U S A. 115(38):9628–9633. https://doi.org/

10.1073/pnas.1722677115. Epub 2018 Sep 5. PubMed PMID: 30185558; PubMed Central PMCID: PMC6156681

Kellogg EH, Hejab NMA, Poepsel S, Downing KH, DiMaio F, Nogales E (2018) Near-atomic model of microtubule-tau interactions. Science 360(6394):1242–1246. https://doi.org/10.1126/science.aat1780. Epub 2018 May 10. PubMed PMID: 29748322

Kurochkina N, Bhaskar M, Yadav SP, Pant HC (2018) Phosphorylation, Dephosphorylation, and multiprotein assemblies regulate dynamic behavior of neuronal cytoskeleton: a mini-review. Front Mol Neurosci 11:373. https://doi.org/10.3389/fnmol.2018.00373. eCollection 2018. PubMed PMID: 30349458; PubMed Central PMCID: PMC6186834

Kurochkina N, Choekyi T (2011) Helix-helix interfaces and ligand binding. J Theor Biol 283:92–102

Leo JC, Lyskowski A, Hattula K, Hartmann MD, Schwarz H, Butcher SJ, Linke D, Lupas AN, Goldman A (2011) The structure of E. coli IgG-binding protein D suggests a general model for bending and binding in trimeric autotransporter adhesins. Structure 19(7):1021–1030. doi: https://doi.org/10.1016/j.str.2011.03.021. PubMed PMID: 21742268

Littler DR, Harrop SJ, Fairlie WD, Brown LJ, Pankhurst GJ, Pankhurst S, DeMaere MZ, Campbell TJ, Bauskin AR, Tonini R, Mazzanti M, Breit SN, Curmi PMG (2004) The intracellular chloride ion channel protein CLIC1 undergoes a redox-controlled structural transition. J Biol Chem *279*:9298–9305

Mynott AV, Harrop SJ, Brown LJ, Breit SN, Kobe B, Curmi PM (2011) Crystal structure of importin-a bound to a peptide bearing the nuclear localisation signal from chloride intracellular channel protein 4. FEBS J 278(10):1662–1675. https://doi.org/10.1111/j.1742-4658.2011.08086.x. Epub 2011 Mar 30. PubMed PMID: 21388519

Oder E, Safo MK, Abdulmalik O, Kato GJ (2016) New developments in anti-sickling agents: can drugs directly prevent the polymerization of sickle haemoglobin in vivo? Br J Haematol 175 (1):24-30. https://doi.org/10.1111/bjh.14264. Epub 2016 Sep 8. Review. PubMed PMID: 27605087; PubMed Central PMCID: PMC5035193

Prota AE, Bargsten K, Zurwerra D, Field JJ, Díaz JF, Altmann KH, Steinmetz MO (2013) Molecular mechanism of action of microtubule-stabilizing anticancer agents. Science 339 (6119):587–590. https://doi.org/10.1126/science.1230582. Epub 2013 Jan 3. PubMed PMID: 23287720

Rudrabhatla P, Albers W, Pant HC (2009) Peptidyl-Prolyl Isomerase 1 regulates protein phosphatase 2A-mediated topographic phosphorylation of neurofilament proteins. J Neurosci 29 (47):14869–14880. https://doi.org/10.1523/JNEUROSCI.4469-09.2009

Schmidt H, Carter AP (2016) Review: structure and mechanism of the dynein motor ATPase. Biopolymers 105(8):557–567. https://doi.org/10.1002/bip.22856

Stella S, Molina R, López-Méndez B, Juillerat A, Bertonati C, Daboussi F, Campos-Olivas R, Duchateau P, Montoya G (2014) BuD, a helix-loop-helix DNA-binding domain for genome modification. Acta Crystallogr D Biol Crystallogr 70(Pt 7):2042–2052. https://doi.org/10.1107/S1399004714011183. Epub e2014 Jun 29. PubMed PMID: 25004980; PubMed Central PMCID: PMC4089491

Ueno H, Yasunaga T, Shingyoji C, Hirose K (2008) Dynein pulls microtubules without rotating its stalk. Proc Natl Acad Sci U S A 105(50):19702–19707. https://doi.org/10.1073/pnas.0808194105. Epub 2008 Dec 8. PubMed PMID: 19064920; PubMed Central PMCID: PMC2604933

van den Heuvel MG, Dekker C (2007) Motor proteins at work for nanotechnology. Science 317 (5836):333–6. Review. PubMed PMID: 17641191.PMID: 17641191

Verdon G, Oh S, Serio RN, Boudker O (2014)Coupled ion binding and structural transitions along the transport cycle of glutamate transporters. Elife 3:e02283. https://doi.org/10.7554/eLife.02283. PubMed PMID: 24842876; PubMed Central PMCID: PMC4051121

Yakovlev S, Makogonenko E, Kurochkina N, Nieuwenhuizen W, Ingham K, Medved L (2000) Conversion of fibrinogen to fibrin: mechanism of exposure for tPA-plasminogen binding site. Biochemistry 39:15730

Wu H, Maciejewski MW, Marintchev A, Benashski SE, Mullen GP, King SM (2000) Solution structure of a dynein motor domain associated light chain. Nat Struct Biol 7(7):575–579. PubMed PMID: 10876244

Zhmurov A, Protopopova AD, Litvinov RI, Zhukov P, Mukhitov AR, Weisel JW, Barsegov V (2016) Structural basis of interfacial flexibility in fibrin oligomers. Structure 24(11):1907–1917. https://doi.org/10.1016/j.str.2016.08.009. Epub 2016 Sep 29. PubMed PMID: 27692965; PubMed Central PMCID: PMC5240993

Zhou GP (2011) The structural determinations of the leucine zipper coiled-coil domains of the cGMP-dependent protein kinase I alpha and its interaction with the myosin binding subunit of the myosin light chains phosphase. Proteins Pept Lett 18:966–978

Zhou GP, Huang RB (2013) The pH-Triggered Conversion of the PrP(c) to PrP(sc.). Curr Top Med Chem 13:1152–1163

Chapter 3
Enzymes

Abstract Enzymes increase rate of chemical reactions. Structure of the enzyme reveals how functional groups and the entire molecule work together to optimize this process. Structural data combined with kinetic data allow propose and refine catalytic mechanisms, understand relationships between structure and function of the enzyme, and apply this knowledge to design new enzymes, inhibitors, modulators, and activators for industrial and biomedical applications. From first steps to modern approaches, studies of enzymes bring new isoforms that work more efficiently and selectively than native molecules.

Keywords Enzyme · Catalysis · Specificity · Stereoselectivity · Active site

Enzymes increase rate of chemical reactions by organizing a framework for specific binding of substrates and cofactors, properly aligning catalytic subunits, and providing environment for stabilizing transition states. Myoglobin and hemoglobin were first enzymes for which detailed structures at atomic resolution were solved (Kendrew et al. 1958; Perutz et al. 1960). Enzyme structure shows how important positioning and movements of functional groups are for the catalytic reaction. Catalytic amino acids and the entire molecule in the enzyme/substrate complex, protonation states of individual amino acids, hydrogen bonds and salt bridges, conformational states involving slight motions of protein groups and ligands or large scale rearrangements of subunits: all are important and can be addressed in high resolution structure. Structural data combined with kinetic data allow propose and refine catalytic mechanisms, understand relationships between structure and function of enzyme, and apply this knowledge to design new enzymes, inhibitors, modulators, and activators for new treatments. From first steps to modern approaches, studies of enzymes brought new isoforms that work more efficiently and selectively than native molecules (Arnold 2017). The flow of catalytic reaction is viewed more as a pathway through ensemble of functional conformations rather than a linear process with sequence of ordered steps. Enzyme as dynamic molecule performs the movements of domains, secondary structure elements, and loops/

N. Kurochkina, *Protein Structure and Modeling*,
https://doi.org/10.1007/978-981-13-6601-7_3

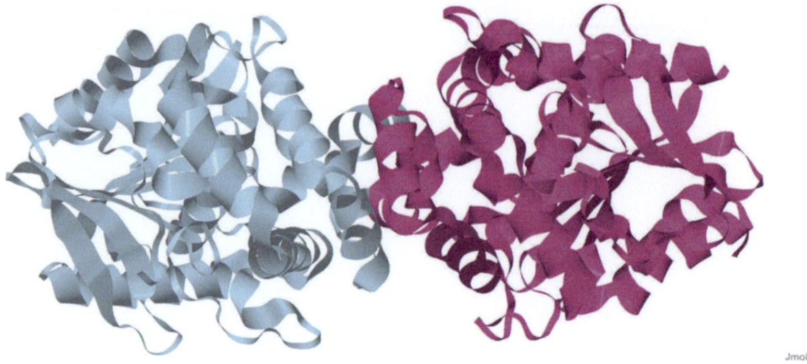

Fig. 3.1 Fluoroacetate dehalogenase dimer in apo form /3r3u/

linkers on time scales from minutes to picoseconds. Protein backbone and side chains undergo breathing/pumping movements (Saleh and Kalodimos 2017).

Fluoroacetate dehalogenase (FAcD) hydrolyzes the strongest covalent bond between carbon and fluorine (Chan et al. 2011a). Freeze trapping X-ray crystallography of FAcD shows about 38 states between substrate binding and product release. Two subunits cooperate and exhibit the major ground state and minor excited state which changes toward growing population of the excited state as substrate binds. One subunit binds substrate and the other reproduces the conformational changes of the first subunit. Substrate binding dynamics in one subunit is allosterically propagated to the unliganded subunit. B factor values that characterize mobility of atoms show increased values for the unliganded protomer during the reaction. The increased motions are accompanied by reduction in bound water molecules indicating that the intrinsic cost of the ligand binding to catalytic subunits is paid by the unliganded subunit (Kim et al. 2017). While performing catalytic reaction, enzyme subunits work in concert (Fig. 3.1). During defluorination reaction, enzyme exhibits main conformational states in the free form, enzyme-fluoroacetate Michaelis complex, covalent intermediate with bound glycolyl, and enzyme-product complex. In the halide pocket of the active site, three hydrogen bonds stabilize the fluoride ion, and provide selectivity towards fluorinated substrates. The crystal structure of the FAcD from *Burkholderia sp* shows typical homodimer of a/ß hydrolase fold. The conserved catalytic triad comprises catalytic Asp110, base His280 and carboxylate of Asp134 located in the domain interface to which substances can get only along the channel 11 Å-long. The active site hosts the aspartate nucleophile. Proximal histidine and tryptophan stabilize the leaving fluoride as suggested by their ability to bind chloride. The proposed mechanism involves an SN2 attack of aspartate nucleophile that results in the fluoride anion abstracted from the substrate. Asp110 Oδ2 atom attacks C2 atom of FAc to break the CF bond, release fluoride and invert stereoconfiguration at C2 via glycolyl-enzyme covalent intermediate. S_N2 reactions require a collinear alignment from the nucleophile Oδ2, electrophile C2 and the leaving group fluoride. Active site and molecular framework dynamics are critical

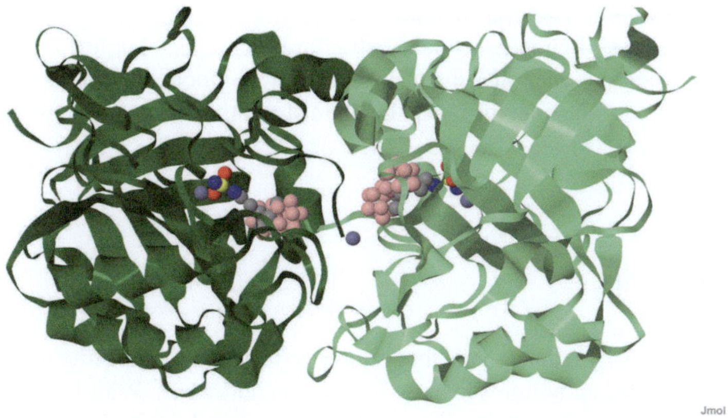

Fig. 3.2 Carbonic anhydrase /5ogn/

for the design of novel defluorinases which process diverse set of fluorinated organic compounds, and can advance many industrial applications.

During defluorination reaction, enzyme exhibits main conformational states in the free form, enzyme-fluoroacetate Michaelis complex, covalent intermediate with bound glycolyl, and enzyme-product complex. In the halide pocket of the active site, three hydrogen bonds stabilize the fluoride ion, and provide selectivity towards fluorinated substrates.

The S_N2 mechanism is also used by other dehalogenases including haloalkane dehalogenases, fluoroacetate dehalogenases, L-2haloacid dehalogenases, D- and DL-2-haloacid dehalogenases, haloalcohol dehalogenases and dichloromethane dehalogenases.

Carbonic anhydrase (CA) catalyzes the reversible reaction of hydration of CO_2 to produce HCO_3^- and a proton. Three distinct forms of CA, CA A, CA B and CA C (or CA III, CA I, and CA II in the alternative nomenclature), have similar structure (Fig. 3.2) and differ in cellular localization and catalytic efficiency. They are subdivided into the α, β, γ, δ, ζ, η, θ, and θ classes.

The ellipsoidal molecule and conically shaped active site host zinc tetrahedrally coordinated to three histidine residues and water molecule. Water molecules in the active site form hydrogen bonds with hydrophobic residues. The monomer comprises seven right-handed α-helices around a 10 stranded ß-sheet core.

Efficiency of CA catalytic reaction as measured by a turnover rate of 1.1–4.4 μs^{-1} is very high. The two-step ping-pong mechanism involves the first step of generating bicarbonate via a nucleophilic attack of CO2 and the second step of the regeneration of the catalytic zinc-bound hydroxyl. Diffusion of CO2 into the hydrophobic pocket of the active site and binding in the proximity of the catalytic zinc-bound hydroxyl at a distance 2.8°A triggers conformational change and initiates a nucleophilic attack from the free pair of electrons on the hydroxyl group onto the partially positive carbon of the CO2 which results in product formation HCO3- anchored to zinc through the same nucleophilic oxygen that performed the attack and product release.

Hyperthermostable bacterial α-type CA is a non-covalent dimer. An increased number of hydrogen bonded secondary structure elements, charges on the surface involved in ionic interactions suggest structural components of thermostability compared to mesophilic enzymes, may be main structural sources of these properties (Fredslund et al. 2018).

Stereoselectivity

Transaminase (TAm) is an enzyme that catalyses the reversible reaction of the amino group transfer from an amino substrate to a ketone/aldehyde. With keto group, this reaction generates a chiral amine (Sayer et al. 2014 4cmd 4chi). TAms are used for the synthesis of chiral amino acid analogues such as *tert*-leucine, homophenylalanine, and glutamate. These enzymes can process most of the 20 naturally occurring amino acids. The (S)- and (R)-selective TAms can catalyse the amino group transfer from substrates containing an amino group distal to the carboxyl group (ωATms) or no carboxyl group (amine transaminases), the enzymes, particularly interesting for biotechnology. D amino acid Tams and branched chain Tams are applied for both the asymmetric synthesis and resolution of chiral amines. The ωTAms from *Vibrio fluvialis* and *Chromobacterium violaceum* have been studied as potential biocatalysts in the production of chiral amines by a stereoselective synthesis. Optically pure compounds are very important for pharmaceutical, agrochemical and medical industrial production. These enzymes have a PLP type I protein structure fold which contains two α/β domains, antiparallel seven stranded and mixed eight stranded. Enzyme branched chain amino acid aminotransferase binds acidic and hydrophobic/aromatic side chains of the substrates at the same active site but binds acidic and basic side chains at different active sites (Goto et al. 2003). *Mesorhizobium* sp. transaminase active site accommodates both α and β amino acids binding in reverse orientation (Wyrenga et al. 2012).

Transaminases, acylases, and tryptophan synthases synthesize tryptophan analogues that are precursors of many medicinal compounds used as anticancer, antimicrobial, immunosuppressive, and phytotoxic agents. Tryptophan synthase (TrpS) produces enantiopure tryptophan derivatives from L serine and indole ring. TrpS two subunits, allosteric regulatory α subunit TrpA and catalytic β subunit TrpB, are subject to optimization by directed evolution. Mutations that improve positioning of 4 indole and serine in the active site eliminating stereoclashes with cofactor pyridoxal phosphate (PLP) make better enzyme (Romney et al. 2017).

CDK5 and its Activators/Inhibitors

Cyclin dependent kinases (CDKs) regulate cell cycle progression, and their function depends on cyclins, activators, and inhibitors. Phosphorylation and dephosphorylation are important signaling events that regulate dynamics of the skeleton (Kurochkina et al. 2018). Primarily active in nervous system, CDKs also function in pancreatic, muscle, and other cells. Involved in cancer, diabetes, inflammation, muscular dystrophy and other pathologies, CDKs are targets of therapeutic intervention and drug design (Reviews: Noble et al. 1997; Contreras-Vallejos et al. 2012; Arif 2012).

Comparison of CDK5/p25, CDK6/Cyclin V, CDK2/Cyclin and other complexes (Tarricone et al. 2001; Schulze-Gahmen, and Kim 2002, Li et al. 2006) shows that each kinase has high specificity for binding of its activator/inhibitor. CDKs other than CDK5 do not bind p25 with high affinity. CDK5 binds cyclins D1, D2, and D3. Only D2 inhibits CDK5/p35. Cyclin E inhibits CDK5 in neuronal cells (Arif, 2012). Cyclin I and G1 binding to CDK5 were shown to activate the kinase (Brinkkoetter et al. 2009; Contreras-Vallejos et al. 2012). Inhibition of CDKs is also very specific (Li et al. 2006). Binding of p16 (ankyrin repeat fold protein) to CDK6, for instance, misaligns N- and C-lobe perturbing active state, affects cyclin D binding site and blocks the catalytic cleft for ATP binding. INK4 proteins bind and inhibit CDK4/6, but not CDK2 in spite of high similarity between CDK2, 4, and 6. Several ATP antagonists bind to the catalytic site and inhibit CDKs (Mapelli et al. 2005). At the site of CDK5 activation, CIP (Zheng et al. 2010a) and small peptides p1- p5 derived from activator p35/p25 also act as inhibitors with p3 and p5 more potent than CIP (Zheng et al. 2010b).

Inhibition of CDK5 at the site of activation by small peptides-inhibitors emerges as source of new therapeutic agents comparable with activators. Mechanisms of their action represents important object of study.

Kinase action. Inactive CDK complexes are first activated by binding of their regulatory subunits cyclins. Complex is further activated by phosphorylation of the threonine in the T-loop (Jeffrey et al. 1995; Schulze-Gahmen, and Kim 2002). The active site of the kinase is located in the cleft between its N-lobe rich in β-sheet and α-helical C-lobe. In inactive conformation, T-loop blocks access to the active site. Binding of cyclin makes it possible to get T-loop out of the active site. Further phosphorylation of threonine in the T-loop is required to achieve active state in which phosphothreonine T160 forms hydrogen bonds with helix α1 (or C), catalytic loop and T-loop, and residue at P + 3 position of the substrate whereas conformational transition of V164 brings it in contact with residue at P + 1 position of the substrate. In phosphorylated state, C-helix moves into the active site and catalytic residues align optimally for the reaction. Phosphorylation of CDK2 T160 in the activation loop, T-loop (D145-E172 carrying DFG and APE motifs), is required for activation of CDK2/Cyclin A. However, activation of CDK5/P25 is independent of phosphorylation of corresponding S159 of CDK5. P39, p35, or P25 contribute to substrate specificity of CDK5 by triggering all conformational changes required for

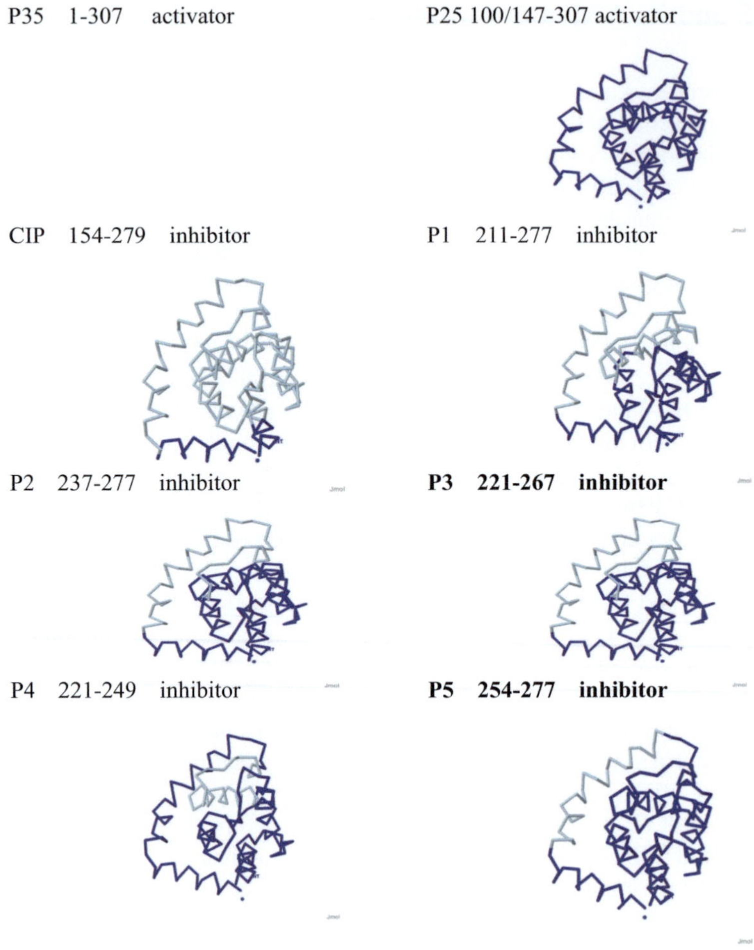

Fig. 3.3 CDK5 activators and inhibitors

CDK5 activation in the absence of activation loop phosphorylation (Tarricone et al. 2001). CIP, 124-residue peptide derived from p25 (aa 154–277; Fig. 3.3), inhibits CDK5 (Zheng et al. 2010b). P35 and p25 molecules trigger conformational changes in CDK5 that are necessary for its activation. The inhibitor CIP, derived from p25, lacks N-terminal (aNT) and C-terminal (α7) helices. Possibly, if CIP binding to CDK5 does not induce the same conformational changes as those induced by p35/p25, CDK5 remains in inactive conformation. Smaller peptides derived from p35/p25, p1-p5, also inhibit the kinase, but only p5 and p3 are more potent inhibitors. Structural modifications (Fig. 3.3) remove all elements from p25 that are not in direct contact with CDK5. In the smallest and most potent inhibitor, p5, the remaining portion comprises C-terminal α-helices α5-α6.

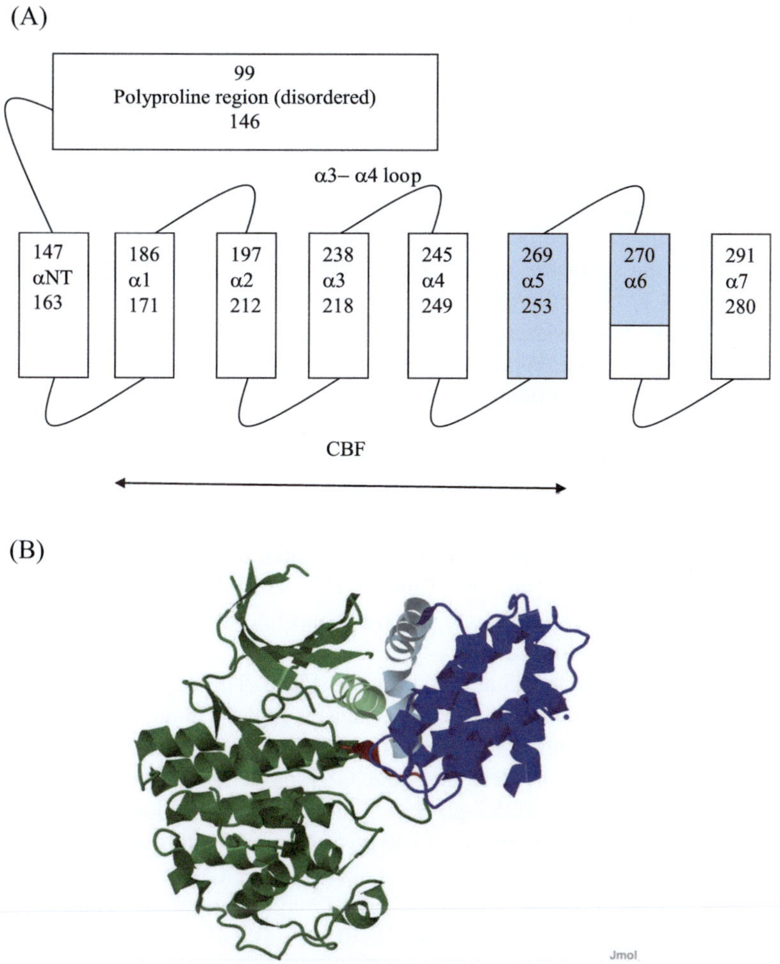

Fig. 3.4 (**A**) Structural elements of P25 (aa 99–307) and P5 (aa 254–277; shaded). PDB designation 1unh (Mapelli et al. 2005). Loop α3-α4 containing E240 contributes to catalytic activity of CDK5. (**B**) Three-dimensional structure of CDK5 (green) and p25 (blue) complex. CDK5 C-helix – light green; activation loop – brown; inhibitor p5 (amino acid residues 254–277) – light blue

Cyclin structure Cyclins consist of two domains, each domain carrying eight α-helices, αNT and α1-α7. Only one of these two domains binds the kinase (Noble et al. 1997). The two domains have very similar structure in spite of low amino acid similarity. The homologous CDK-binding site of cyclins, named a Cyclin box fold (CBF), consists of helices α1-α5.

Activators/inhibitors P25 comprises residues 99–307 of p35 (Fig. 3.4) (Mapelli et al. 2005). In crystallographic structure of CDK5/P25 complex (Tarricone et al. 2001), fragment 147–293 forms a globular domain with eight α-helices. Its structure is similar to the structure of Cyclin A and α-helices are named as αNT and α1-α7 in both proteins in which α1-α5 belong to cyclin box fold (CBF) motif. Region 99–146 of p25 is disordered.

CDK5/CIP CDK inhibitory peptide (CIP) 125 residue long (aa 154–279) inhibits p25/CDK5 while retains intact p35/CDK5 activity *in vivo* (Sundaram et al. 2013). P35 is phosphorylated by CDK5/p35 at Ser8 and Thr138 (Kamei et al. 2007). The inhibitor CIP, derived from p25, lacks N-terminal (αNT) and C-terminal (α7) helices (Fig. 3.3). Although structure of CIP is not solved, its inability to activate CDK5 suggests that it might undergo conformational transition. Removal of two α-helices can destabilize the structure of CIP. These α-helices are not in direct contact with CDK5 (Table 3.1) but influence conformation of the remaining helices. αNT also contains Thr138, the residue that is phosphorylated by CDK5/p35 and carries important regulatory role. Whether conformational transition upon removal of two helices, lack of Thr138, or other reason converts activator to inhibitor is a subject of further studies.

Modeling of CDK5/p5 interactions. Structural modifications (Fig. 3.3) that remove all elements from p25, which are not in direct contact with CDK5, produce a set of inhibitors. In the smallest and most potent inhibitor, p5, the remaining portion comprises C-terminal a-helices α5-α6. Contacts of p25 and p5 with CDK5 are subdivided into three groups (Table 3.1). First group of activator/inhibitor residues contacts C-helix of CDK5, second group interacts with catalytic site, and third group – with the N-terminus.

The p25 region between α3 and α4 contains E240. This amino acid is necessary for CDK5 activity (Tarricone et al. 2001). E240 is present in all inhibitors, GST-p1 (Glu^{211}-Ala^{277}), 67 amino acid residues; GST-p2 (Met^{237}-Ala^{277}), 41 residues; GST-p3 (Glu^{221}-Leu^{267}), 47 residues; and GST-p4 (Glu^{221}-Leu^{249}), 29 residues; however, it is not present in GST-p5 (Lys^{254}-Ala^{277}), 24 residues.

In this set of inhibitory peptides, p3 and p5 are more effective inhibitors than CIP. All peptides except p5 contain E240 that is necessary for CDK5 activity. Inability of p5 to process substrate in the absence of E240 may contribute to its high potency. However, p3 is an exception. It contains E240 but misses C-terminus. The question remains whether residues 267–277 in the C-terminal region of p3 might be needed for the reaction or for binding (as shown by molecular dynamics for residues Glu^{255}, Trp^{259}, and Ala^{233}). The fragment of p5 residues 267–277 interacts with C-helix of CDK5 (Table 3.1). Although many factors that influence activity of the inhibitory peptides can be explained by their interactions with CDK5, further studies will provide data to highlight factors necessary for inhibition and activation.

CDK5/p5 interactions with peptide substrate. Structures of CDK2/Cyclin A alone and in complex with peptide substrate $S_0P_{+1}R_{+2}K_{+3}$, CDK2/Cyclin /substrate complex (Brown et al. 1999), and CDK5/p25 complex (Tarricone et al. 2001) are determined. They show how CDK5/P5 could interact with its substrate. Loop α3-α4 of P5, particularly amino acid E240, is a candidate for interaction with the lysine at +3 position of the peptide substrate (Fig. 3.4B). Impaired ability of E240Q and E240A mutants to phosphorylate substrate shows important role of p25 for CDK5 activity.

Table 3.1 Contact residues of CDK5 and Activator/Inhibitor (P5, P25/P35)

CDK5		Activator/Inhibitor	
Amino acid	Position	Amino acid	Position
H71	N-lobe; β4	E255	P5, p25 α5
L37		W258	P5, p25
P45	N-lobe; C-terminus		
Val 69		L262	P5, p25 α5
His 71			
K56	C-helix	N266	P5, p25 α5
C53	C-helix	S269	P5, p25 α5
K56	C-helix		p25
E57	C-helix		p25
E57		S270	P5, p25 α5
E57		L273	P5, p25 α6
V122			p25
F151	Catalytic site	N276	P5, p25 α6
G152	Catalytic site		p25
I153	Catalytic site		p25
N121		A277	P5, p25 α6
		S242	p25 α3 α4
R50	C-helix	I241	p25 α3 α4
E161			p25
R149	Catalytic site	G238	p25 α3
R50	C-helix	Y236	p25 α3
L54			p25
R149	Catalytic site	M237	p25 α3
I153	Catalytic site		p25
S46	C-helix	S235	p25 α3
R50	C-helix		p25
P45	C-helix	Y231	p25 α3
S46	C-helix		p25
R156	Catalytic site	Q193	p25 α1 α2

Abbreviations: CDK5 **C-helix**: G43-K56 P^{45}SSALREICL^{54} Inhibitor/Activator, **p5**, spanning CIP residues Lys^{245}-Ala^{277} (Zheng et al. 2010a) P25 (Mapelli et al. 2005); CIP- residues 154–279 of p35; GST-p1 (Glu^{211}-Ala^{277}), 67 amino acid residues; GST-p2 (Met^{237}-Ala^{277}), 41 residues; GST-p3 (Glu^{221}-Leu^{267}), 47 residues; GST-p4 (Glu^{221}-Leu^{249}), 29 residues; and GST-p5 (Lys^{254}-Ala^{277}), 24 residues (Zheng et al. 2010b).

Interactions of CDK5/p35/p5 and CDK5/p25/p5. P5 competes with P25 for binding to CDK5. When P5 displaces P25 from CDK5, absence of loop α3-α4 in P5 possibly does not allow efficient processing of substrate by CDK5/P5 complex (inhibition). Since contacts with P67 mainly occur in P10 fragment, P67 protects P35/CDK5 complex better then it protects P25/CDK5. P5 competes with P35 only at high concentrations of P5. Then P35/CDK5 complex is not inhibited by p5 since it

```
                                                                          ↓
P35       101   PPPAQPP-APPASQLS--GSQT GSSSVKKAPHPAVTSAGTPK /1unh/ disordered
                P PA PP   PP    LS   GS+T
Ack1      771   PGPASPPRVPPREPLSPQGSRT PSPLVPPGSSPLPPRLSSSPGKTMPTTQSFASDPKYATP
MIG6      311   PLGSRPPKVPPREPLSPSNSRT  /2rf9/disordered
Synapsin25      LQRPQPP-PPPPSAAS--PGAT  /1pk8/rat synapsin I disordered
```

Fig. 3.5 Amino acid sequence alignment of CDK5 inhibitor activator p25 and MIG6/ ACK1 EGFR-binding regions. In Ack, EGFR-binding region is part of proline-rich region; it was shown that this region binds its SH3 domain; F820 (green) located C-terminal to this region is a candidate for interaction with SH3 W424 (Galisteo et al. 2006). Proline-rich region is subdivided into 3 subregions p1, p2, and p3. The fragment shown here is p3. It binds Grb2 (Chan et al. 2011) and is located N-terminal to 138 T phosphorylation site (arrow). Synapsin also shows homology; it is CDK5 substrate, and it binds Grb2 and Src (Onofri et al. 2000); its N-terminal region that is homologous to p25 is involved in association with synaptic vesicles regulated by phosphorylation (Brautigam et al. 2004)

contains CDK5 catalytic site and α3-α4 loop of P5 and is protected by P67 from replacement by P5.

P25N-terminal region. Region 99–146 of p25 that is disordered in crystallographic structure of the CDK5/p25 complex (Tarricone et al. 2001), does not show homology to Cyclin A but does show homology to the N-terminal proline-rich regions of MIG6 and Ack, inhibitors of Epidermal Growth Factor Receptor (EGFR). EGFR-binding and CDK-binding domains of MIG6/ACK and Cyclin/p25 differ, but regions N-terminal to the kinase-binding domain are similar and may indicate presence of common ligands (Fig. 3.5). For Ack, this polyproline segment was shown to bind GRB2, most likely its SH3 domain (Chan et al. 2011b). SH3 domains are modules of protein-protein interactions that specifically recognize polyproline sequences (Review: Kurochkina and Guha 2013). Studies of CDK5 function have shown that postmitotic neurons do not migrate normally into cortical layers in $CDK5^{-/-}$ mice (Ohshima et al. 1996; Ayala et al. 2007; Zheng et al. 2010a). Inhibition of CDK5 activity results in reduced migration of lung cancer cells (Demelash et al. 2012). This provides evidence of more similarities in the mode of interactions of the two kinases with their partners that can be related to their common function in protein migration.

Amino acid sequence similarities of CDK5 activators/inhibitors CDKs show high degree of homology to each other whereas cyclins are homologous only in the region of the cyclin box fold (CBF) spanning approximately 100 amino acids comprising N-terminal α-helix aNt and α-helices α1-α7 (Jeffrey et al. 1995). CDK5 activator p35/p25 is homologous to several cyclins and carries cyclin box. CDK5 inhibitor p5 spans α5-α7.

Fragment α1-α4 of p35 shows 30% identity 45% homology to Cyclin G1 (aa 178–250). Short p35 fragment of the N-terminal helix (aa 147–173) is 33% identical and 51% homologous to Cyclin I. Both of these homologous fragments lack p5 residues (Fig. 3.6).

Inhibition by CIP and p1-p5 has two sites of action: (1) inhibitory peptide displaces p35 or p25 from the CDK5/p35 or CDK5/p25 complex; for CIP or p1-p5, this results in binding to CDK5 but not triggering conformational changes

(A)

```
                 α1                     α2                    α3
P35     178   LRSVDRSLLLQGWQDQGFITPANVVFLYMLCRDVISSEVGSDHELQAVLLTCLYLSYSYM
              LR  +   LL   Q  GF T    + + +L R +   +V   H L  V L+C YL+   +
CycG1    51   LRDFEVKDLLSLTQFFGFDTETFSLAVNLLDRFLSKMKVQPKH-LGCVGLSCFYLAVKSI

                 α4
P35     238   GNEISYPLKPFLV
                E + PL   L+
CycG1   110   EEERNVPLATDLI
```

(B)

```
                     αNt
P35     147   STSELLRCLGEFLCRRCYRLKHLSPTD
              S+SE+LR     L +  + L   +P D
CycI    123   SSSEILRMERIILDKLNWDLHTATPLD
```

Fig. 3.6 Amino acid sequence alignment of p35 and cyclins (**A**) G1 and (**B**) I

necessary for its activation; additionally, p5, would not be not able to process substrate in the absence of E240; (2) inhibitory peptide binds to CDK5/p35 or CDK5/p25 and acts as pseudosubstrate; for CIP or p1- p5 binding to the kinase complex would prevent processing of other substrates. Inhibition of kinases by pseudosubstrates is described for protein kinase C (PKC) (Wang et al. 2012), p21-activated kinases (PAK) (Haa et al. 2012), and other proteins. This mechanism is widely used for the design of new inhibitors.

CDK and CDC42 inhibitors/activators CDK inhibitors subdivide into two classes: inhibitors of CDK4 (INK4) and $p21^{Cip1}$ (Sherr and Roberts 1999). CDK4/6 interacts with ankyrin domain of INK4 (Krzywda et al. 2003). Cdc42 interacts with TPR domain of $p67^{phox}$ (Lapouge et al. 2000) and with Ack CRIB (Mott et al. 1999). Protein crystallographic structures of CDK5 and its complexes in the Protein Data Bank (PDB) (Bernstein et al. 1977) include CDK5 in complex with its activator p25 / 1unh/; CDK5/p25 in complex with inhibitors binding to ATP site roscovitine, indirubin, and aloisine /1unl 1ung/; CDK5/p25 in complex with ATP analog / 3o0g/; CDK2/Cyclin A in complex with peptide substrate /1qmz/; CDK6 in complex with inhibitor P16INK4A /1bi7/ and P19INK4D /1bi8/.

CDK5 activity can depend not only on p25/p35, but also on other proteins. For example, 15-kDa CDK4- and CDK5-binding protein stimulates phosphorylation of one group of substrates but not the other. Among 24 substrates tested, only the chromosomal high mobility group proteins HMG I/Y and P1 and myelin basic protein (MBP) were phosphorylated by the $pl5^{CDK\text{-}BP}$-bound kinase (Azzi et al. 1994). So some fragments of the activator are needed to bind one group of substrates and other fragments to bind another substrate.

Role of phosphorylation in activation of kinase was shown, for instance, for PKB (Yang et al. 2002). Phosphorylation of activation loop results in ordering of αC and αB helices.

Activation effects of phosphorylation include increase of activity in CDK5 but decrease in CDK2 upon Tyr 15 phosphorylation by Abl facilitated by binding to Cables. Inhibitory effects of phosphorylation were demonstrated for CDK5 as phosphorylation at Thr 14 (Zukerberg et al. 2000).

In Hck, Lck, Abl, and Src, intramolecular binding of SH3 to proline rich peptides of a linker region provides mechanism of inhibition of kinase activity; consequently, release of peptide by SH3 results in kinase activation (Sicheri et al. 1997).

CDK4 inhibitors in INK4 class have ankyrin repeat fold (Yuan et al. 1999). Substrate of CDK5, TRPV1, is phosphorylated in the ankyrin repeats domain (Pareek et al. 2006). Ankyrin repeats of TRPA1 (Hall et al. 2018) are phosphorylated at the same structurally equivalent positions within anryrin repeat characteristic sequence **Thr/Ser** Pro Leu His.

GNE

The devastating neuromuscular disorder hereditary inclusion body myopathy (HIBM) and sialuria characterized by increased sialic acid levels are caused by (primarily missense) mutations in the bifunctional enzyme uridine diphosphate-*N*-acetylglucosamine 2-epimerase/*N*-acetylmannosamine kinase (GNE/MNK) that catalyzes the first two committed, rate-limiting steps in the biosynthesis of *N*-acetylneuraminic acid (sialic acid).

Uridine diphosphate (UDP)-*N*-acetylglucosamine (GlcNAc) 2-epimerase /*N*-acetylmannosamine (ManNAc) kinase (GNE/MNK) is a bifunctional enzyme of sialic acid biosynthesis encoded by the *GNE* gene (Hinderlich et al. 1997; Eisenberg et al. 2001). One domain (GNE) carries out epimerase function; another domain (MNK) is responsible for kinase activity. *GNE* mutations can result in a human disorder hereditary inclusion body myopathy (HIBM) (Huizing and Krasnewich 2009) and sialuria (Huizing 2005).

HIBM is an autosomal recessive disorder; patients harbor two *GNE* mutations outside the enzyme's allosteric site, in either the GNE and/or MNK coding domains (Eisenberg et al. 2001). HIBM mutations lead to decreased GNE and MNK enzymatic activities (Effertz et al. 1999; Sparks et al. 2005) and, in some patients, decreased sialylation of glycoproteins (Huizing et al. 2004; Tajima et al. 2005). Mutations in one enzymatic domain also affect the activity of the other domain (Sparks et al. 2005). GNE/MNK exists in two major oligomeric states, hexamers and dimers, which are in a dynamic interplay with monomers and higher aggregates. The fully functional tetrameric state of GNE/MNK is stabilized by ligands of the GNE domain, UDP-N-acetylglucosamine and CMP-N-acetylneuraminic acid (Ghaderi et al. 2007).

Two novel isoforms of the GNE/MNK enzyme were recently identified; isoform *GNE1* encodes the GNE/MNK protein described in all previous studies, while *GNE2* and *GNE3* encode proteins with extended and deleted N-termini, respectively (Reinke et al. 2007; Reinke et al. 2009). Isoforms GNE1 and GNE2 possess both epimerase and kinase activities, whereas isoform GNE3 exhibits only kinase activity (Reinke et al. 2009).

Structure studies of the active and allosteric sites of GNE/MNK contribute to further understanding of GNE/MNK function and its ligands. Modeling of the mutations associated with HIBM reveals why these mutations contribute to decreased/inhibited enzymatic activities. One of the goals is to reveal ligands (with a similar or different structure to ManNAc) that may be good candidates for rescuing the effects of mutated GNE/MNK. Newly discovered ligands may be powerful candidate drugs for the treatment of HIBM and sialuria.

The N-terminal GNE domain is homologous to several prokaryotic 2-epimerases, for which a three-dimensional model was determined. The C-terminal *N*-acetylmannosamine kinase (MNK) domain belongs to the sugar kinases superfamily and its crystallographic structure is solved at 2.84 Å. So far, no detailed description of the complete three-dimensional structure of the GNE/MNK enzyme has been provided.

All available structural data of GNE/MNK homologs together with assessment of interhelical angles were used to model the structure and active sites of the GNE/MNK enzyme and effects of all reported missense mutations to date associated with HIBM on interhelical arrangements, substrate binding, and enzyme action (Kurochkina et al. 2009). Modeling of human GNE/MNK enzyme with Dr. M. Huizing laboratory (NHGRI/NIH) identified structure of the active site which was later confirmed by new crystallographic structures of MNK (Fig. 3.7A) in complex with substrate, UDP and inhibitors (Martinez et al. 2012). ROK family members include *E.coli* (2aa4) and *L. monocytogenes* (4htl) enzymes. The zinc binding site is essential for the active site function. Recently determined structure of *F. nucleatum* MNK, however, (Caing-Carlsson et al. 2017) shows a functional active site in the absence of zinc. Molecular modeling studies of the structure, arrangement of domains and location of the allosteric and active sites of the GNE/MNK enzyme as well as modeling the effects of therapeutic substrates and/or candidate drugs on enzyme conformation are subjects of future studies.

We identified that all mutations mapped on the three-dimensional structure/model are located inside or in the proximity of the active sites and at the secondary structure interfaces (Kurochkina et al. 2009). We studied the Persian-Jewish HIBM GNE/MNK founder mutation M712 T in more detail. Helix-helix interfaces form

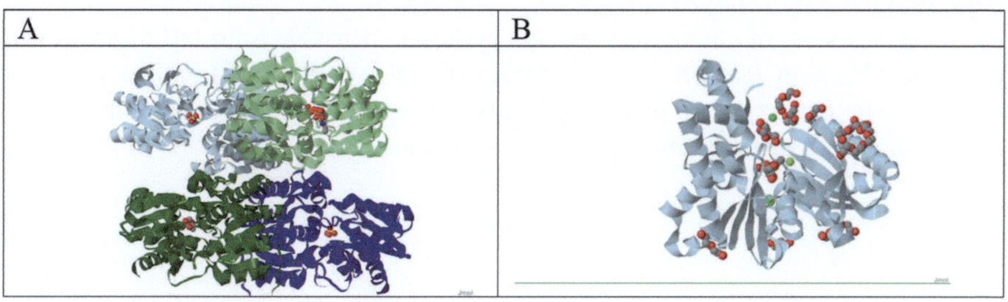

Fig. 3.7 GNE (**A**) epimerase and (**B**) kinase domains

structural frames to help position protein atomic groups for their interactions with ligands. Interhelical angles influence the ability of molecules to bind ligands. Amino acids at key interior positions of helix-helix interfaces are important for the orientation of helices (Kurochkina 2007 and 2008). Mutations away from the active site of the enzyme may affect catalytic activity via changes in the composition of secondary structure interfaces (Kurochkina et al. 2009).

Comparison of the three isoforms *GNE1*, *GNE2* and *GNE3* shows that only the epimerase domain exhibits differences in all three isoforms. The 31-residue extension at N-termini of GNE2 isoform possibly adopts an alpha-helical conformation and shows amino acid similarities with ATP-binding fragments of some kinases and helicases. The 14-residue extension at the N-terminus of GNE3 isoform is similar to proteins of the Rossmann fold L-lactate dehydrogenase, amidase.

GNE enzymatic activity is inhibited by the product CMP/Neu5Ac. Crystal structure of the human GNE epimerase domain tetramer (Fig. 3.7B) contains UDP and CMP/Neu5Ac in the active site. UDP/GlcNAc binding mode in hydrolyzing and bacterial nonhydrolyzing enzymes suggests Arg113 and Ser302 involvement in product hydrolysis and Arg263/Arg266 in sialuria. Structure reveals interactions of the active sites via channel for ManNAc transport (Chen et al. 2016). Recently determined structure of *Acinetobacter baumannii* epimerase (Fig. 3.7B) shows that similarly to epimerase domain of the human GNE it forms dimers and tetramers; however, unlike closed only protomer of GNE, it exhibits both open and closed protomers (Ko et al. 2018).

Building on these data, further modeling is now being performed, involving secondary structure interfaces with low degree of similarity among GNE/MNK enzymes and its homologs, domains arrangement and inter-subunit interactions, human allosteric site and structural effects of the mutations, sites of conformational transitions, and more detailed interactions of the active sites.

Enzymatic Glucose Biosensor

Electrochemical sandwich-type glucose biosensor uses carbon nanotubes (CNT), mucin and albumin as the scaffold of the enzymatic matrix with embedded glucose oxidase (GO). GO catalyzes the glucose oxidation in the presence of oxygen with consequent release of gluconolactone and hydrogen peroxide followed by hydrogen peroxide oxidation that produces a current (Comba et al. 2018). GO monitoring using mucin in enzymatic matrix provides the enzyme stability and the stable analytical performance as an amperometric biosensor.

Ligand Binding

Proteins interact with many ligands: nucleic acids, lipids, carbohydrates and their components.

Helix-helix interfaces and ligand binding Helix-helix parallel and antiparallel interfaces are characterized by certain combinations of amino acids, which repeatedly occur at core positions **a** and **d** (leucine zipper nomenclature) in similar and nonsimilar proteins. Amino acids at positions **a** and **d** correlate with interhelical angle. They are responsible for mutual orientation of helices. Therefore, they to large extent influence interactions of helices with numerous ligands. Interhelical angles can be predicted for various combinations of amino acids. In a set of proteins comprising glutathione S-transferase, intracellular chloride channel and annexin molecules from various sources, correct relationship between amino acids combination and interhelical angle is found in 31 out of 34 proteins. Also, interhelical angle correlates with conformation of the ligand: glutathione S-transferase ligands glutathione, s-hexylglutathione, glutathione sulfonic acid, and glutathione-s-dinitrobenzene are good examples (Kurochkina and Choekyi 2011). New structures of GSTs and DHARs show consistency with previously obtained data (Table 3.2).

Thioredoxin fold is one of the most conserved (Fig. 1.18). Structure of seven Precambrian enzymes from species ~4 billions years old shows high preservation of fold βαβαββα: pleated β-sheet of three parallel and two antiparallel β-strands and four α-helices and its characteristics such as number of hydrogen bonds and salt bridges, surface charge, accessible surface area. The length of α-helix α1 is increased: bacterial enzymes contain shorter helix (Ingles-Prieto et al. 2014). Gluthathione S transferase N-terminal domain also has theoredoxin fold (Fig. 1.18).

Two parallel interacting α-helices of GST/thioredoxin present a good structural unit to address important questions how amino acids at interface positions relate to mutual orientation of helices, how changes of amino acids result in changes of structure, and what effect structural changes have on the function of the molecule.

Amino acids at helix-helix interfaces appear to be important determinants of the protein three-dimensional fold. Interior protein residues form hydrophobic core and are involved in interactions, which are very sensitive to changes. In the families of similar proteins, correlated mutations lead to changes in the packing of the core. Groups of interactions that are responsible for structure conservation and structural changes migrate from protein to protein and conserved groups of interacting residues, or amino acid combinations, are seen in distantly related or even unrelated proteins. Structural modifications can preserve or alter the function such as ligand binding.

Glutathione S-transferase, cytoplasmic chloride channels and annexin molecules compose a group of proteins used to predict amino acid composition of helix-helix interfaces in similar and distantly related proteins and to find possible link to their ligand binding properties (Kurochkina and Choekyi 2011).

Glutathione S-transferase catalyses conjugation of reduced glutathione (GSH) to xenobiotic compounds making them more soluble and more easily excreted. This

Table 3.2 Examples of repeating sequence combinations at helix-helix interfaces of Glutathione S-transferase, DHAR (helices α1, α3) and annexin molecules (helices B, E). Interface formed by amino acids at positions **a, d** of one helix and **a', d'** of another helix is shown as $\binom{d\ d'}{a'a}$. Side chain rotamer of each amino acid is given by +, t(*trans*), or – dihedral angles. (**A**) Change of one amino acid (magenta) at the interface correlates with change of interhelical angle (Kurochkina and Choekyi 2011); newly added combinations are shaded yellow; (**B**) with the same amino acids at the interface, ligand binding /unbinding and consequent change in side chain(s) rotamers correlates with change of interhelical angle

(**A**)

Atomic groups	Amino acids at positions a and d	Interhelical angle, degrees	Atomic groups	Amino acids at positions a and d	Interhelical angle, degrees
δδ γγ	$L_{-t}I_{-t}$ / $I_{-t}V_{-t}$; $I_{-t}L_{tt}$ / $T_{-}I_{tt}$; $L_{-t}I_{-t}$ / $I_{-t}I_{-t}$; $L_{-t}L_{-t}$ / $I_{-t}I_{-t}$; $L_{-t}I_{-t}$ / $I_{-t}I_{-t}$	10°–36°	δβ γγ	$I_{-t}A$ / $V_{t}T_{+}$; $I_{-t}A$ / $V_{t}V$; $I_{-t}A$ / $V_{t}V$; $I_{-t}A$ / $T_{t}V$	−4°– 1°
δδ γδ	$L_{-+}L_{t+}$ / $I_{-t}L_{-t}$; $L_{-t}L_{t+}$ / $I_{-t}L_{-t}$; $I_{-t}L_{t+}$ / $I_{-t}L_{-t}$; $L_{-t}I_{-t}$ / $I_{-t}I_{-t}$; $I_{-t}L_{t+}$ / $T_{-}L_{-t}$; $L_{+t}I_{-t}$ / $L_{-t}I_{t+}$; $L_{-t}I_{-+}$ / $L_{-t}I_{-t}$; $I_{-t}I_{t-}$ / $T_{t}L_{-t}$	5°–49°	δβ γδ	$L_{t}A$ / $C_{-}L_{-t}$; $L_{t}A$ / $V_{t}I_{-+}$	−11°
GST *Plasmodium falciparum* /2aaw/			GST *Arabidopsis thaliana* /1gnw/		
GST *Schistosoma japonicum* /1gne/ /1m9b/			GST *Glycine max* /2vo4/		
Annexin VI *Bos taurus* /1avc/			GST *Chlamydomonas reinhardtii* /5xft/		
Annexin A8 *Homo sapiens* /1w3w/			GSTU23 *Arabidopsis thaliana* /6ep7/		
Annexin A2 *Homo sapiens* /2hyw/ /1xjl/					
GST4–4 *Mus musculus*/1guk/			GST Oryza sativa /1oyj /		
GST *Caenorhabtidis elegans* /1yq1/			GST Aegilops tauschii /1gwc /		
GST *Homo sapiens* /1gtu/					
GSTM2–2 *Homo sapiens* /1hna/					
GST *Homo sapiens* /1gsd/					
Annexin V Homo sapiens /1avr/					
Annexin V Rattus norvegicus /2ran/					
Annexin XII Hydra attenuata /1 dm5/					
Annexin A8 Homo sapiens /1w3w/					
Annexin IV Bos taurus /1ann/					
Annexin III Homo sapiens /1axn/					
Annexin VI Bos taurus /1avc/					
Annexin III Homo sapiens /1axn/					
Annexin V Homo sapiens /1avr/					
Annexin V Gallus gallus /1ala/					
GSTA1–1 *Homo sapiens* /5lcz/					

(continued)

Table 3.2 (continued)

(B)			
Atomic groups	Amino acids at positions a and d	Interhelical angle, degrees	Protein
δγ	$L^{80}_{-t}\,T^{29}_{+}$ $A^{26}V^{77}_{-}$	–3	DHAR1 *Arabidopsis thaliana* /5el8/ Apo
αγ	$L^{80}_{-t}\,T^{29}_{+}$ $A^{26}V^{77}_{-}$	–19	DHAR1 *Arabidopsis thaliana* /5ela/ Sulfonic acid
	$L^{80}_{-t}\,T^{29}_{+}$ $A^{26}V^{77}_{-}$	22	DHAR1 *Arabidopsis thaliana* /5elg/ GSH

detoxifying enzyme is present almost in all organisms. In plants, GSTs can be activated by pathogens, dehydration, or stress due to salt, ozone, or heavy metals levels. It has important implications in design of catalysts with new improved properties and green chemistry biocatalysts (Valenzuela-Chavira et al. 2017). Annexin is known as an important cancer biomarker (Chin and Gray 2008) involved in inflammatory pathways, regulation of apoptosis signaling, and cell proliferation (Lim and Pervaiz 2007). Proteins contain both parallel and antiparallel orientation of helices. They participate in multiple pathways involving interactions with various ligands, play significant physiological role and exhibits complex genotype-phenotype relationships (Rubin et al. 2006).

Glutathione S-transferase can be subdivided into into alpha, mu, pi, theta, zeta, tau, and omega classes based on substrate/inhibitor specificity, amino acid sequence similarity, structure similarity, and immunological identity (Sheehan et al. 2001). Some members, such as glutathione S-transferase from *Plasmodium falciparum* do not belong to any class (Fritz-Wolf et al. 2003). Sequence alignment of representatives of alpha, mu, and pi classes and their biochemical characterization show diverse catalytic properties of the enzyme (Vuilleumier 1997). Glutathione S-transferase crystallographic structures were determined for the members of each class which show very similar fold although amino acid similarity between members can be only 10%. Pdb contains 316 GSTs, 40 of them from plants (Valenzuela-Chavira et al. 2017).

GST consists of two domains: N-terminal thioredoxin-like and C-terminal as series of four helix bundles. In N-terminal domain of glutathione S-transferase of βαβαββα thioredoxin-like fold, the N-terminal (α1) and C-terminal (α3) parallel helices on one side of the β-sheet interact with each other and with C-terminal α-helical domain; middle helix (α2) is on the opposite side of the β-sheet. Conserved glutathione binding G-site (domain 1) and less conserved accommodating various ligands xenobiotic substrate binding H-site (domain 2) and structural communication between the active sites of the GST subunits are important for the enzyme function (Karpusas et al. 2013). GSH is conjugated with various hydrophobic compounds containing an electrophilic atom: arene oxides, organic halides, epoxides, organic thiocyanates, esters, and α,β-unsaturated carbonyls. Catalytic role is carried out by N-terminal amino acid residue tyrosine in most GST classes, serine in Theta and Zeta or cysteine in Omega and Beta classes (Sheehan et al. 2001). GST domain is also found in translational factors such as aminoacyl-tRNA synthetases, and as a part of Elongation factor 1B from *Saccharomyces cerevisiae* (Jeppesen et al. 2003). Presence of GST domain in methionyl-, valyl, glutaminyl, cysteinyl-tRNA synthetases suggests its roles in protein folding, assembly and catalysis (Cho et al. 2015).

Novosphingobium aromaticivorans GST cuts glutathione from (R)- and (S)-GS-HPV, β-glutathionyl-γ-hydroxypropiovanillone, at very high rate following the reaction of breaking the β-aryl ether bond of gualacylglycerol-β-gualacyl ether (Kontur et al. 2018). This enzyme shows potential for applications in the field of new sources of fuels and chemicals obtained from depolymerization of lignin, component of lignocellulosic plant biomass available in great abundance.

Mutations in *H. sapiens* GST A1–1 do not disrupt tertiary structure significantly but destabilize α/β interface in I71V, I71A (Achilonu et al. 2010), and α1/α6/α8 interdomain interface in W21 L mutants (Balchin et al. 2010; 3l0h) also diminishing catalytic activity. Mutations at helix interfaces in *Litopenaeus vannamei* GST C-terminal domain affect stability and ligand binding to H site (Juarez-Martinez et al. 2017).

H. sapiens tetrameric GST forms multiprotein complex with multi-tRNA synthetase complex (MSC). Four MSC components, methionyl-tRNA synthetase, glutaminyl-prolyl-tRNA synthetase, AIMP2, AIMP3, are organized in a dynamic complex that carries important role in protein synthesis (Cho et al. 2015). AIMPs are nonenzymatic factors that are involved in immune response hormonal regulation of glucose metabolism, angiogenesis, wound healing, response to DNA damage. AIMP3 tertiary structure is very similar to that of GST. Mutations of AIMP3, including those located at helix interfaces and proximal to ligand binding sites, are associated with chronic myeloid leukemia (Kim et al. 2008).

Binding of drugs chlorambucil (CBL) to GST shows partial overlap of the active site residues interacting with GSH and CBL as well as coordination of subunits important for the enzyme function. As a site of enzyme regeneration, GST contains noncatalytic ligand binding pocket named L site binding fatty acids, bile acids, drugs which are large lipophilic compounds (Brock et al. 2013). Chlorambucil induces mutations of the juvenile congenital polycystic kidney disease and affects five other genes including Ank3 (Bryda et al. 1996). GST inhibitors are used as anticancer and anti-inflammatory agents (Ramkumar et al. 2016).

Soluble chloride channels CLICs belong to GST structural family, consist of the N terminal thioredoxin like and C terminal all helical domains and for some members show similarities in ligand binding properties. CLIC1 covalently binds GSH (Board et al. 2004). CLIC6 do not bind GSH in solution (Ferofontov et al. 2018).

DHAR catalyses the reaction of dehydroascorbate GSH dependent reduction. Although DHAR and GST active sites are similar and overall structure of DHAR is similar to those of GST and CLIC, only few GSTs and CLICs carry out this reaction (Bodra et al. 2017).

Positions at helix-helix interface are designated by letters **a** to **g** according to leucine zipper nomenclature. Hydrophobic core is formed by positions **a** and **d** located in protein interior. Identical amino acid combinations can be found at interfaces with similar values of interhelical angles in homologous and nonhomologous proteins (Kurochkina 2007, 2008). In spite of low sequence similarity among glutathione S-transferase, DHAR, annexin, and intracellular chloride channel, amino acids at positions **a** and **d** of these proteins carry identical and/or similar sequence combinations at helix-helix interfaces. Similarities in three dimentional sequence patterns are reflected by similarities in structural units, arrangements of pairs of helices. Two types of interfaces differ by combinations at the interfaces. Interfaces with positive interhelical angles show similarity to those of coiled coil leucine zipper, whereas interfaces with negative interhelical angles group with proteins of TIM-barrel fold (triose phosphate isomerase, 2-keto-3-deoxy-6-phosphogluconate aldolase, mandelate racemase, tagatose 1, 6-bisphospate aldolase, L-ALA-D/L-GLU epimerase), isocitrate dehydrogenase, acetylcholine esterase, glutathione reductase, acetylcholine receptor M2, ankyrin repeats. They also contain similar topological arrangement of rows of parallel α-helices. Interhelical angle correlate with conformation of the ligand in glutathione S-transferase and DHAR.

As previously found and supported by new data, parallel interface between helices α1 and α3 exhibits both positive and negative interhelical angles in homologous proteins and distinct amino acid combinations corresponding to each type of the interface (Table 3.3).

Twenty crystallographic structures of GST from various organisms previously identified fourteen α1α3 interfaces with positive and six with negative interhelical angles. In addition, fourteen annexin molecules show only positive interhelical

Table 3.3 GST structures subdivided into α1α3 interfaces with positive and negative interhelical angles

Positive interhelical angles	Negative interhelical angles
Rattus norvegicus, pi and theta class	*Glycine max*
Mus musculus, pi and mu class	*Chlamydomonas reinhardtii*
Homo sapiens	*Ochrobactum anthropi*
Gallus gallus	*Anopheles dirus*
Caenorabtidis elegans	*Anopheles gambiae*
Drosophila melanogaster	*Aegilops tauschii*
Litopenaeus vannamei	*Otyza sativa*
	Burkholderia xenovorans
	Arabidopsis thaliana
	Homo sapiens theta and omega class
	Mus musculus zeta class

angles. In all, 31 out of 34 proteins of GST and annexin combined exhibit combinations of amino acids at positions **a** and **d** consistent with characteristic combinations of leucine zipper type (positive interhelical angles or left handed) and TIM barrel type of interface (negative interhelical angles or right handed). Three combinations, which do not show amino acid combination characteristic for the type of interface, have distinct side chain rotamers. All these structures contain aliphatic or hydrogen bonded polar amino acids at positions **a** and **d**.

Leucine zipper coiled coil combination $\binom{\delta\delta}{\gamma\gamma}$ is a reference point of interfaces with positive interhelical angles; it is present in five GST proteins which also show positive interhelical angles. TIM barrel protein is a reference point of interfaces with negative interhelical angles; it is present in a large number of GST proteins which also show negative interhelical angles with alanine present at positions **a** and **d** in $\binom{\delta\beta}{\gamma\gamma}$, $\binom{\gamma\beta}{\beta\gamma}$, $\binom{\delta\beta}{\beta\delta}$ and $\binom{\delta\beta}{\gamma\delta}$ combinations.

Second important group of $\binom{\delta\delta}{\gamma\delta}$ combinations is present in GSTs and annexins at helix-helix interfaces with positive interhelical angles. One amino acid change $\binom{\delta\beta}{\gamma\delta}$ also produces interfaces with negative interhelical angles.

Some combinations repeatedly occur at interfaces with positive angles: $\binom{\delta\delta}{\gamma\varepsilon}$ is found in four unrelated proteins *Rattus rattus* glutathione S-transferase, *Homo sapiens* annexin III, *Homo sapiens* fibrinogen and *Xenopus laevis* olfactory cyclic-nucleotide-gated channel complexed with calmodulin.

Although amino acids at positions **a** and **d** are the most conserved and most important for the fold, amino acids at positions **e** and **g** conformation adjusts one helix relative to another as observed, for example, in DHAR1. Enzyme apo (5ela), sulfonic acid bound (5elg) and GSH bound (5el8) conformations (Table 3.2B) exhibit conformational transitions, particularly in dihedral angles of interface side chains. The largest changes in dihedral angles occur for side chains of Ser23 (position **e**), Leu30 (**e**), Glu32 (**g**), and backbone of Gly78 and Glu82.

Therefore, it is possible to distinguish two types of interfaces from amino acid composition at positions **a** and **d** for this group of homologous and distantly related proteins (Kurochkina and Choekyi 2011). Newly determined structures expand described groups as *G. max* GST (2ov4 4top) or form new groups of amino acid combinations (Table 3.2A).

Since the basis for preference of a left handed or right handed helix-helix arrangement is in position of the side chain methyl groups, the question arises whether given amino acid sequence of the protein it would be possible to determine positive or negative interhelical angle between helices α1 and α3. A good correlation exists between the position of side chain methyl groups and the interhelical angle in a set of 200 proteins (Kurochkina 2008). This set of proteins was used to calculate average position of the vector S, which reflects displacement of methyl groups at positions **a**, **a'** relative to methyl groups at positions **d**, **d'**. Such correlation also exists in glutathione S-transferase and annexin if position of vector S is calculated for each interface. The correlation coefficient between the X coordinate of vector S and the interhelical angle is high for all combinations (R = −0.71). Position of the vector S for three types of interfaces $\binom{\delta\delta}{\gamma\gamma}$, $\binom{\delta\beta}{\gamma\gamma}$, and $\binom{\delta\beta}{\gamma\delta}$ of glutathione S-transferase and annexin in XY plane indicates link to interhelical angle or handedness of the interface.

Hydrophobic residues in protein interior are in contacts with each other and less mobile compared to residues located at the surface. Many contacts remain intact when secondary structure elements or subunits move as rigid bodies. If we estimate conformational states using side chain dihedral angles, we can see that few core residues change rotameric states in response to protein movements which are necessary for its function. Molecules move in response to changes in surrounding solvent. Movements of secondary structure elements or subunits occur with changes in rotameric states.

Approximately several hundred glutathione S-transferase and other proteins which possess glutathione S-transferase activity are listed in gene ontologies of cellular component and molecular function (Ashburner et al. 2000) and more and more are newly added from genome studies. Sequences of proteins with glutathione S-transferase activity, when aligned at the site of two interacting helices (Table 3.4), show conservation of the atomic group combinations. Proteins which contain aliphatic amino acids at positions **a** and **d** form groups of combinations that are clearly associated with handedness. In alignment, it is also important to include positions **d** preceeding interface positions **a** and **d** because they are involved in catalysis and ligand binding. Within each group, we can see conservation of both interface positions **a** and **d** and ligand binding amino terminal or carboxy terminal positions. Conserved interface positions **a** and **d** allow to derive important structural characteristics, such as interhelical angle and location of catalytic residues, from the members for which three-dimensional structure is known and apply them to other members of the group. Some groups with conserved amino acids at the interface positions show diverse set of organisms carrying the same characteristics. Therefore, it is possible to trace evolutionary relationships and potential origin of similarities and differences between species.

Aliphatic amino acids are main contributors to the layers of hydrophobic core. We can compare contribution of four aliphatic amino acids alanine, valine, isoleucine, and leucine, to positions **a** and **d** of the interface with contribution of the extended set which includes alanine, valine, isoleucine, leucine, proline, cysteine, and threonine. Preference for interacting atomic groups to form interface can be estimated as difference between experimental frequencies and expected values. It appears that this difference is larger for the set of seven types compared to the set of four types of amino acids. It occurs because aliphatic only residues occur at positions **a** and **d** of the interface much more frequently than polar residues. Expected values estimated for the two sets, one as ($n = 4$, $N = n^2 = 16$), the second as ($n = 7$, $N = 49$), show low probability of certain combination to be at the interface by chance but higher probability of the same combination in the test data set of 200 proteins (Kurochkina and Choekyi 2011). Expected values compared with experimental results show highest increase in experimental frequency compared to

Table 3.4 Examples of amino acid sequence alignment for ligand-binding helices α1 and α3 of glutathione S-transferase, and similar proteins. Grooping according to similar atomic contacts and interhelical angles (Table 3.2) allows to compare proteins, for which crystallographic structure was determined (PDB accession code of proteins), and proteins with known amino acid sequence only. Positions **a** and **d** of helix-helix interface are in bold

Protein name, source and pdb code	Helix α1 d---**a**--**d**----	Helix α3 d---**a**--**d**----
Xx γγ		
δδ γγ	----γ--δ----	----γ--δ---
GST P1 *C. elegans*	LAEP**I**RL**L**LADK *D*	SGAI**I**RH**L**ARL *D*
GST *S. japonica* /1gne/	LVQP**T**RL**L**LEYL *D*	SMAI**I**RY**I**ADK *D*
GST *P.falciparum* /2aaw/	KAEL**I**RL**I**FLYA *D*	SQAI**V**RY**L**SKK *D*
GST Q8ILQ7 *P.falciparum*	KAEL**I**RL**I**FAYL *D*	SQAI**V**RY**L**SKK *D*
GST ABA40394 *P. vivax*	KAEL**I**RL**I**FAYL*D*	SQAI**V**RY**L**SKK
δβ γγ	----γ--β----	----γ--δ---
GST *A. thaliana* /1gnw/	ATRR**V**LI**A**LHEK	SRAI**T**QY**I**AHR
GST *G. max* /2vo4 4top/	FGMR**V**RI**A**LAEK *D*	SLIA**V**QY**I**EEV *D*
GST F2/02520.1 *A. thaliana*	ATRR**V**LI**A**LHEK *D*	SRAI**T**QY**I**AHR *D*
GST F3 *A. thaliana*	STRR**V**LI**A**LHEK *D*	SRAI**T**QY**I**AHR *D*
GST Z2 *A. thaliana*	CAHR**V**RI**A**LTLK *D*	SFAI**I**MY**L**DDK *D*
GST Z1 *A. thaliana*	CAHR**V**RI**A**LALK *D*	SFAI**I**MY**L**DEK *D*
GST T2 *S. cereviciae*	YPAR**V**RI**A**LAEK	CTAI**T**EY**I**DAL
GST *B. xenovarans*	YCQK**V**LT**A**LYEN	SSII**I**EY**L**GLY
Maleylacetoacetate isomerase *H. sapiens*	CSWR**V**RI**A**LALK	SLAI**I**EY**L**EE-

expected value for atomic group combination $\binom{xx}{\delta\delta}$. This difference increases when type of atomic group contributing by amino acid is considered.

There is also a strong correlation between the interhelical angle of the interface α1α3 and the virtual dihedral angle N_1-$C\alpha_1$-C_1-N_2 for glutathione ligands. Glutathione (and s-hexylglutathione, glutathione sulfonic acid, glutathione-s-dinitrobenzene) binds glutathione S-transferase mainly by the N-termini of both helices. The highest value of correlation coefficient $R = -0.8$, and interfaces with positive interhelical angles bind ligand in more extended conformation than those with negative interhelical angles.

References

Achilonu I, Gildenhuys S, Fisher L, Burke J, Fanucchi S, Sewell BT, Fernandes M, Dirr HW (2010) The role of a topologically conserved isoleucine in glutathione transferase structure, stability and function. Acta Crystallogr Sect F Struct Biol Cryst Commun 66(Pt 7):776–780. https://doi.

org/10.1107/S1744309110019135. Epub 2010 Jun 23. PubMed PMID: 20606271; PubMed Central PMCID: PMC2898459
Arif A (2012) Extraneuronal activities and regulatory mechanisms of the atypical cyclin-dependent kinase Cdk5. Biochem Pharmacol 84:985–993
Ashburner M, Ball CA, Blake JA, Botstein D, Butler H, Cherry JM, Davis AP, Dolinski K, Dwight SS, Eppig JT, Harris MA, Hill DP, Issel-Tarver L, Kasarskis A, Lewis S, Matese JC, Richardson JE, Ringwald M, Rubin GM, Sherlock G (2000) Gene ontology: tool for the unification of biology. The gene ontology consortium. Nat Genet 25:25–29. http://www.ncbi.nlm.nih.gov/pubmed/10802651?ordinalpos=3&itool=EntrezSystem2.PEntrez.Pubmed.Pubmed_ResultsPanel.Pubmed_RVDocSum
Azzi L, Meijer L, Ostvold A-C, Lew J, Wang JH (1994) Purification of a 15-kDa cdk4- and cdk5-binding Protein. J Biol Chem 269:13279–13288
Ayala R, Shu T, Tsai LH (2007) Trecking across the brain: the journey of neuronal migration. Cell 128:29–43
Balchin D, Fanucchi S, Achilonu I, Adamson RJ, Burke J, Fernandes M, Gildenhuys S, Dirr HW (2010) Stability of the domain interface contributes towards the catalytic function at the H-site of class alpha glutathione transferase A1-1. Biochim Biophys Acta 1804(12):2228–2233. https://doi.org/10.1016/j.bbapap.2010.09.003. Epub 2010 Sep 15. PubMed PMID: 20833278
Bernstein FC, Koetzle TF, Williams GJB, Meyer EF Jr, Brice MD, Rodgers JR, Kennard O, Shimanouchi T, Tasumi M (1977) The protein data bank: a computer-based archival file for macromolecular structures. J Mol Biol 112:535
Board PG, Coggan M, Watson S, Gage PW, Dulhunty AF (2004) CLIC-2 modulates cardiac ryanodine receptor Ca2+ release channels. Int J Biochem Cell Biol 36(8):1599–1612. PubMed PMID: 15147738
Bodra N, Young D, Astolfi Rosado L, Pallo A, Wahni K, De Proft F, Huang J, Van Breusegem F, Messens J (2017) Arabidopsis thaliana dehydroascorbate reductase 2: Conformational flexibility during catalysis. Sci Rep 7:42494. doi: 10.1038/srep42494. Erratum in: Sci Rep 7:46896. PubMed PMID: 28195196; PubMed Central PMCID: PMC5307357
Brinkkoetter PT, Olivier P, Wu JS, Henderson S, Krofft RD, Pippin JW, Hockenbery D, Roberts JM, Shankland SJ (2009) J Clin Invest 119:3089
Brock J, Board PG, Oakley AJ (2013) Structural insights into omega-class glutathione transferases: a snapshot of enzyme reduction and identification of a non-catalytic ligandin site. PLoS One 8 (4):e60324. https://doi.org/10.1371/journal.pone.0060324. Print 2013. PubMed PMID: 23593192; PubMed Central PMCID: PMC3621891
Brown NR, Noble MEM, Endicott JE, Johnson LN (1999) The structural basis for specificity of substrate and recruitment peptides for cyclin-dependent kinases. Cell biology 1:438–443
Bryda EC, Ling H, Rathbun DE, Burmeister M, Flaherty L (1996) Fine genetic map of mouse chromosome 10 around the polycystic kidney disease gene, jcpk, and ankyrin 3. Genomics 35 (3):425–430. PubMed PMID: 8812475
Caing-Carlsson R, Goyal P, Sharma A, Ghosh S, Setty TG, North RA et al (2017) Crystal structure of N-acetylmannosamine kinase from Fusobacterium nucleatum. Acta Crystallogr Sect F Struct Biol Cryst Commun 73(Pt 6):356–362. https://doi.org/10.1107/S2053230X17007439
Chan W, Sit ST, Manser E (2011a) The Cdc42-associated kinase ACK1 is not autoinhibited but requires Src for Activation. Biochem J. (2011) 435:355–364
Chan PW, Yakunin AF, Edwards EA, Pai EF (2011b) Mapping the reaction coordinates of enzymatic defluorination. J Am Chem Soc 133(19):7461–7468. https://doi.org/10.1021/ja200277d. Epub 2011 Apr 21. PubMed PMID: 21510690; PubMed Central PMCID: PMC3101105
Chin L, Gray JW (2008) Translating insights from the cancer genome into clinical practice. Nature 452:553–563
Cho HY, Maeng SJ, Cho HJ, Choi YS, Chung JM, Lee S, Kim HK, Kim JH, Eom CY, Kim YG, Guo M, Jung HS, Kang BS, Kim S (2015) Assembly of multi-tRNA Synthetase complex via heterotetrameric glutathione transferase-homology domains. J Biol Chem 290

(49):29313–29328. https://doi.org/10.1074/jbc.M115.690867. Epub 2015 Oct 15. PubMed PMID: 26472928; PubMed Central PMCID: PMC4705937

Comba FN, Romero MR, Garay FS, Baruzzi AM (2018) Mucin and carbon nanotube-based biosensor for detection of glucose in human plasma. Anal Biochem 550:34–40. doi: 10.1016/j.ab.2018.04.006. [Epub ahead of print] PubMed PMID: 29649474

Contreras-Vallejos E, Utreras E, Gonzalez-Billault C (2012) Going out of the brain: Non-nervous system physiological and pathological functions of Cdk5. Cellular Signalling 24(2012):44–52

Demelash A, Rudrabhatla P, Pant HC, Wang X, Amin ND, McWhite CD, Xu N, Ilona Linnoila R (2012) Achaete-scute homologue-1 (ASH1) stimulates migration of lung cancer cells through Cdk5/p35 pathway. Mol Biol Cell 23:2856–2866

Effertz K, Hinderlich S, Reutter W (1999) Selective Loss of either the Epimerase or Kinase Activity of UDP-*N*-acetylglucosamine 2-Epimerase/*N*-Acetylmannosamine Kinase due to Site-directed Mutagenesis Based on Sequence Alignments. J Biol Chem 274:28771

Eisenberg I, Avidan N, Potikha T, Hochner H, Chen M, Olender T, Barash M, Shemesh M, Sadeh M, Grabov-Nardini G, Shmilevich I, Friedmann A, Karpati G, Bradley WG, Baumbach L, Lancet D, Asher EB, Beckmann JS, Argov Z, Mitrani-Rosenbaum S (2001) The UDP-N-acetylglucosamine 2-epimerase/N-acetylmannosamine kinase gene is mutated in recessive hereditary inclusion body myopathy. Nat Genet 29:83

Ferofontov A, Strulovich R, Marom M, Giladi M, Haitin Y (2018) Inherent flexibility of CLIC6 revealed by crystallographic and solution studies. Sci Rep 8(1):6882. https://doi.org/10.1038/s41598-018-25231-z. PubMed PMID: 29720717; PubMed Central PMCID: PMC5931990

Fredslund F, Borchert MS, Poulsen JN, Mortensen SB, Perner M, Streit WR, Lo LL (2018) Structure of a hyperthermostable carbonic anhydrase identified from an active hydrothermal vent chimney. Enzyme Microb Technol 114:48–54. https://doi.org/10.1016/j.enzmictec.2018.03.009. Epub 2018 Mar 30. PubMed PMID: 29685353

Fritz-Wolf K, Becker K, Rahlfs S, Harwaldt P, Schirmer RH, Kabsch W, Becker K (2003) X-ray structure of glutathione S-transferase from the malarial parasite Plasmodium falciparum. 100 (24):13821–13826

Ghaderi D, Strauss HM, Reinke S, Cirak S, Reutter W, Lucka L, Hinderlich S (2007) Evidence for dynamic interplay of different oligomeric states of UDP-N-acetylglucosamine 2-epimerase/N-acetylmannosamine kinase by biophysical methods. J Mol Biol 369:746–758

Goto M, Miyahara I, Hayashi H, Kagamiyama H, Hirotsu K (2003) Crystal structures of branched-chain amino acid aminotransferase complexed with glutamate and glutarate: true reaction intermediate and double substrate recognition of the enzyme. Biochemistry 42 (13):3725–3733. PubMed PMID: 12667063

Haa BH, Matthew J, Davisa B, Chena C, Loua HJ, Jia Gaoa C, Zhanga R, Krauthammerd M, Halabane R, Schlessingera J, Turka BE, Boggon TJ (2012) Type II p21-activated kinases (PAKs) are regulated by an autoinhibitory pseudosubstrate. Proc Natl Acad Sci 109:16107–16112

Hall BE, Prochazkova M, Sapio MR, Minetos P, Kurochkina N, Binukumar BK, Amin ND, Terse A, Joseph J, Raithel SJ, Mannes AJ, Pant HC, Chung MK, Iadarola MJ, Kulkarni AB (2018) Phosphorylation of the Transient Receptor Potential Ankyrin 1 by Cyclin-dependent Kinase 5 affects Chemo-nociception. Sci Rep 8(1):1177. https://doi.org/10.1038/s41598-018-19532-6. PubMed PMID: 29352128; PubMed Central PMCID: PMC5775258

Hinderlich S, Stasche R, Zeitler R, Reitter W (1997) A bifunctional enzyme catalyzes the first two steps in N-acetylneuraminic acid biosynthesis of rat liver. Purification and characterization of UDP-N-acetylglucosamine 2-epimerase/N-acetylmannosamine kinase. J Biol Chem 272:24313–24318

Huizing M (2005) Disease mechanisms associated with mutations of the GNE gene. Drug discovery today: Disease mechanisms 2:519–527

Huizing M, Krasnewich DM (2009) Hereditary inclusion body myopathy: a decade of progress. Biochim Biophys Acta 1792:881

Huizing M, Rakocevic G, Sparks SE, Mamali I, Shatunov A, Goldfarb L, Krasnewich D, Gahl WA, Dalakas MC (2004) Hypoglycosylation of alpha-dystroglycan in patients with hereditary IBM due to GNE mutations. Mol Genet Metab 81:196

Jeffrey PD, Russo AA, Polyak K, Gibbs E, Hurwltz J, Massague J, Pavletich NP (1995) Mechanism of CDK activationrevealed by the structure of a cyclinA-CDK2 complex. Nature 376:313–320

Jeppesen MG, Ortiz P, Shepard W et al (2003) The Crystal Structure of the Glutathione *S*-Transferase-like Domain of Elongation Factor 1B from *Saccharomyces cerevisiae*. J Biol Chem 278:47190–47198

Kamei H, Saito T, Ozawa M, Fujita‡ Y, Asada‡ A, Bibb§ JA, Takaomi C, Saido HS, Hisanaga S-i (2007) Cleavage of the CDK5 activator p35 to p25 by site-specific phosphorylation. J Biol Chem *282*:1687–1694

Karpusas M, Axarli I, Chiniadis L, Papakyriakou A, Bethanis K, Scopelitou K, Clonis YD, Labrou NE (2013) The interaction of the chemotherapeutic drug chlorambucil with human glutathione transferase A1-1: kinetic and structural analysis. PLoS One 8(2):e56337. https://doi.org/10.1371/journal.pone.0056337. Epub 2013 Feb 27. PubMed PMID: 23460799; PubMed Central PMCID: PMC3584069

Kendrew JC, Bodo G, Dintzis HM, Parrish RG, Wyckoff H, Phillips DC (1958) A three-dimensional model of the myoglobin molecule obtained by x-ray analysis. Nature 181 (4610):662–666

Kim KJ, Park MC, Choi SJ, Oh YS, Choi EC, Cho HJ, Kim MH, Kim SH, Kim DW, Kim S, Kang BS (2008) Determination of three-dimensional structure and residues of the novel tumor suppressor AIMP3/p18 required for the interaction with ATM. J Biol Chem 283 (20):14032–14040. https://doi.org/10.1074/jbc.M800859200. Epub 2008 Mar 14. PubMed PMID: 18343821

Kim TH, Mehrabi P, Ren Z, Sljoka A, Ing C, Bezginov A, Ye L, Pomès R, Prosser RS, Pai EF (2017) The role of dimer asymmetry and protomer dynamics in enzyme catalysis. Science 355 (6322). pii: eaag2355. doi:10.1126/science.aag2355. PubMed PMID: 28104837

Ko TP, Lai SJ, Hsieh TJ, Yang CS, Chen Y (2018) The tetrameric structure of sialic acid-synthesizing UDP-GlcNAc 2-epimerase from Acinetobacter baumannii: A comparative study with human GNE. J Biol Chem. 293(26):10119–10127. doi: https://doi.org/10.1074/jbc.RA118.001971. Epub 2018 May 15. PubMed PMID: 29764940; PubMed Central PMCID: PMC6028959

Kontur WS, Bingman CA, Olmsted CN, Wassarman DR, Ulbrich A, Gall DL, Smith RW, Yusko LM, Fox BG, Noguera DR, Coon JJ, Donohue TJ (2018) Novosphingobium aromaticivorans uses a Nu-class glutathione S-transferase as a glutathione lyase in breaking the ß-aryl ether bond of lignin. J Biol Chem 293(14):4955–4968. doi: https://doi.org/10.1074/jbc.RA117.001268. Epub 2018 Feb 15. PubMed PMID: 29449375; PubMed Central PMCID: PMC5892560.

Kurochkina N, Bhaskar M, Yadav SP, Pant HC (2018) Phosphorylation, Dephosphorylation, and multiprotein assemblies regulate dynamic behavior of neuronal cytoskeleton: a mini-review. Front Mol Neurosci 11:373. https://doi.org/10.3389/fnmol.2018.00373. eCollection 2018. PubMed PMID: 30349458; PubMed Central PMCID: PMC6186834

Kurochkina N (2007) J Theor Biol 247:110–121

Kurochkina N (2008) J Theor Biol 255:188–198

Kurochkina N, Choekyi T (2011) Helix-helix interfaces and ligand binding. J Theor Biol 283:92–102

Kurochkina N, Guha U (2013) SH3 domains: modules of protein-protein interactions. Biol Rev 5:29–39

Lapouge K, Smith SJM, Walker PA, Gamblin SJ, Smerdon SJ, Rittinger K (2000) Structure of the TPR Domain of p67phox in Complex with Rac·GTP. Molecula Cell 6:899–907

Lim LH, Pervaiz S (2007) Annexin 1: the new face of an old molecule. Faseb J *21*:968–975

Mapelli M, Massimiliano L, Crovace C, Seeliger MA, Tsai L-H, Meijer L, Musacchio A (2005) Mechanism of CDK5/p25 binding by CDK inhibitors. J Med Chem 2005(48):671–679

Martinez J, Nguyen LD, Hinderlich S, Zimmer R, Tauberger E, Reutter W et al (2012) Crystal structures of N-Acetylmannosamine Kinase provide insights into enzyme activity and inhibition. J Biol Chem 287(17):13656–13665. https://doi.org/10.1074/jbc.M111.318170

Mott HR, Owen D, Nietlispach D, Lowe PN, Manser E, Limk L, Laue ED (1999) Structure of the small G protein Cdc42 bound to the GTPase-binding domain of ACK. Nature 399:384–388

Noble ME, Endicott JA, Brown NR, Johnson LN (1997) The cyclin box fold: protein recognition in cell-cycle and transcription control. Trends Biochem Sci 22:482–487

Ohshima T, Ward JM, Huh CG, Longenecker G, Veeranna Pant HC, Brady RO, Martin LJ, Kulkarni AB (1996) Targeted disruption of the cyclin-dependent kinase 5 gene results in abnormal corticogenesis, neuronal pathology and perinatal death. Proc Natl Acad Sci USA 93:11173–11178

Pareek TK, Keller J, Kesavapany S, Agarwal N, Kuner R, Pant H, Iadarola MJ, Brady RO, Kulkarni AB (2006) Cyclin-dependent kinase 5 modulates nociceptive signaling through direct phosphorylation of transient receptor potential vanilloid 1. Proc Natl Acad Sci 104:660–665

Perutz MF, Rossmann MG, Cullis AF, Muirhead H, Will G, North AC (1960) Structure of haemoglobin: a three-dimensional Fourier synthesis at 5.5-A. resolution, obtained by X-ray analysis. Nature 185(4711):416–422. PubMed PMID: 18990801

Ramkumar K, Samanta S, Kyani A, Yang S, Tamura S, Ziemke E, Stuckey JA, Li S, Chinnaswamy K, Otake H, Debnath B, Yarovenko V, Sebolt-Leopold JS, Ljungman M, Neamati N (2016) Mechanistic evaluation and transcriptional signature of a glutathione S-transferase omega 1 inhibitor. Nat Commun 7:13084. https://doi.org/10.1038/ncomms13084. PubMed PMID: 27703239; PubMed Central PMCID: PMC5059489

Reinke SO, Eidenschink C, Jay CM, Hinderlich S (2009) Biochemical characterization of human and murine isoforms of UDP-N-acetylglucosamine 2-epimerase/N-acetylmannosamine kinase (GNE). Glycoconj J 26:415

Romney DK, Murciano-Calles J, Wehrmüller JE, Arnold FH (2017) Unlocking reactivity of TrpB: a general biocatalytic platform for synthesis of tryptophan analogues. J Am Chem Soc 139 (31):10769–10776. https://doi.org/10.1021/jacs.7b05007. Epub 2017 Jul 28. PubMed PMID: 28708383; PubMed Central PMCID: PMC5589443

Rubin DL, Lewis SE, Mungall CJ, Misra S, Westerfield M, Ashburner M, Sim I, Chute CG, Solbrig H, Storey MA, Smith B, Richter JD, Noy NF, Musen MA (2006) The national center for biomedical ontology: advancing biomedicine through structured organization of scientific knowledge. OMICS 10:185–198

Saleh T, Kalodimos CG (2017) Enzymes at work are enzymes in motion. Science 355 (6322):247–248. https://doi.org/10.1126/science.aal4632. PubMed PMID: 28104853

Sayer C, Martinez-Torres RJ, Richter N, Isupov MN, Hailes HC, Littlechild JA, Ward JM (2014) The substrate specificity, enantioselectivity and structure of the (R)-selective amine: pyruvate transaminase from Nectria haematococca. FEBS J 281(9):2240–2253. https://doi.org/10.1111/febs.12778. Epub 2014 Apr 7. PubMed PMID: 24618038; PubMed Central PMCID: PMC4255305

Schulze-Gahmen U, Kim S-H (2002) Structural basis for CDK6 activation by a virus-encoded cyclin. Nature Str Biol 9:177–181

Sheehan D, Meade G, Foley VM, Dowd CA (2001) Structure, function and evolution of glutathione transferases: implications for classification of non-mammalian members of an ancient enzyme superfamily. Biochem J 360:1–16

Chen S-C, Huang C-H, Lai S-J, Yang CS, Hsiao T-H, Lin C-H, Fu P-K, Ko T-P, Chen Y (2016) Mechanism and inhibition of human UDP-GlcNAc 2-epimerase, the key enzyme in sialic acid biosynthesis. Sci Rep 6:23274. https://doi.org/10.1038/srep23274

Sherr CJ, Roberts JM (1999) CDK inhibitors: positive and negative regulators of G1-phase progression. Genes Dev 13(12):1501–1512. Review. PubMed PMID: 10385618

Sicheri F, Moarefi I, Kuriyan J (1997) Crystal structure of the Src family tyrosine kinase Hck. Nature 385:602–609

Sparks SE, Ciccone C, Lalor M, Orvisky E, Klootwijk R, Savelkoul PJ, Dalakas MC, Krasnewich DM, Gahl WA, Huizing M (2005) Use of a cell-free system to determine UDP-N-acetylglucosamine 2-epimerase and N-acetylmannoseamine kinase activities in human hereditary inclusion body myopathy. Glycobiology 15:1102

Sundaram JR, Poore CP, Sulaimee NHB, Pareek T, Asad ABMA, Rajkumar R, Cheong WF, Wenk MR, Dawe GS, Chuang K-H, Pant HC, Kesavapan S (2013) Specific inhibition of p25/Cdk5 activity by the Cdk5 inhibitory peptide reduces neurodegeneration *in vivo*. J Neurosci 33 (1):334–343

Tajima Y, Uyama E, Go S, Sato C, Tao N, Kotani M, Hino H, Suzuki A, Sanai Y, Kitajima K, Sakuraba H (2005) Distal myopathy with rimmed vacuoles: impaired O-glycan formation in muscular glycoproteins. Am J Pathol 166:1121

Tarricone C, Dhavan R, Peng J, Areces LB, Tsai L-H, Musacchio A (2001) Structure and regulation of the CDK5-p25nck5a complex. Mol Cell 8:657–669

Valenzuela-Chavira I, Contreras-Vergara CA, Arvizu-Flores AA, Serrano-Posada H, Lopez-Zavala AA, García-Orozco KD, Hernandez-Paredes J, Rudiño-Piñera E, Stojanoff V, Sotelo-Mundo RR, Islas-Osuna MA (2017) Insights into ligand binding to a glutathione S-transferase from mango: Structure, thermodynamics and kinetics. Biochimie 135:35–45. https://doi.org/10.1016/j.biochi.2017.01.005. Epub 2017 Jan 17. PubMed PMID: 28104507; PubMed Central PMCID: PMC5346462.

Vuilleumier S (1997) Minireview. Bacterial glutathione *S*-transferases: What are they good for? J Bact 179:1431–1441

Wang C, Shang Y, Yu J, Zhang M (2012) Substrate recognition mechanism of atypical protein Kinase Cs revealed by the structure of PKCi in complex with a substrate peptide from Par-3. Structure 20:791–801

Yang J, Cron P, Good VM, Thompson V, Hemmings BA, Barford D (2002) Crystal structure of an activated Akt/protein kinase B ternary complex with GSK3-peptide and AMP-PNP. Nat Struct Biol 9(12):940–944. PubMed PMID: 12434148

Yuan C, Li J, Selby TL (1999) Byeon I-J L and Tsai M-D (1999) tumor suppressor INK4: comparisons of conformational properties between p16INK4A and p18INK4C J. Mol Biol 294:201–211

Zheng Y-L, Li B-S, Rudrabhatla P, Shukla V, Amin ND, Maric D, Kesavapany S, Kanungo J, Pareek TK, Takahashi S, Grant P, Kulkarni AB, Pant HC (2010a) Phosphorylation of p27Kip1 at Thr187 by Cyclin-dependent Kinase 5 modulates neural stem cell differentiation. Mol Biol Cell 21:3601–3614

Zheng Y-L, Amin ND, Hu Y-F, Rudrabhatla P, Shukla V, Kanungo J, Kesavapany S, Grant P, Albers W, Pant HC (2010b) A 24-residue Peptide (p5), derived from p35, the Cdk5 neuronal activator, specifically inhibits Cdk5-p25 hyperactivity and tau hyperphosphorylation. J Biol Chem 285(44):34202–34212

Zukerberg LR, Patrick GN, Nikolic M, Humbert S, Wu CL, Lanier LM, Gertler FB, Vidal M, Van Etten RA, Tsai LH (2000) Cables links Cdk5 and c-Abl and facilitates Cdk5 tyrosine phosphorylation, kinase upregulation, and neurite outgrowth. Neuron 26(3):633–646. PubMed PMID: 10896159

Chapter 4
Channels

Abstract Recent advancements in structure elucidation of channels answer many important questions regarding ligand induced conformational transitions which accompany activation and desensitization, allosteric regulation by ions, involvement of regulatory proteins in their function, and many other. Membrane and intracellular proteins such as ion channels, amino acid and lipid transporters and enzymes carry out important metabolic and signaling cellular functions. Structural flexibility, fluctuations and dynamics of channels are important for ligand recognition and function regulation. Selectivity of channels is determined by interactions of the groups in protein interior.

Keywords Ion channel · Transporter · Current · Potential · Permeability · Pore · Cellular signal transduction · Diffusion

Ionotropic Glutamate Receptors

Ionotropic glutamate receptors can be found in a wide variety of organisms from bacteria to plants and animals. This suggests their potential role in transition from single cell to multicellular organism and as an object of evolutionary studies (Oswald 2004). Structural aspects of agonist binding control of channel gating are particularly important (Erreger et al. 2004). Mechanisms of activation and desensitization of ionotropic glutamate receptors are extensively studied. Structure of the ligand binding site gives possibility for the rational design of subtype-specific ligands (Johansen et al. 2007; Gardoni and Luca M. 2006).

Advancements in crystallographic analysis of ligand binding extracellular domain of the receptor complexed with full and partial agonists and antagonists reveal possible mechanisms of their action. Crystallographic structures of NMDA, AMPA, and kainate subtypes of ionotropic glutamate receptors with respect to mechanisms of ligand induced conformational changes, which accompany activation and desensitization, allosteric regulation by ions, and involvement of regulatory

N. Kurochkina, *Protein Structure and Modeling*,
https://doi.org/10.1007/978-981-13-6601-7_4

proteins in channel function demonstrate progress in our understanding of structure/function relationship.

Mechanisms of action of ionotropic glutamate receptors as major neurotransmitters represent particular interest. With high-resolution crystallographic data obtained for the ligand binding portion of the receptor it became possible to extend our knowledge and understand principles of ligand-receptor interactions leading to activation and desensitization. Review articles devoted to NMR and high resolution crystallographic structures (McFeeters and Oswald 2004; Mayer 2005) emphasize that significantly larger part of the receptor molecule is involved in ligand recognition then could be found on the basis of mutagenesis and homology models. Data obtained for binding modes of agonists, partial agonists, and antagonists to GluR2, GluR5/6, NR1 and NR2A subunits show how they are influenced by steric constraints, cavity size and water-bridged interactions (Chen and Wyllie 2006). Structural flexibility, fluctuations and dynamics are important for ligand recognition, activation and desensitization (Kubo and Ito 2003).

Receptor Structure

There are three major subtypes of ionotropic glutamate receptor N-methyl-D-aspartate (NMDA), α-amino-3-hydroxy-5-methyl-4-isoxazole propionate (AMPA), and kainate receptors. This categorization is based on the compound, which activates the receptor. They are homologous in their amino acid sequence and share similar three-dimensional structure.

Ionotropic glutamate receptor is a tetramer assembled as an arrangement of dimers, which form homomeric or heteromeric subunit associations (Song et al. 2018). Each of the four subunits contains transmembrane ion channel pore region and soluble extracellular ligand binding region. Transmembrane region consists of three transmembrane domains M1, M3, and M4 combined with M2 domain organized similar to potassium channel reentrant P loop. Ligand binding region forms S1S2 domain (Fig. 4.1a). Segment S1 precedes transmembrane domain M1. Segment S2 contains extracellular domain between transmembrane domains M3 and M4. Together S1 and S2 segments form agonist binding site as S1S2 domain, which exists in soluble form separately from the transmembrane part. Crystallographic structures were determined for the S1S2 ligand binding region in apo form and in complexes with agonists and antagonists. These two domains of the receptor subunit are preceded by the N-terminal domain which shows sequence similarity to bacterial leucine/isoleucine/valine binding proteins. At the C-terminus of the subunit, intracellular domain is located. For some subunits, it is structured when bound to calmodulin in the presence of Ca^{2+} (Fig. 4.1b). Totally four domains compose the entire receptor subunit.

C-terminal intracellular domain structure can be seen for NR1 subunit of NMDA receptor when it is bound to calmodulin. Calmodulin binds C0 and C1 regions of NR1 subunit via tetrad FLMM of the N-domain residues, which form wrap-around

(a) (b)

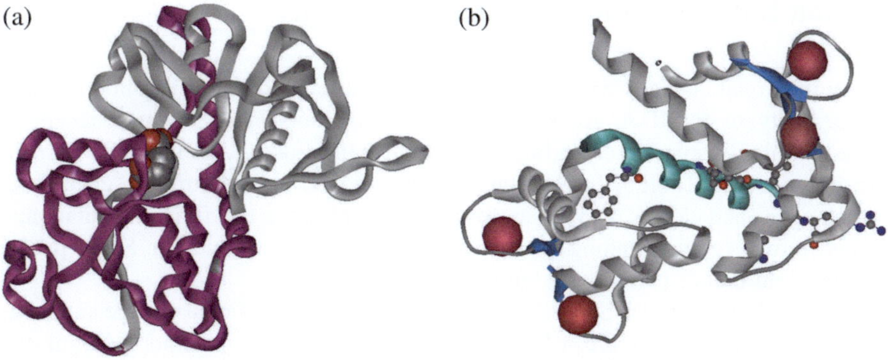

Fig. 4.1 Domains of NMDA receptor. (**a**) Ligand binding domain S1S2 (S1 grey, S2 magenta) of the NR2A subunit of NMDA receptor in complex with glutamate (CPK model). (**b**) Fragment of the C-terminal intracellular domain of NMDA receptor (cyan) in complex with calmodulin (grey and blue). Amino acids Phe880, Thr886, Ser890, Arg893–895 are shown as ball and stick models, Ca^{2+} ions as red balls

conformation of CaM-ligand, so that F880 contacts CaM C-pocket whereas T886 contacts its N-pocket (Ataman et al. 2007). Important residue affected by binding is S890. Its phosphorylation regulates membrane localization. C-terminal to the fragment, endoplasmic reticulum retention sequence R893-R895 is located (Fig. 4.1b).

Transmembrane pore region of glutamate receptors show high degree of homology to K+ channels (Wood 1995) although membrane topology differs. Crystallographic structure of K+ channel pore region was determined (Jiang et al. 2003).

Extracellular ligand binding S1S2 domain exhibits properties of the entire receptor molecule in terms of its response to ligand binding and leads to the conclusion that its structure is not distorted when being extracted. Comparison of the S1S2 region of NR1 subunit of NMDA receptor, GluR2 subunit of AMPA receptor, and GluR5 and GluR6 subunits of kainate receptor in their response to ligand binding shows many common charactiristics although it is possible to identify subtype-specific features (Furukawa and Gouaux 2003; Nanao et al. 2004; Hald et al. 2007).

Amino terminal domain is involved in subunit assembly and receptor desensitization. It influences binding of modulators such as protons, polyamines and Zn^{2+} (Wong et al. 2005). Since in ionotropic and metabotropic glutamate receptors these domains are homologous to bacterial leucine/isoleucine/valine binding protein, the structure of NMDA receptor amino terminal domain could be modeled (Hugging and Grant 2005) on the basis of crystallographic structure determined for extracellular dimer of metabotropic glutamate receptor (Kunishima et al. 2000). Conserved cysteine residues form a disulfide bond, which holds a four-helix bundle structure located between subunits. Zn-binding site in the cavity of the NR2A subunit stabilizes the open state.

Ligand Induced Conformational Transitions

Conformational changes which characterize activation and desensitization of three major subtypes (NMDA, AMPA, and kainate) of ionotropic glutamate receptors based on high-resolution crystallographic and NMR structures of the S1S2 domain are described in Table 4.1. Domain S1S2 is bilobed and resembles "clamshell" which opens and closes. "Open" conformation is present in apo form and antagonist bound form. Agonist binding leads to active "closed" conformation. Some partial agonists can induce partially closed conformation.

Agonist binding to the S1S2 domain results in a conformational change and leads to closed bilobed structure which causes opening of ion channel pore of the transmembrane domain.

Both glycine binding to NR1 subunit and glutamate binding to NR2A-D subunit are required for activation of NMDA receptor. Cocrystal structures of NR1 S1S2 ligand-binding core with the agonists glycine and D-serine (DS), the partial agonist D-cycloserine (DCS) and the antagonist 5,7-dichlorokynurenic acid (DCKA) were solved (Furukawa and Gouaux 2003). The cleft of the S1S2 "clamshell' is open in the presence of the antagonist DCKA and closed in the glycine, DS and DCS complexes.

Domains of the S1S2 ligand binding core of GluR2 (AMPA) receptor are expanded in apo form and contract in ligand-bound form with domain separation decreasing from apo to DNQX, kainite and glutamate/AMPA (Armstrong and Gouaux 2000). AMPA and glutamate also induce 180 degrees flip of a *trans* peptide bond in the ligand binding site.

In GluR2 structure of the complex with antagonist (S)-NS1209 in one protomer and the endogenous ligand (S)-glutamate in the other, ligand-induced hyperextension of the binding domains has been observed. Novel binding mode of (S)-NS1209 shows new larger GluR2 areas with even not conserved among receptor subtypes residues involved in contacts (Kasper et al. 2006).

Extension of binding pocket can be seen in an experiment of designing a new agonist compound 2-Me-Tet-AMPA, which discriminates GluR2–4 receptors from GluR1 by more than ten-fold preference (Vogensen et al. 2007).

CNQX is AMPA receptor antagonist. However, in AMPA receptor bound to TARP, CNQX acts as partial agonist. Transmembrane AMPA regulatory protein TARP is AMPA receptor auxiliary subunit. Complex of CNQX with AMPA shows that ligand induces domain closure. Binding of TARP possibly strengthens coupling between domain closure and channel opening and increases agonist efficacy (Menuz et al. 2007).

In kainate receptors GluR5 and GluR6, domain opening of 11 and 13 degrees correspondingly is observed in domoic acid (partial agonist) bound state compared to glutamic acid (agonist) bound state. Larger opening can be seen upon antagonist binding with 28 degrees in ATPO and 29–30 degrees in UBP302 or UBP310 complexes of GluR5 (Hald et al. 2007).

Table 4.1 Conformational changes of the ionotropic glutamate receptors accompanying ligand binding

Ligands	Ligand activity	PDB code	Conformational change
NMDA receptor *Rattus norvegicus*			
NR1-NR2A/ Glycine, glutamate	Agonist	2a5t	
NR2A/glutamate	Agonist	2a5s	
NR1/glycine	Agonist	1pb7	Closed
NR1/D-serine	Agonist	1pb8	Closed
NR1/D-cycloserine	Partial agonist	1pb9	No change of domain orientation compared to agonist form but binding affinity is lower
NR1/DCKA	Antagonist	1pbq	24° more open than with glycine /1pb7/
AMPA receptor GluR2 *Rattus norvegicus*			
Apo		1fto	Open 19° compared to glutamate bound /1ftj/
Glutamate	Agonist	1ftj	Closed 19° compared to apo form /1fto/
AMPA	Agonist	1ftm	Closed 21° compared to apo form /1fto/
CPW	Partial agonist	1syh	Partially opened 6° compared to glutamate bound /1ftj/
Kainate	Partial agonist	1fw0 1ftk	Partially opened 8° compared to glutamate bound /1ftj/
DNQX	Antagonist	1ftl	Open, apo-like, 19° compared to glutamate bound /1ftj/
ATPO	Antagonist	1n01	Open, apo-like, 19° compared to glutamate bound /1ftj/
S-glutamate	Agonist	2cmo	More open conformation of the D1, D2 domains compared to apo form, hyperextension of contacts
(S)-NS1209	Antagonist		
CNQX	Antagonist	3b7d	Partial domain closure not tranduced into linker separation
	Partial agonist in GluR2/ TARP		
2-me-Tet-Ampa	Agonist	2p2a	Extension of ligand contacts
			Discrimination GluR2–4 from GluR1 by ten-fold preference
S729C mutant/ glutamic acid	Agonist	2i3w	Compensation for the 21° glutamate induced domain closure
G725C mutant/ glutamic acid	Agonist	2i3v	
L483Y mutant/ AMPA	Agonist	1 lb8	Stabilization of intradimer interface
L483Y mutant/ DNQX	Antagonist	1 lb9	Stabilization of intradimer interface
Fluoro-willardiine	Positive modulators	2al5	Stabilization of agonist bound closed-cleft conformation, slowing of agonist unbinding and deactivation
Aniracetam			

(continued)

Table 4.1 (continued)

Ligands	Ligand activity	PDB code	Conformational change
Kainate receptors GluR5–7 *Rattus norvegicus*			
GluR5/glutamate	Agonist	2f36	Closed 25° relative to GluR2 apo form /1fto/
		1ycj	
GluR5/domoic acid	Partial agonist	2pbw	Open 11° compared to glutamate bound / 2f36/
GluR5/ATPO	Antagonist	1vso	Open 28° compared to glutamate bound / 2f36/
GluR5/UBP302	Antagonist	2f35	Open 29° compared to glutamate bound / 2f36/
GluR5/UBP310	Antagonist	2f34	Open 30° compared to glutamate bound / 2f36/
GluR6/glutamate	Agonist	1 s50	Closed
		1s7y	
GluR6/domoic acid	Partial agonist	1yae	Open 13° compared to glutamate bound / 1 s50/
			Contacts to lobe I only
GluR6/2 s, 4r-4-Methylglutamate	Agonist	1sd3	Closed 26.4° relative to GluR2 apo form / 1fto/
GluR6/quisqualate	Agonist	1s9t	Closed 26.2° relative to GluR2 apo form / 1fto/
GluR6/L-glutamate	Agonist	1ii5	Closed 26.6° relative to GluR2 apo form / 1fto/
GluR6/kainate	Partial agonist	1tt1	Closed 23.3° relative to GluR2 apo form / 1fto/
GluR6 ELKQ mutant/ glutamic acid	Agonist	2i0b	Introduction of 4 conserved AMPA receptor residues into kainate receptor to create non-desensitizing phenotype
GluR6 Y490C, L752C mutant/ glutamic acid	Agonist	2i0c	Disulfide bond

Structures of kainate receptor in complex with antagonists UBP302 and UBP310 show novel mechanism of binding different from that described for AMPA receptors with hyperextension of the ligand-binding core and a 22 A extension of the ion channel linkers compared to glutamate-bound forms (Mayer et al. 2006).

Deactivation and Desensitization

Receptor after activation by bound agonist can be either deactivated if agonist unbinds from receptor returning to apo state or desensitized if agonist stays bound to the inactive receptor.

Heterodimeric interface NR1-NR2A of NMDA receptor is involved in modulation of the rate of ion channel deactivation via participation of Tyr535 of NR1. From the comparison of heterodimer NR1-NR2A S1S2 and homodimer GluR2 S1S2, it can be seen that there are two regions (mainly helix D of one subunit and helix J of another subunit) that are involved in modulating receptor desensitization. Another region with key amino acids P532 (NR1) and P527 (NR2A) defines binding site of small molecules such as aniracetam that modulate receptor deactivation by stabilizing closed-cleft activated conformation. Closed-cleft conformation is stabilized either by interdomain hydrogen bonds or by locking the "clamshell" hinge as a result of modulator binding (Furukawa et al. 2005).

Conformational changes, which characterize desensitized state can be seen in AMPA receptor mutants L483Y (Sun et al. 2002), S729C and G725C (Armstrong et al. 2006). Stabilization of the intradimer interface by either mutations or allosteric modulators reduces desensitization (Sun et al. 2002). Desensitization involves the rupture of an interface between the N terminal domains of the two-fold related glutamate-binding subunits. This compensates 21 degrees of domain closure induced by glutamate binding. The rupture of the domain interface causes ion channel to close. Therefore, it provides explanation how agonist binding is decoupled from ion channel gating upon desensitization of the receptor (Armstrong et al. 2006).

Release of glutamate from presynaptic terminals and binding to postsynaptic AMPA receptors result in conformational changes, which lead to rapid and near-synchronous ion channel opening to provide efficient transmission. Glutamate must dissociate from the receptor to enable discrimination of closely spaced in time presynaptic action potentials. Therefore, the binding domain must open. Structures of GluR2 ligand binding core T686 s and T686a mutants give evidence that binding domain closing preceedes channel opening (Zhang et al. 2008).

Non-desensitizing receptor was designed for GluR5, GluR6, and GluR7 with mutant Y490C and L752C forming intermolecular disulfide bond (Weston 2006), which is equivalent to GluR2 L483Y mutant.

Receptor Modulation by Ions

AMPA and kainate receptors are activated and antagonized by the same family of compounds and have similar structure, amino acid composition, and properties of ion permeability. Among characteristics which distinguish between the two receptors is their sensitivity toward ions. In contrast to GluR3 (AMPA), kainate receptors response is reduced by the absence of extracellular sodium ions. With other monovalent ions used, sensitivity of the kainate receptor-mediated current to the removal of the ion decreases in order Na^+, Li^+, K^+, Cs^+. This effect is not caused by the change of the receptor affinity toward glutamate due to interaction with ion. Kainate receptor-specific gating mechanism is associated with amino acid residue M770 of the S2 fragment. Its mutation to lysine, present at equivalent position in AMPA

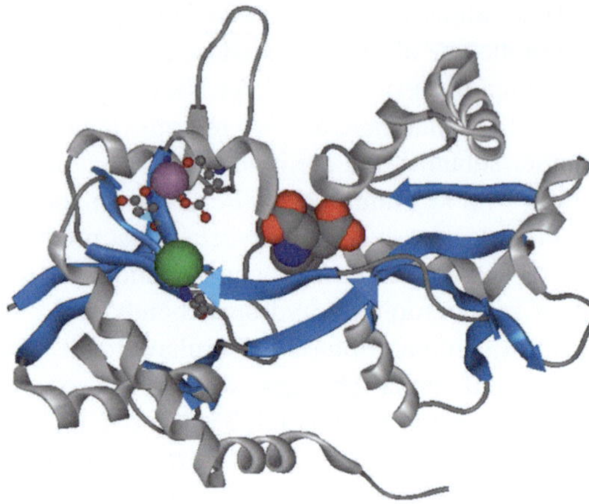

Fig. 4.2 Domain S1S2 of kainate receptor GluR5 subunit (helices grey, β-sheet blue) in complex with Na^+ (magenta) and Cl^- (green) ions. Amino acids coordinating ions Glu569, Ile572, Asp573, Lys576 are shown as ball and stick models

receptor, is sufficient to remove sensitivity. Nonpolar nature and size of methionine (or leucine/isoleucine) are important to retain this effect. In S1 segment, amino acid D528K or D528C replacements also abolish dependence of the receptor on ions (Paternain et al. 2003). Activation of kainate receptors by glutamate requires presence of Na^+ and Cl^- ions in the extracellular solution. Cl^- is essential structural component of the receptor bound in the cavity formed at the interface between two subunits. Absence of Cl^- results in reduced dimer stability, increased rate of desensitization and drop in the portion of the receptors competent for activation by glutamate (Plested and Mayer 2007). Since introducing positively charged amino acids in the proximity of position 770 abolishes effect of anions and cations, Na^+ and Cl^- ions coactivate GluR6 receptors by establishing a dipole. Using charged amino acids as tethered ions, researchers demonstrated that the docking order was fixed with cations binding first, followed by anions. Divalent ions Ca^{2+} and Mg^{2+}, but not H^+ are equivalent to Na^+ in coactivation of the kainate receptors. This study demonstrates a mechanism by which allosteric anions and cations regulate ligand-binding dimer stability and therefore the rate of dimerization and receptor availability for gating by glutamate (Wong et al. 2007). Crystallographic structures of complexes show that allosteric anions and cations bind at physiologically discrete sites (Fig. 4.2) with strong electric fields (Plested et al. 2008).

Receptor Dynamics

NMR experiment with GluR2 ligand binding core bound to the 5-substituted willardiine partial agonists demonstrates that the protein backbone amid chemical shift deviations correlate mainly with extent and efficacy of desensitization for those amino acid residues of S1S2 domain that are located near the putative hinge axis of lobe closure (Fenwick and Oswald 2008). The ligands show correlation of the

5-position substituent size with ligand efficacy and extent of receptor desensitization. Substituent electronegativity correlates with ligand potency.

Amino Acid Transporters

Enzymes and membrane transporters maintain cell homeostasis providing amino acid processing (synthesis and degradation) and flow (absorption, distribution and excretion). Alanine, serine, cysteine transporters (ASCT), such as ASCT1 and ASCT2, transport neutral amino acids. ASCT2, in fact, preferentially bidirectionally transports glutamine. ASCT2 tertiary structure (6gct) comprises a trimer in which each monomer contains fixed and elevator domains. ASCT link to proliferative pathways and growth makes it target of anticancer therapies (Scalize et al. 2018).

Glutamate/ Glutamine transporters Glutamatergic excitatory neurotransmission in the central nervous system involves the following processes: loading of glutamate into synaptic vesicles at presynaptic terminals by glutamate transporters, release of glutamate into synaptic clefts as action potential propagates to presynaptic terminals, activation of glutamate receptors on synaptic membranes, clearance of glutamate and preparation for the next signal. Glutamate/ glutamine cycle includes conversion to glutamine by glutamine synthetase in astrocytes and back to glutamate by glutaminase in neurons after being transferred from astrocytes to neurons (Hayashi 2018) (Chap. 1). The transporters of the glutamate/glutamine cycle are transmembrane proteins that use electrochemical gradient of various ions between the membranes: plasma membrane glutamate transporters EAAT and SNAT on the plasma membrane of presynaptic terminals and astrocyte processes or vesicular glutamate transporter vGLuT on synaptic vesicles.

Trimeric glutamate transporters EAAT1 bind substrate by a pair of helical hairpins (Fig. 4.3). Each hairpin is a helix/turn/helix substructure. Helix adjacent to the substrate binding hairpin coordinates sodium. Three sodium ions and one

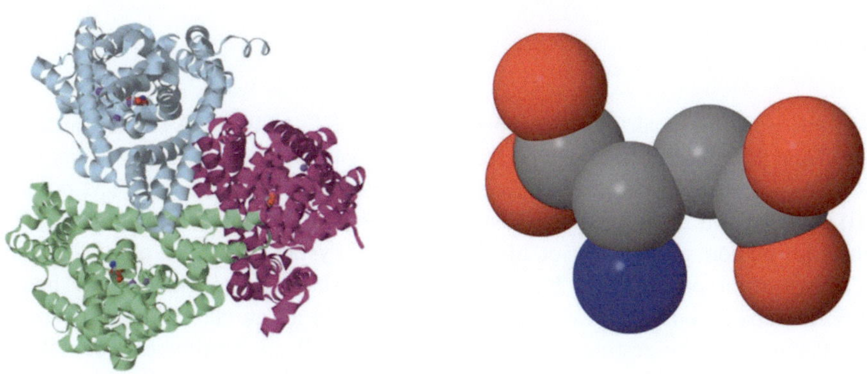

Fig. 4.3 Glu transporter and its substrate /5e9s/

proton enter the cell and one potassium ion leaves the cell with glutamate transport. Lipids modulate the transporter activity, particularly movement of its transport domain, by binding at the site between the trimerization and transport domains.

ASCT1, ASCT2, SLC1A4, AND SLC1A5 transport alanine, serine, and cysteine.

GABA and Glycine Receptors

Studies on excitatory circuits of cellular and molecular mechanisms that regulate synaptic transmission, wiring patterns of circuits, and specific propagation of connections significantly advanced. In contrast, similar work with respect to inhibitory circuits mediated by GABA and glycine is slower.

Only recently new approaches started to address parts of the CNS that show developmental changes in inhibitory circuits involving transmitter or receptor composition at the inhibitory synapses. Some new results show that disturbances in the same circuit in the development or maintenance of one type of inhibitory connection influences other inhibitory synapse types. Particularly important for the development of innovative therapeutic approaches is understanding inputs that co-release both neurotransmitters, GABA and glycine.

Scaffolding and intracellular signaling proteins play an important role in organizing dynamic network interactions of the receptor subunits. Postsynaptic scaffold protein gephyrin binds both glycine and GABA receptors at overlapping binding sites selectively targeting GABA or glycine receptor subunits. Inhibitory signaling protein collybistin, binds gephyrin, subunits NL2, and NL4, and affects both GABAergic and glycinergic synapses. Specific patterns of wiring and mechanisms of connections between GABAergic interneurons in the cerebellum involve various cell adhesion molecules that target the AIS and GABAergic Purkinje cells, Sema3A and semaphorin receptor neuropilin-1 (NRP1), neurofascin, and ankyrin G (Gamlin et al. 2018).

ATPase

Sodium potassium adenosine triphosphate (Na^+,K^+–ATPase) ATPase transports Na + and K+ through the plasma membrane to provide electrochemical gradients as essential energy source for transport of substances. It consists of TM domain (10 helices), and three cytoplasmic domains: the activator, nucleotide binding, and phosphorylation (Gregersen et al. 2016). The crystal structures of of Na^+,K^+-ATPases from porcine and bovine kidney and shark rectal gland show bound BeF3, ouabain and Mg^{2+} ions. ATPases function in all animal cells is to support maintenance of the resting membrane potential, cell excitability and activity of transporters. The cycle consists of the extrusion of three sodium ions and taking in of two potassium ions transported into the cell.

Cardiac glycosides, natural compounds and Na,K-ATPases ligands, are attractive drug candidates against cancer (Schneider et al. 2018). Natural cardenolide digitoxigenin monodigitoxoside (DGX) is one of the candidates against non-small cell lung cancer.

Ouabain is a cardiotonic steroid. Ouabain binding involves an extensive hydrogen bonding network of the ß-surface of its steroid core and the side chains of αM1, αM2, and αM6 and requires the presence of Mg^{2+}; high-affinity binding also requires autophosphorylation of the Na^+,K^+-ATPase. The presence of monovalent cations inhibits ouabain binding to the enzyme.

Three cation transport sites I, II, and III, bind cations. Site II can be occupied by Mg^{2+}, Rb^+ or Mn^{2+}, but not Na^+. Ouabain binding pocket accommodates deeply bound ouabain with long-range interactions between its polarized five-membered lactone ring and the Mg^{2+}. If K^+ binds the same site causing unwinding of a turn of αM4 and stimulating Ile318-Val325 interactions with the cation site, ouabain deep binding is hindered as we can see in E2P–ouabain complex of the homologous porcine Na^+,K^+-ATPase; porcine E2P–bufalin Na^+,K^+–ATPase structure: E2P–CTS structures of the porcine kidney enzyme (Laursen et al. 2013).

Transient Receptor Potential, TRP

TRP superfamily comprises seven subfamilies. Structure of TRP channels contains a tetramer in which each subunit has six-transmembrane helices organized around a cation-permeable pore. They carry out important functions in all organisms (Latorre et al. 2009). Ca^{2+} pumps, ion exchangers, and TRP channels are main molecules for Ca^{2+} entrance into the cell. Regulation of calcium ion homeostasis by TRP channels of TRPV (TRPV6) and TRPM (TRPM1, TRPM7, TRPM8) subfamilies is explored in design of new therapies targeting proliferation pathways in cancer cells (Dou et al. 2013; Chen et al. 2014; Chen et al. 2016; Zsombok and Derbenev 2016). Cellular calcium maintenance is important for proliferation, migration, and carcinogenesis. Ca^{2+}, calmodulin (CaM), CaM kinase, cyclin dependent kinases, Cdks, and their activators and inhibitors are key molecules regulating cell cycle. Rb phosphorylation by cyclin dependent kinase triggers progression of G1 to S cycle of cell division (Krzywda et al. 2004).

Members of TRP are extensively studied by CryoEM and resolution rapidly improves (Table 4.2). After first successful crystallographic structures of sodium/potassium voltage-gated channels (Jiang et al. 2003), more and more newly determined structures appear. TRPV1 belongs to vanilloid subfamily. It is activated by such stimuli as capsaicin, anandamide, heat, phosphatidylinositol 4,5-biphosphate ($PI(4,5)P_2$), and protons and binds numerous ligands as we observe in TRPV1/capsaicin, TRPV1/resiniferatoxin, TRPV1/DkTx (spider double knot toxin) and TRPV1/ATP complexes. TRPV1/$PI(4,5)P_2$ complex shows that phospholipid binding site partially overlaps with capsaicin binding site: current induced by $PI(4,5)P_2$ is smaller if capsaicin is added after $PI(4,5)P_2$ (Poblete et al. 2015). TRPV1 has

Table 4.2 Structure of TRP channels

Channel/ subfamily	Source	Resolution, Angstrom	PDB entry
TRPV1	*Rattus norvegicus*	3.275	2nyj 3j5p
TRPV2	*Orictolagus cuniculus*	2.2 3.9	2eta 6bwm
TRPV3	*Mus musculus*	4.3	6dvw
TRPV4	*Gallus gallus*	1.9	3w9f 6bbj
TRPA1	*Homo sapiens*	4.24	3j9p
TRPML	*Callithrix jacchus*	2.94	5w3s
NOMPC/TRPN	*Drosophila melanogaster*	3.55	5vkq

potential for drug development for its role in the suppression of seizure (Gonzalez-Reyes et al. 2013). ATP and PIP2 sensitize TRPV1 (Lishko et al. 2007). TRPV1 desensitization is calcium-dependent (Numazaki et al. 2003). The Ca^{2+}-CaM / TRPV1 interaction was found to reduce channel open probability (Rosenbaum et al. 2004). TRPV1-ARD binds CaM, and this interface is shared with triphosphate nucleotides. Interaction with CaM is crucial to inactivation of the channel upon repeated stimulation. Ankyrin repeats domain (ARD) of TRPA1, similarly to that of TRPV, is the site of interaction of several ligands (Lishko et al. 2007; Nilius et al. 2011). Ankyrin repeats in TRPs are implicated in tethering proteins to membrane, binding wide variety of ligands, spring behavior and mechanotransduction (Gaudet 2008). Intracellular PPPi that bind ARD are important for channel activation, stabilize the channel and increase channel conductance (Nilius et al. 2011). ARD is necessary for plasma membrane localization of TRPA1 (Pareek et al. 2007). Expressed in nociceptive, or pain sensitive, neurons, TRPA1 responds to electrophilic (acrolein in smoke, allylisothiocyanate in mustard oil, cinnamaldehyde, formaldehyde) and nonelectrophilic (menthol, carvacrol, nicotine) chemical irritants, protons, and noxious cold. Electrophilic compounds activate TRPA1 through covalent modification of three cysteines. TRPA1 is targeted as candidate for analgesics and treatments of airway diseases (Iadarola et al. 2018). Agonist induced sensitization (increase in currents passing through the channel) occurs through a longlasting and reversible change in structure; sensitization is followed by shift of voltage dependent activation as gain of function. Other mechanisms suggest relief of voltage dependent pore block by calcium or time dependent pore dilation enhanced by the absence of calcium (Meents et al. 2016).

Hot sensing and cold sensing properties of TRP ion channels reside in the same molecule. The rationale for TRP tempetature sensitivity is provided by a difference in heat capacity of the open and closed states (Clapham and Miller 2011).

Successful crystallization and structure determination of membrane proteins brings new molecules.

NOMPC (NO Mechanoreceptor Potential C isoform L) is a mechanotransduction channel involved in hearing and touch sensation. Recently determined electron cryomicroscopy structure of *Drosophila melanogaster* channel at 3.6 Å contains large ankyrin repeats domain which tether the channel to microtubules and cytoskeleton (Jin et al. 2017). NOMPC 29 ankyrin repeats form a helical spring

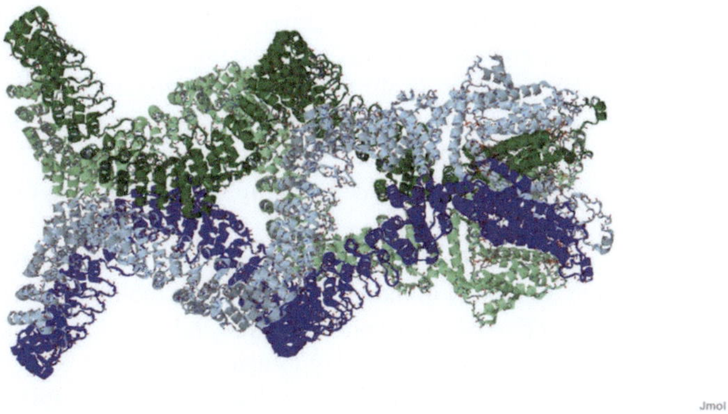

Fig. 4.4 NOMPC /5vkq/

(Fig. 4.4). Another mechanotransduction channel, *Xenopus* TRPN1, contains N terminal cytoplasmic domain with 28 ankyrin repeats interacting with penta EF hand protein peflin (Wiemuth et al. 2011). *Nectria haematococca* Ca2+ activated lipid scramblase NHTMEM16 and chloride channel NHTMEM16A structures contain ten transmembrane helices with hydrophilic interior cavity and shows how mutations of calcium coordinating residues influence lipid binding and ion conduction (Brunner et al. 2018).

Channelopathies

Channelopathies are mutations in ion channel or related proteins associated with more than 55 inherited human diseases. The drugs that act through targeting ion channels are widely used for treatment of channelopathies and many of these molecules are in developmental stages.

SCN5A gene encodes the pore-forming α-subunit of the cardiac Na+ channel (NaV1.5). Patients with **cardiac syndromes** and **electrocardiogram abnormalities** carry multiple mutations in this gene. Many of these mutations exhibit multiple disease phenotypes mainly due to changes in the current generated by NaV1.5 channels. SCN5A mutations also can be associated with extracardiac phenotypes such as gastrointestinal diseases and epilepsy (Verkerk et al. 2018).

Long-QT syndrome, a cardiac arrhythmia, is caused by mutations in human ether-a-go-go-related gene17 (hERG) which encodes a voltage gated K^+ channel that produces cardiac IKr currents These mutations or action of some drugs disrupt currents and cell excitability (Jiang et al. 2017). Mutations in another gene that is K^+ channel, KCNK4, result in cardiac arrhythmias, altered neurotransmission, endocrine dysfunction, and disturbances in development (Bauer et al. 2018). Many channelopathies are linked to mutations in CA^{2+} channels (Pancaroglu et al. 2018).

The astrocytic glutamate transporters-1 (GLT-1) remove excess of synaptic glutamate to prevent its accumulation to toxic levels. GLT-1 and glutamine synthetase are key proteins that maintain healthy extracellular glutamate concentrations. **Transient global cerebral ischemia** (tGCI) causes excessive release of glutamate from neurons (Li et al. 2018). Channels such as connexins important for building gap junction hemichannels in astrocytes and cellular-Src (c-Src) regulate expression of GLT-1.

Charcot-Marie-Tooth (CMT) disease, a motor and sensory neuropathy is a channelopathy linked to mutations in more than 120 genes including TRP channels. It is characterized by degeneration of peripheral nerves and muscle weakness (Juneja et al. 2018) and continuous motor unit activity (Vasilescu 1984). CMT gene testing on 2 unrelated families with CMT identifies 2 novel heterozygous missense mutations, p.Arg237Gly and p.Arg237Leu, in the TRPV4 ankyrin repeats domain, mainly convex face of the TRPV4 tetramer. The mutations do not change TRPV4 subcellular localization but cause elevations of cytosolic Ca2+ levels and cytotoxicity (Sullivan et al. 2015). Studies of other genes involved in CMT show mutations in glycyl-tRNA synthetase (Xie et al. 2007) and phosphoinositide phosphatase (Begley et al. 2003).

Spinocerebellar ataxia is one of channelopathies (Sullivan et al. 2015). KChIP3 is high-affinity Ca^{2+} sensor that prevents oversecretion of mucin (Cantero-Recasens et al. 2018). Skeletal muscle channelopathies **non-dystrophic myotonia** and **periodic paralysis** are currently studied for the presence of mutations (Phillips et al. 2018). In patients suffering from type 1 hypokalaemic periodic paralysis the disease is linked to missense mutations in the a1 subunit of the dihydropyridine receptor which resides in the tubular membrane of skeletal muscle fibres which regulate the Ca^{2+} release from sarcoplasmic reticulum. Movement of monovalent cations through the mutated Ca^{2+} channels and depolarization provide explanation of the pathological processes (Allard et al. 2018).

Type 2 Diabetes TRP ion channels such as TRPA1 and TRPV1 are expressed in nonsensory pancreatic beta cells, stimulate insulin secretion, and are implicated in channelopathies/diabetes (Cao et al. 2012; Malenczyk et al. 2017). TRPA1-mediated depolarization and synchronized inhibition of TRPA1 channel induces insulin release. TRPA1 agonists, endogenous and exogenous, trigger Ca2+ influx and stimulate insulin release from pancreatic beta cells (Cao et al. 2012). As Ca^{2+}-sensor proteins, TRPA1 mediate insulin release via SNARE interactions. Many pathways involving TRP channels are potential sites of antidiabetic drug design (Ahlstrom et al. 2014).

Inside Protein Molecule

All types of biological macromolecules, proteins, nucleic acids, carbohydrates, and lipids, contain carbohydrate chains. Repeating units of various types as building blocks fold into regular three dimensional structure in which arrangement of atomic groups in space is governed by molecular forces, electrostatic, London dispersion, van der Waals and hydrogen bonding interactions. Molecular packing is very important characteristic of protein structure. According to the model of protein interior proposed by Kauzmann (1959), molecule is similar to an organic liquid or "oil-drop." Packing density of lysozyme and ribonuclease (Richards 1977) is approximately 0.75 which indicates that the protein interior is similar to an organic crystal. Amino acids of the same type occupy the same volume inside the protein core (Chothia 1975). With protein interior so well packed and resembling more solid than liquid, the best description is molecular crystal. Protein interior is heterogeneous with values of packing efficiency from .66 to .84 (Richards 1977). Mobile regions near active sites are less dense. If packing efficiency is addressed for atomic group types, we can see that low density regions are mainly located at the "knob/hole" interfaces where side chains of one secondary structure element packs against the backbone of the second structure element. We can conclude that grooves of hydrogen bonded secondary structure elements host specific channels protected from the solvent exposure by side chain which protrudes from the ridges and packs into the grooves. The carbonyl oxygens, -CH3 groups of aliphatic side chains, and -CH- groups of aromatic side chains have lower than average packing density value. In the structure, these groups come together in one place and form a channel. Each of these channels has a uniform structure: hydrogen bonded resonance polar groups of the peptide group on one side and packed nonpolar methyl groups of side chains on the other side (Kurochkina and Privalov 1996). Besides, packed nonpolar methyl groups provide a structure of carbon lattice hosting carbon/hydrogen covalent bonds and hydrogen/hydrogen van der Waals bonds, which are also sites of low density/high density combination. Presence of such regularly organized low density and high density regions allows to explain many processes inside protein molecule such as diffusion of ions and molecules through protein interior, channeling of substrates and products during catalytic reaction, and breathing/pumping motions of the molecule.

References

Ahlstrom M, Roxhed N, Nordquist L (2014) Intradermal insulin delivery: a promising future for diabetes management. J Diabetes Sci Technol 8(3):453–457. https://doi.org/10.1177/1932296814530060. Epub 2014 Apr 10. Review. PubMed PMID: 24876605; PubMed Central PMCID: PMC4455430

Armstrong N, Gouaux E (2000) Mechanisms for activation and antagonism of an AMPA-sensitive glutamate receptor: crystal structures of the GluR2 ligand binding core. Neuron 28:165–181

Armstrong N, Jasti J, Beich-Frandsen M, Gouaux F (2006) Measurement of conformational changes accompanying desensitization in an ionotropic glutamate receptor. Cell 127:85–97

Ataman ZA, Gakhar L, Sorensen BR, Hell JW, Shea MA (2007) The NMDA receptor NR1 C1 region bound to calmodulin: structural insights into functional differences between homologous domains. Structure 15:1603–1617

Bauer CK, Calligari P, Radio FC, Caputo V, Dentici ML, Falah N, High F, Pantaleoni F, Barresi S, Ciolfi A, Pizzi S, Bruselles A, Person R, Richards S, Cho MT, Claps Sepulveda DJ, Pro S, Battini R, Zampino G, Digilio MC, Bocchinfuso G, Dallapiccola B, Stella L, Tartaglia M (2018) Mutations in KCNK4 that Affect Gating Sullivan R, Yau WY, O'Connor E, Houlden H. Spinocerebellar ataxia: an update. J Neurol. https://doi.org/10.1007/s00415-018-9076-4. [Epub ahead of print] PubMed PMID: 30284037

Begley MJ, Taylor GS, Kim SA, Veine DM, Dixon JE, Stuckey JA (2003) Crystal structure of a phosphoinositide phosphatase, MTMR2: insights into myotubular myopathy and Charcot-Marie-Tooth syndrome. Mol Cell 12(6):1391–1402. PubMed PMID: 14690594

Cantero-Recasens G, Butnaru CM, Valverde MA, Naranjo JR, Brouwers N, Malhotra V (2018) KChIP3 coupled to Ca(2+) oscillations exerts a tonic brake on baseline mucin release in the colon. Elife. 7. pii: e39729 Phillips L, Trivedi JR. Skeletal Muscle Channelopathies. Neurotherapeutics https://doi.org/10.1007/s13311-018-00678-0. [Epub ahead of print] Review. PubMed PMID: 30341599

Cao DS, Zhong L, Hsieh TH, Abooj M, Bishnoi M, Hughes L, Premkumar LS (2012) Expression of transient receptor potential ankyrin 1 (TRPA1) and its role in insulin release from rat pancreatic beta cells. PLoS One 7(5):e38005. https://doi.org/10.1371/journal.pone.0038005. Epub 2012 May 31. PubMed PMID: 22701540; PubMed Central PMCID: PMC3365106

Chen PE, Wyllie DJA (2006) Pharmacological insights obtained from structure–function studies of ionotropic glutamate receptors. BJ Pharm 147:839–853

Chothia C (1975) Structural invariants in protein folding. Nature 254:304–308

Clapham DE, Miller C (2011) A thermodynamic framework for understanding temperature-sensing by TRP channels. Proc Natl Acad Sci USA 108:19492–19497

Dou Y, Li Y, Chen J, Wu S, Xiao X, Xie S, Tang L, Yan M, Wang Y, Lin J, Zhu W, Yan G (2013) Inhibition of cancer cell proliferation by midazolam by targeting transient receptor potential melastatin 7. Oncol Lett 5(3):1010–1016. Epub 2013 Jan 11. PubMed PMID: 23426784; PubMed Central PMCID: PMC3576402

Erreger K, Chen PE, Wyllie DJ, Traynellis SF (2004) Glutamate receptor gating. Crit Rev Neurobiol 16:187–124

Fenwick MK, Oswald RE (2008) NMR spectroscopy of the ligand-binding core of ionotropic glutamate receptor 2 bound to 5-substituted willardiine partial agonists. J Mol Biol. 378:673–685

Furukawa H, Gouaux E (2003) Mechanisms of activation, inhibition and specificity: crystal structures of the NMDA receptor NR1 ligand-binding core. Embo J 22:2873–2885

Furukawa H, Singh SK, Mancusso R, Gouaux E (2005) Subunit arrangement and function in NMDA receptors. Nature 438:185–192

Gamlin CR, Yu WQ, Wong ROL, Hoon M (2018) Assembly and maintenance of GABAergic and Glycinergic circuits in the mammalian nervous system. Neural Dev 13(1):12. https://doi.org/10.1186/s13064-018-0109-6. Review. PubMed PMID: 29875009; PubMed Central PMCID: PMC5991458

Gaudet R (2008) A primer on ankyrin repeat function in TRP channels and beyond. Mol Biosyst 4:372–379

Gonzalez-Reyes LE, Ladas TP, Chiang CC, Durand DM (2013) TRPV1 antagonist capsazepine suppresses 4-AP-induced epileptiform activity in vitro and electrographic seizures in vivo. Exp Neurol 250:321–332. https://doi.org/10.1016/j.expneurol.2013.10.010. Epub 2013 Oct 18. PubMed PMID: 24145133; PubMed Central PMCID: PMC4104534

Gregersen JL, Mattle D, Fedosova NU, Nissen P, Reinhard L (2016) Isolation, crystallization and crystal structure determination of bovine kidney Na(+),K(+)-ATPase. Acta Crystallogr F Struct

Biol Commun 72(Pt4):282–287. https://doi.org/10.1107/S2053230X1600279X. Epub 2016 Mar 16. PubMed PMID: 27050261; PubMed Central PMCID: PMC4822984

Hald H, Naur P, Pickering DS, Sprogoe D, Madsen U, Timmermann DB, Ahring PK, Liljefors T, Schousboe A, Edebjerg J, Gajhede M, Kastrup JS (2007) Partial agonism and antagonism of the Ionotropic Glutamate Receptor iGLuR5. J Biol Chem 282:25726–25736

Hayashi MK (2018) Structure-Function Relationship of Transporters in the Glutamate-Glutamine Cycle of the Central Nervous System. Int J Mol Sci 19(4):pii: E1177. https://doi.org/10.3390/ijms19041177. Review. PubMed PMID: 29649168; PubMed Central PMCID: PMC5979278

Iadarola MJ, Sapio MR, Raithel SJ, Mannes AJ, Brown DC (2018) Long-term pain relief in canine osteoarthritis by a single intra-articular injection of resiniferatoxin, a potent TRPV1 agonist. Pain 159(10):2105–2114. https://doi.org/10.1097/j.pain.0000000000001314. PubMed PMID: 30015705

Jiang Y, Lee A, Chen J, Ruta V, Cadene M, Chait BT, MacKinnon R (2003) X-ray structure of a voltage-dependent k+ channel. Nature 423:33–41

Jiang YY, Hou HT, Yang Q, Liu XC, He GW (2017) Chloride channels are involved in the development of atrial fibrillation–A transcriptomic and proteomic study. Sci Rep 7(1):10215. doi: 10.1038

Jin P, Bulkley D, Guo Y, Zhang W, Guo Z, Huynh W, Wu S, Meltzer S, Cheng T, Jan LY, Jan YN, Cheng Y (2017) Electron cryo-microscopy structure of the mechanotransduction channel NOMPC. Nature. 547(7661):118-122. https://doi.org/10.1038/nature22981. Epub 2017 Jun 26. PubMed PMID: 28658211; PubMed Central PMCID: PMC5669069.

Juneja M, Burns J, Saporta MA, Timmerman V (2018) Challenges in modelling the Charcot-Marie-Tooth neuropathies for therapy development. J Neurol Neurosurg Psychiatry. pii: jnnp-2018-318834. doi: 10.1136/jnnp-2018-318834. [Epub ahead of print] PubMed PMID: 30018047

Kasper C, Pickering OS, Mirza O, Olsen L, Kristensen AS, Greenwood JR, Liljefors T, Schousboe A, Watjen F, Gajhede M, Siqurskjold BW, Kastrup JS (2006) The structure of a mixed GluR2 ligand-binding core dimer in complex with (S)-glutamate and the antagonist (S)-NS1209. J Mol Biol 357:1184–1201

Kauzmann W (1959) Some factors in the interpretation of protein denaturation. Adv Protein Chem 14:1–63

Krzywda S, Brzozowski AM, Higashitsuji H, Fujita J, Welchman R, Dawson S, Mayer RJ, Wilkinson AJ (2004) The Crystal Structure of Gankyrin, an Oncoprotein Found in Complexes with Cyclin-dependent Kinase 4, a 19 S Proteasomal ATPase Regulator, and the Tumor Suppressors Rb and p53. J Biol Chem *279*:1541–1545

Kunishima N, Shimada Y, Tsuji Y, Sato T, Yamamoto M, Kumasaka T, Nakanishi S, Jingami H, Morikawa K (2000) Structural basis of glutamate recognition by a dimaric metabotropic glutamate receptor. Nature 407:971–977

Laursen M, Yatime L, Nissen P, Fedosova NU (2013) Crystal structure of the high-affinity Na+K+-ATPase ouabain complex with Mg2+ bound in the cation binding site. Proc Natl Acad Sci U S A 110(27):10958–10963. https://doi.org/10.1073/pnas.1222308110. Epub 2013 Jun 17. PubMed PMID: 23776223; PubMed Central PMCID: PMC3704003

Li K, Zhou H, Zhan L, Shi Z, Sun W, Liu D, Liu L, Liang D, Tan Y, Xu W, Xu E (2018) Hypoxic preconditioning maintains GLT-1 against transient global cerebral ischemia through upregulating Cx43 and inhibiting c-Src. Front Mol Neurosci 11:344. https://doi.org/10.3389/fnmol.2018.00344. eCollection 2018. PubMed PMID: 30323740; PubMed Central PMCID: PMC6172853

Lishko PV, Procko E, Jin X, Phelps CB, Gaudet R (2007) The ankyrin repeats of TRPV1 bind multiple ligands and modulate channel sensitivity. Neuron 54:905–918

Mayer ML (2005) Glutamate receptor ion channels. Curr Opin Neurobiol 15:282–288

Mayer ML, Ghosal A, Colman NP, Jane DE (2006) Crystal structures of the kainate receptor GluR5 ligand binding core dimer with novel GluR5-selective antagonists. J. Neurosci 26:2852–2861

McFeeters RL, Oswald RE (2004) Emerging structural explanations of ionotropic glutamate receptor function. Faseb J 18:428–438

Meents JE, Fischer MJ, McNaughton PA (2016) Agonist-induced sensitisation of the irritant receptor ion channel TRPA1. J Physiol 594(22):6643–6660. https://doi.org/10.1113/JP272237. Epub 2016 Aug 4. PubMed PMID: 27307078; PubMed Central PMCID: PMC5108891

Menuz K, Stroud RM, Nicoll RA, Hays FA (2007) TARP auxiliary subunits switch AMPA receptor antagonists into partial agonists. Science 318:815–817

Nanao MH, Green T, Stern-Bach Y, Heinemann SF, Choe S (2004) Structure of the kainate receptor subunit GluR6 agonist-binding domain complexed with domoic acid. Proc Natl Acad Sci 102:1708–1713

Nilius N, Prenen J, Owsianik G (2011) Irritating channels: the case of TRPA1. J Physiol 589 (7):1543–1549

Oswald RE (2004) Ionotropic glutamate receptor recognition and activation. Adv Protein Chem 68:313–349

Paternain AV, Cohen A, Stern-Bach V, Lerma J (2003) A role for extracellular Na+ in the channel gating of native and recombinant kainate receptors. J Neurosci 23:8641–8648

Plested AJ, Mayer ML (2007) Structure and mechanism of kainate receptor modulation by anions. Neuron 53:829–841

Plested AJ, Vijauan R, Biggin PC, Mayer ML (2008) Molecular basis of kainate receptor modulation by sodium. Neuron 58:720–735

Poblete H, Oyarzún I, Olivero P, Comer J, Zuñiga M, Sepulveda RV, Báez-Nieto D, González Leon C, González-Nilo F, Latorre R (2015) Molecular determinants of phosphatidylinositol 4,5-bisphosphate (PI(4,5)P2) binding to transient receptor potential V1 (TRPV1) channels. J Biol Chem 290(4):2086–2098. https://doi.org/10.1074/jbc.M114.613620. Epub 2014 Nov 25. PubMed PMID: 25425643; PubMed Central PMCID: PMC4303662

Richards FM (1977) Areas, volumes, packing, and protein structure. Ann Rev Biophys Bioeng 6:151–176

Rosenbaum T, Gordon-Shaag A, Munari M, Gordon SE (2004) Ca2+/calmodulin modulates TRPV1 activation by capsaicin. J Gen Physiol 123(1):53–62. PubMed PMID: 14699077; PubMed Central PMCID: PMC2217413

Schneider NFZ, Cerella C, Lee JY, Mazumder A, Kim KR, de Carvalho A, Munkert J, Pádua RM, Kreis W, Kim KW, Christov C, Dicato M, Kim HJ, Han BW, Braga FC, Simões CMO, Diederich M (2018) Cardiac glycoside glucoevatromonoside induces cancer type-specific cell death. Front Pharmacol 9:70. https://doi.org/10.3389/fphar.2018.00070. eCollection 2018. PubMed PMID: 29545747; PubMed Central PMCID: PMC5838923

Chen S-C, Huang C-H, Lai S-J, Yang CS, Hsiao T-H, Lin C-H, Fu P-K, Ko T-P, Chen Y (2016) Mechanism and inhibition of human UDP-GlcNAc 2-epimerase, the key enzyme in sialic acid biosynthesis. Sci Rep 6:23274. https://doi.org/10.1038/srep23274

Song X, Jensen MØ, Jogini V, Stein RA, Lee CH, Mchaourab HS, Shaw DE, Gouaux E (2018) Mechanism of NMDA receptor channel block by MK-801 and memantine. Nature 556 (7702):515–519. https://doi.org/10.1038/s41586-018-0039-9. Epub 2018 Apr 18. PubMed PMID: 29670280; PubMed Central PMCID: PMC5962351

Sullivan JM, Zimanyi CM, Aisenberg W, Bears B, Chen D-H, Day JW et al (2015) Novel mutations highlight the key role of the ankyrin repeat domain in TRPV4-mediated neuropathy. Neurology: Genetics 1:e29

Sun Y, Olson R, Horning M, Armstrong N, Mayer M, Gouaux E (2002) Mechanism of glutamate receptor desensitization. Nature 417:245–253

Verkerk AO, Amin AS, Remme CA (2018 Oct 1) Disease Modifiers of Inherited SCN5A Channelopathy. Front Cardiovasc Med. 5:137. https://doi.org/10.3389/fcvm.2018.00137. eCollection 2018. Review. PubMed PMID: 30327767; PubMed Central PMCID: PMC6174200

Wong AYC, Maclean DM, Bowie D (2007) Na+/Cl- dipole couples agonist binding to kainate receptor activation. J Neurosci 27:6800–6809

Wong E, Ng F-M, Yu C-Y, Lim P, Lim L-H, Traynelis SF, Low CM (2005) Expression and characterization of soluble amino-terminal domain of NR2B subunit of N-methyl-D-aspartate receptor. Prot Sci 14:2275–2283

Xie W, Nangle LA, Zhang W, Schimmel P, Yang XL (2007) Long-range structural effects of a Charcot-Marie-Tooth disease-causing mutation in human glycyl-tRNA synthetase. Proc Natl Acad Sci USA 104(24):9976–9981. Epub 2007 Jun 1. PubMed PMID: 17545306; PubMed Central PMCID: PMC1891255

Zhang W, Cho Y, Lolis E, Howe JR (2008) Structural and single channel results indicate that the rates of ligand binding domain closing and opening directly impact AMPA receptor gating. J Neurosci 28:932–943

Zsombok A, Derbenev AV (2016) TRP Channels as Therapeutic Targets in Diabetes and Obesity. Pharmaceuticals (Basel) 9(3). pii: E50. https://doi.org/10.3390/ph9030050. Review. PubMed PMID: 27548188; PubMed Central PMCID: PMC5039503

Chapter 5
Helical Assemblies

Abstract Repeating structural motifs in proteins exist in many forms as groups of amino acids, secondary structure elements, domains or molecules. Protein assemblies of ankyrins, ARM/HEAT, tetratricopeptide and other repeats contain a repeating unit for which major part represents a pair of antiparallel alpha-helices. These repeating units are packed parallel to each other in a super-helix shaped as solenoid or barrels and rings. Ankyrin repeats form a left-handed spiral whereas ARM and HEAT repeats assemble into a right-handed spiral. Ankyrin repeats contain 24 whereas tetratricopeptide 8 repeats per turn. Repeats such as leucine-rich repeats and TIM-barrel contain one external row of alpha-helices and one internal row of beta-strands. Most of the repeats function as modules for binding other proteins. Amino acid sequence, specific interactions, and distribution on the surface of the secondary structure elements of amino acids determine structure of the assembly.

Keywords Helix · Assembly · Helix interface · Chirality · Enantioselectivity

Repeating character of structural motifs is property of many proteins including leucine zipper (O'Shea et al. 1991), myohemerithrin (Hendrickson and Ward 1977) and TIM-barrel (Wierenga 2001). Helical assemblies represent large group of protein interactions domains containing repeat units (Andrade et al. 2001b). This group includes toroids/spirals: ankyrin, ARM/HEAT, Sel1 like repeat (SLR), tetratricopeptide (TPR), pentatricopeptide (PPR), leucine-rich repeats (LRR), and barrels: 6α/6α, 8α/8β, rings of connexins and ATPases. The repeat units of ankyrin, ARM/HEAT, PUM, LRR, TAL, TPR, PPR, SLR, and other types comprising several secondary structure elements stacked parallel to each other form a toroid, helix of helices. Helical toroids and barrels are mostly scaffold proteins. Many repeats (ankyrin, ARM/HEAT, TPR) consist of two antiparallel α-helices stacked parallel to each other so that one helix forms inner (concave) row whereas another helix forms outer (convex) row. Repeats such as LRR (Kobe and Deisenhofer 1996; Kajava 1998) and TIM-barrel (Wierenga 2001) contain one external row of alpha-helices and one internal row of beta-strands. Most of the repeats function as modules for binding other proteins (Smith 2004). Conformational transitions accompany

N. Kurochkina, *Protein Structure and Modeling*,
https://doi.org/10.1007/978-981-13-6601-7_5

binding of ligands to superhelical structure in HEAT repeats of importin β (Lee et al. 2000; Forwood et al. 2010), CRM1 (Monecke et al. 2012), ARM (Huber et al. 1997), TPR (Allan et al. 2011) and other motifs. Ankyrin repeats possess elastic properties important in mechanotransduction (Lee et al. 2006). Measurements of strength of intermolecular interactions were performed for some repeats of helical assemblies such as those of clathrin (Jin et al. 2013). Consensus sequences, derived for ankyrins (Michaely et al. 2002; Mosavi et al. 2002), four groups of ARM/HEAT (Andrade et al. 1995; Malik et al. 1997), several groups of TPR (Blatch and La¨ ssle 1999; Kajander et al. 2006; Perry et al. 2006; Han et al. 2007; Sawyer, N. et al. 2013), and other repeats, help to design repeats that would be representatives of groups of proteins and carry new properties. Designed ankyrin repeats (Binz et al. 2003) have shown to have potential applications in biotechnology. Chiral properties of helical polymers are applied in the design of new materials for electronic devices (Yang, et al. 2013).

Amino acid combinations at helix-helix interfaces (Kurochkina 2007; Kurochkina 2008; Kurochkina and Choekyi 2011) and location of amino acids at helical edges (Kurochkina 2010) are important determinants of the fold that characterize positive and negative interhelical angles (or handedness) and overall shape of the helical assembly. Distribution of contacts on the edges of helical surfaces and characteristic amino acid combinations of helix-helix interfaces in ankyrin and HEAT/ARM repeat assemblies are unique. For both left-handed and right-handed assemblies, involvement of edges of helical surfaces in the formation of helix-helix interfaces follow characteristic pattern demonstrating chiral character of van der Waals interactions. Consistent with presence of negative interhelical angles is the fact that amino acid combinations of parallel ankyrin interfaces are identical or homologous to those of TIM-barrel. Amino acid combinations of parallel HEAT/ARM interfaces are homologous to those present at helix-helix interfaces with positive interhelical angles.

Assembly Shape and Helix Edges

Protein structural motifs such as four-α-helix bundle (Hendrickson et al. 1975), TIM-barrel 8α/8β fold (Alber et al. 1981), Rossmann dinucleotide binding fold (Rao and Rossmann 1973), helical rod or spiral (Andrade et al. 2001) exhibit a unique type of helix-helix interactions (Chap. 1). In each type of fold, edges of the helix surface, which participate in the formation of helix-helix contacts with surrounding helices (Fig. 1.26), determine direction and specific arrangement of the assembly. The association of the edge orientation of the helix surfaces with circular arrangements of the four, eight, and sixteen α-helices, which are found in the four-α-helical motif, TIM-barrel 8α/8β fold, and helical rod of $16.\bar{3}$ helices per turn correspondingly exists (Kurochkina 2010). Edges (i, i + 1) – (i + 1, i + 2) of the helix surface are central for the interhelical contacts in a four-α-helix bundle. Edges (i, i + 1) – (i + 2, i + 3) are involved in the assembly of four-α-helix subunits into helical rod of a tobacco mosaic virus and a three-helix fragment of a Rossmann fold. In 8α/8β TIM-barrel fold, edges (i, i + 1) – (i + 5, i + 6) are involved in the octagon arrangement. A good correlation exists between polygon interior angles and angles formed by the edges of helix surfaces; therefore, an approximation of a cross section of each motif with a polygon (n-gon, n = 4, 8, 16) suggests basis for overall shape of each assembly. Edges of the helix are also critical for assembly of spiral rods of helical repeats.

Regularly positioned secondary structure elements comprise protein molecule. For the majority of proteins, secondary structure elements compose almost 80% of the protein molecule; interactions of secondary structure elements are considered as major contributors to the fold of the molecule. The arrangement of contacts between α-helices in each of the structural motifs shows that there exists a relationship between the intrinsic properties of α-helix and the types of tertiary structure motifs. A TIM-barrel protein contains eight-stranded β-barrel surrounded by eight parallel α-helices. First discovered in triose phosphate isomerase (Alber et al. 1981), this structural motif is also a structural unit of many enzymes: pyruvate kinase, malate synthase, xylose isomerase, fructose-1, 6-bisphosphate, 2-keto-3-deoxy-6-phosphogluconate and D-2-deoxyribose-5-phosphate aldolase. In proteins of the Rossmann fold, alternating α-helices and β-strands with βαβ ADP-binding structural unit carry specific sequence patterns of nucleotide binding (Rao and Rossmann 1973). This fold contains five, six, or seven parallel β-strands surrounded by α-helices, in which similarly to TIM-barrel proteins, the number of strands approximately matches number of helices. However, β-structure is almost flat and flanked by three or more α-helices on each side in Rossmann fold whereas it is circular in the TIM-barrel. Rossmann fold is a structural unit of lactate dehydrogenase, malate dehydrogenase, uridine-diphosphate galactose and uridine-diphosphate-N-acetylglucosamine 4-epimerases, pyridoxal phosphorylase B, glycosyltransferases, and other proteins. Classic βαβ dinucleotide-binding (Rossmann) fold has many variants such as nucleotide-binding domain (NBD) and catalytic domain of the D-lactate dehydrogenase; NBD is widely conserved among NAD-dependent dehydrogenases 6-stranded parallel β-sheet with α-helices packed on each side and GxGxxG sequence motif; the catalytic domain has a 5-stranded parallel β-sheet flanked on each side by α-helices and lacking a characteristic nucleotide-binding sequence motif. Some variants such as L-lactate dehydrogenase show deletion of the third β-strand. Other, such as D-lactate dehydrogenase, has an addition of one α-helix/β-strand pair at the N-terminus (Stoll et al. 1996).

In α-helix, one turn of the helix (3.6 in globular proteins or 3.5 residues in fibrous coiled coil proteins) brings each residue close to every third and forth residue so that (i, i + 3) and (i, i + 4) residues form hydrogen and van der Waals bonds. The peptide group atoms C_{α}^{i}, C^{i}, N^{i+1}, C_{α}^{i+1}, and C_{β}^{i+1} located between the two adjacent α-carbon atoms make a central part of the pocket with protruding side chains that interact with side chains of the same helix and with side chains of the neighboring secondary structure elements. On the surface of each helix, edges (i, i + 1), (i, i + 3), (i, i + 4) can be seen bordering “knobs” and “holes” of the interacting helices (Fig. 1.21).

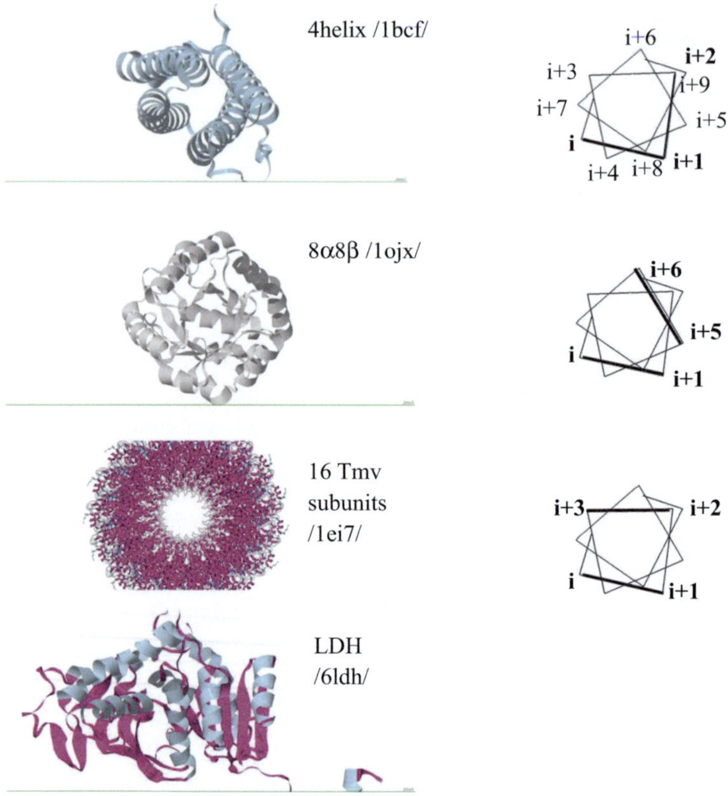

Fig. 5.1 Four, eight, and sixteen helix arrangements and polygon crosssections

Similarly to contacts on the surface of the α-helix in a leucine zipper coiled coil, contacts on all other types of helices follow the same regular pattern. Amino acid residues at the interface of the two α-helices organize a heptad (3–4) periodic repeat. For each residue of the helix, we can estimate its contribution to the formation of the interface. The interface area change of amino acid residues of each helix that is observed in helix-helix interactions with the surrounding helices represents this contribution (Appendix). Comparison of the interfaces in various proteins shows distinct pattern of contacts characteristic for the particular type of fold in four helix bundle myohemerythrin, 8α8β barrel triose phosphate isomerase, and Rossmann fold L-lactate dehydrogenase (Kurochkina 2010).

If we calculate the angles between helical edges and compare them with the angles of the polygons formed at a cross section of each motif (See Appendix A), we can see that circular arrangements of the four, eight, and sixteen α-helices, which comprise the four-α-helical motif, TIM-barrel 8α/8β fold, and helical rod of $16.\bar{3}$ helices per turn (Fig. 5.1) correspondingly, can be associated with the orientation of the edges of the helix surfaces. For each helix, there is one preceding and one following in amino acid sequence contacting helices.

In the four-helix bundle, if edge (i, i + 1) is central for the contacts with the preceding α-helix then edge (i + 1, i + 2) is central for the contacts with the following α-helix. Angle θ between the two planes, one containing edge (i, i + 1), another containing adjacent edge (i + 1, i + 2) shows the average value of approximately 100° ± 2.6. This arrangement of helices is in good agreement with a quadrilateral, which on the plane would be a square with an angle of 90°. Since in three-dimensional space the vertices of a quadrilateral are located on a helix, we see the larger angle between the edges.

Similarly, angle θ calculated between the two planes, one containing edge (i, i + 1), another containing edge (i + 2, i + 3) shows the average value of the angle between these two planes θ = 159° ± 2.8. It is in good agreement with a value of an interior angle of a 16-gon (157°). In the assembly of the four-helix bundle subunits of a tobacco mosaic virus, these two edges are central for the circular arrangement of 49 subunits forming a disc as 3-turn helix. Therefore, each ring of $16.\bar{3}$ subunits makes one circular turn, or a 16-gon. The value of the angle between the planes containing edges (i, i + 1) and (i + 2, i + 3) correlates with the value of ideal angle of a cross section approximated by a 16-gon. Also, L-Lactate dehydrogenase, representative of Rossmann fold, contains a seven-stranded almost flat β-sheet flanked by three α-helices on each side. The three helices arrangement is different from 4-gon or 8-gon and is more similar to an arc of a 16-gon.

A cross section of the TIM-barrel eight α-helices is an octagon. The angle θ = 140° ± 3.1 between the planes containing edges (i, i + 1) and (i + 5, i + 6) is in good agreement with the value to the interior angle of 135° in an ideal octagon.

To summarize, edges (i, i + 1) – (i + 1, i + 2) of the helix surface are central for the interhelical contacts in a four-helix bundle; edges (i, i + 1) – (i + 5, i + 6) in the 8α/8β TIM-barrel fold, and edges (i, i + 1) – (i + 2, i + 3) in helical rod of a tobacco mosaic virus and a three-helix fragment of a Rossmann fold. There exists a relationship between the intrinsic properties of α-helix and the types of tertiary structure motifs that involve helix-helix interactions. Conservation of the enzyme shape as is observed in the enzymes with TIM-barrel or Rossman fold originates in distribution of helix contacting residues at the helix surfaces. The function of each enzyme differs, but common shape of the motif is conserved and is important for binding of similar ligands. Although some protein families, for example, globins, phycocyanins and toxins, have similar tertiary structure but do not exhibit preferences for the common ligands, these shared architectural features may reflect distant evolutionary relationship (Holm and Sander 1993).

Possible arrangements of α-helices originate from structural properties of the interacting helix surfaces. Ability of secondary structure elements to utilize various combinations of interacting edges gives rise to various shapes of assemblies. Presence of limited number of helix arrangement patterns helps to better understanding of such complex processes as crystal growth which depends on satisfying strict requirements of the lattice contacts or biofilm formation which is based on molecular recognition and assembly. Also, similar structural relationships can be found in many other types of packing of secondary structure elements.

Determinants

Helix structure reflects periodicity of contacts. One turn of the α-helix spans 7 amino acids or 10 A along the helix axis. Intrinsic properties of α-helix determine its geometry and modes of interactions with other secondary structure elements. To describe various structural arrangements and tertiary folds, a set of parameters is used. For any repeats molecule such as ankyrin repeats (Fig 5.2A), curvature, twist, and lateral bending (Fig 5.2B) describe orientation of one repeat relative to another. Also, comparison of the repeats in the same coordinate system (Fig. 5.2C, D) shows similarities and differences in their characteristics.

Curvature, twist, and lateral bending characterize orientation of one repeat relative to another (Forwood et al. 2010) using three perpendicular principal axes in consecutive repeats: Axis P1 is the axis perpendicular to the helices in the plane of the repeat; axis P2 is the axis parallel to the helices in the plane of the repeat in individual repeats; and axis P3 is the axis perpendicular to axes P1 and P2. Three angles between axes of the two consecutive repeats describe the orientation of one repeat relative to another: curvature (P1 axes), twist (P2 axes), and lateral bending (P3 axes) (Fig 5.2B).

Coordinate system of the layer (Fig 5.2C) places amino acid at position **a** of the helix at the origin and aligns peptide group with XZ plane. Then interacting helix is along the Y axis. When several pairs of helices are compared in this coordinate system, we can see similarities and differences in the core amino acid positions and direction and shape of the assembly (Fig 5.2D).

Types of Helical Assemblies

Ankyrin repeats Ankyrins are proteins that mediate associations between cytoskeleton and proteins involved in ion exchange, cell adhesion, and membrane transport (Krzywda et al. 2004; Michaely et al. 2002; Lishko et al. 2007; Lumpkin et al. 2007; Deng et al. 2012; Kaszas et al. 2012). Ankyrin membrane binding domain consists of stacked repeat units and binds many different proteins. Another domain binds spectrin molecules of cytoskeleton. Ankyrin repeat unit consists of a pair of antiparallel α-helices and a β-hairpin (Michaely et al. 2002; Mosavi et al. 2002). Ankyrin repeat sequence adopts a fold of two antiparallel α-helices (helix-loop-helix) repeated as a unit and bound in a parallel manner to other similar two-helical units. Pairs of adjacent repeat elements form a four-helix bundle. Together, they assemble in a spiral, which has inner row and outer row of parallel α-helices with antiparallel arrangement of helices between the rows. Gankyrin, a protein overexpressed in human hepatocellular carcinomas, utilizes a concave surface of ankyrin repeats to interact with its partners such as cyclin-dependent kinase 4, retinoblastoma protein, and E3 ubiquitin ligase involved in p53 degradation. Gankyrin-protein complexes are important components of gene expression and apoptotic

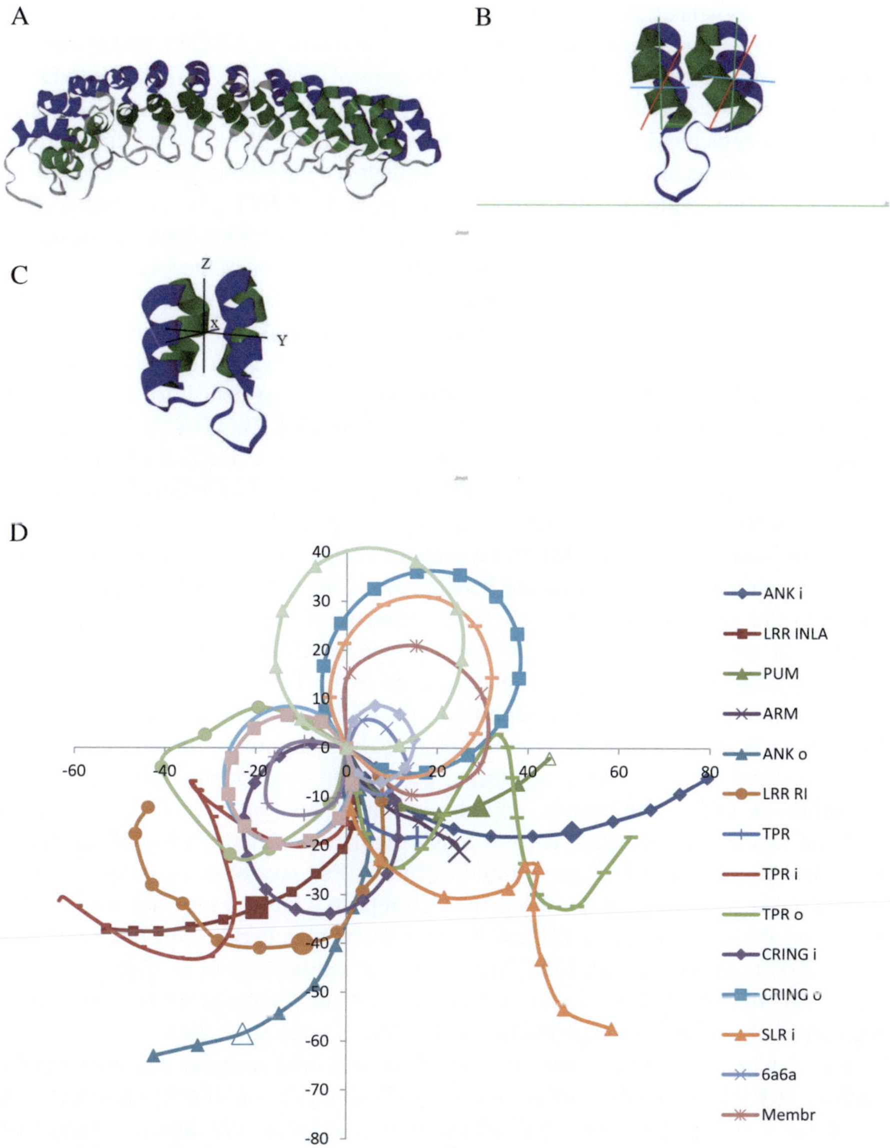

Fig. 5.2 Comparison of toroids and barrels. (**A**) The structure of ankyrin R: ribbon diagram; inner row (green); outer row (blue). (**B**) Three perpendicular principal axes in consecutive repeats: Axis P1 is the axis perpendicular to the helices in the plane of the repeat (blue); axis P2 is the axis parallel to the helices in the plane of the repeat in individual repeats (cyan); and axis P3 is the axis perpendicular to axes P1 and P2 (red). Three angles between axes of the two consecutive repeats describe the orientation of one repeat relative to another: curvature (P1 axes), twist (P2 axes), and lateral bending (P3 axes) (Forwood et al. 2010). Comparison of helical assemblies in the same coordinate system (Kurochkina 2008) and Fig. 1.23 for (**C**) Ankyrin repeats; and (**D**) All types of assemblies superimposed

pathways (Krzywda et al. 2004). Ankyrin repeats function to recognize targets of the complex between breast and ovarian cancer associated protein, BRCA1, and associated RING domain protein BARD1. This complex controls cell cycle in response to DNA damage (Edwards et al. 2008). Huntingtin Interacting Protein 14 (HIP14) and Huntingtin Interacting Protein 14 like (HIP14L) are the only palmitoyl acyltransferases (PATs) that have an ankyrin repeat domain, mediator of the interaction between HIP14 and HTT which exhibits CAG expansion in exon 1 of *HTT* gene as pathological modification (Sanders, et al. 2014). A series of ankyrin repeat proteins was designed using a derived consensus sequence followed by determination of the x-ray structure of the fragments containing three and four (Mosavi et al. 2004) or four to six (Kohl et al. 2002) repeat units that retain ankyrin characteristic fold. Ankyrin consensus sequence shows that all well conserved positions of amino acids are located in the interior of the repeat (Mosavi et al. 2002). Arrangement of ankyrin parallel α-helices to some extent is similar to the arrangement observed in TIM-barrel proteins. This large group includes triosephosphate isomerase, pyruvate kinase, malate synthase, fructose-1,6-biphospate aldolase, xynalase and other proteins. TIM-barrels are 8α/8β proteins with circular arrangement of secondary structure elements (Wierenga et al. 1991). Pairs of parallel α-helices in both ankyrin and TIM-barrel proteins exhibit negative interhelical angles. However, curvature direction of the helical row is opposite in the two proteins. In addition, ankyrin has an antiparallel arrangement of α-helices carrying positive interhelical angles. Some of ankiryn repeat proteins such as gankyrin also contain a row of β-hairpins which resembles inner β-barrel of the TIM-barrel.

Ankyrin repeats are mainly found in eukaryotic proteins although they also function in viruses such as poxvirus (Li et al. 2010). As protein- protein interaction modules, they are involved in anchoring proteins to cytoskeleton and membrane, assembly of large multiprotein complexes, and mediating signaling pathways. Ankyrin repeat proteins gankyrin and INK4 regulate phosphorylation of substrates by CDK4/6, important process in tumorogenesis (Padmanabhan et al. 2004). As a part of the regulatory module of leucine rich repeats kinase 2 together with ARM, LRR, and WDD repeats, ankyrin repeats are implicated in Parkinson's disease (Mills et al. 2012). Important functions include transcription activation (Ferreiro et al. 2007; Zweifel et al. 2003), antiviral defense (Praditwongwan et al. 2014), and regulation of channel activity (Phelps et al. 2008; Kaszas et al. 2001).

Ion channels for sensing noxious stimuli of transient receptor potential family such as TRPV1 and TRPV2 are regulated by phosphorylation of ankyrin repeats, an event that turns on agonist-induced calcium influx, and represent important targets of pharmacological design (Clapham and Miller 2011; Pareek et al. 2006; Lishko et al. 2007; Nilius et al. 2011; Kaszas et al. 2012). In TRPA1, ankyrin repeats are key elements of sensitivity to thermal and chemical stimuli (Cordero-Morales et al. 2011; Nagatomo et al. 2010). Ankyrin repeat domains are important for the assembly of the functional TRPC4, TRPC5, (Lepage et al. 2006; Clapham and Miller 2011), and TRPV6 channels (Erler et al. 2006).˙ Genetic modifications of ankyrin repeats occur in tumor suppressors, transcription regulators, and proteins involved in cell development (Li et al. 2006).

Ankyrin repeats exhibit a variety of interaction patterns with other proteins. DARPins, designed ankyrin repeats, bind proteins via β-hairpin (Pecqueur et al. 2012). Other proteins such as Yar1 (Holzer et al. 2013) arrange two α-helices of ankyrin repeats to bind two α-helices of the partner. Spectrin arranges its contacts with ankyrin on the long connecting helix that hosts two repeats with 180 degrees turn. Ankyrin repeats bind ATP (Lishko et al. 2007), lipids (Kim et al. 2014), and sugars (Rogowski et al. 2015).

Sequences of ankyrin-binding proteins such as the erythrocyte anion exchanger, the amiloride-sensitive sodium channel, the voltage-sensitive sodium channel, the Na^+/Ca^{2+} exchanger, CD44, neurofascin, αNa/K Atpase (Zhang et al. 1998), and IP3-R do not show homology (Dautant et al. 2010). Although sequences of proteins binding to AnkG: autoinhibitory carboxyterminal AnkG and AnkR fragments, neurofascin, Nav1.2, KCNQ2, KCNQ3, Cav1.3, and betaDG, are diverse and use different combinations of the binding sites, there are common features of the peptides such as position of homologous charges combined with hydrophobic residues (Wang et al. 2010).

Malarial invasion strongly affects red cell membrane protein distribution resulting in diffusion of material from the multilamellar body into the cavity in the host cell. Early stage of invasion exhibits absence of spectrin, ankyrin, and trans-membrane protein band 3 from *P. falciparum* and *P. knowlesi* (Dluzewski et al. 1989).

Ankyrin repeats enhance function of enzymes such as cellulases (Cunha et al. 2016), and catalases (Pan et al. 2018). *Xanthomonas oryzae pv. oryzae* and *Xanthomonas oryzae pv. oryzicola* AnkB affects the gene expression of catB, catalase activity, and sensitivity to H2O2. In *Xanthomonas spp*., not only amino acid sequence of AnkB but also the locations of ankB and catB genes are highly conserved. AnkB conserved role in the defense against oxidative stress is well described in prokaryotes.

Armadillo (ARM)/HEAT repeats Protein assemblies of ankyrins, ARM/HEAT, and tetratricopeptide (TPR) contain a repeating unit for which major part, as in ankyrins, represents a pair of antiparallel α-helices. These repeating units are packed parallel to each other in a super-helix or solenoid (Andrade et al. 2001a). Ankyrin repeats form a left-handed spiral whereas ARM and HEAT repeats assemble into a right-handed spiral, both types with on average 24 repeats per turn although the number of repeats per turn can vary. Ankyrin and HEAT repeat unit consists of two antiparallel α-helices, A and B (Andrade et al. 2001b). ARM repeat contains three α-helices H1, H2, and H3 in which H2 and H3 form an antiparallel interface and shorter H1 is almost perpendicular to H2-H3. Tetratricopeptide, a right-handed super-helix, carries 8 repeats per turn (Blatch and La¨ ssle 1999; Allan and Ratajczak 2011; Kajander et al. 2007).

To make a complete turn, each of 24 Ankyrin or ARM/HEAT repeats tilt approximately15°. Parameters characterizing packing, curvature, twist, and lateral bending of Ankyrin (Michaely et al. 2002), HEAT Groves et al. 1999), and ARM (Madhurantakam et al. 2012) repeats show consistency with shape of the assemblies. Binding of ligands to superhelical structure of HEAT repeats of importin β (Lee et al.

Table 5.1 Chiral Helical Assemblies. Arrangement of contacts on the helix surfaces

Protein	Inner row	Outer row	Example
Ank	Prev (i, i + 1)	Prev (i, i + 1)	Ankyrin-R /1n11/
	Outer (i + 1, i + 2)	Inner (i + 1, i + 2)	
	Next (i + 2, i + 3)	Next (i-2, i-1)	
ARM/HEAT	Prev (i, i + 1)	(i, i + 1)	Importin α / 1bk5/
	Outer (i + 1, i + 2)	(i + 1, i + 2)	
	Next (i + 2, i + 3)	(i-2, i-1)	
PUM	Prev (i, i + 1)	(i, i + 1)	Pumilio-homology domain /1ib2/
	Outer (i + 1, i + 2)	(i + 1, i + 2)	
	Next (i + 2, i + 3)	(i-2, i-1)	
Tal	Prev (i + 5, i + 6)	(i-3, i-2)	TAL effector PTHXO1 /3ugm/
	Outer (i, i + 1)	(i + 1, i + 2)	
	Next (i + 3, i + 4)	(i-1, i)	
LRR	β structure	(i, i + 1)	Ribonuclease inhibitor /2bnh/
		(i + 1, i + 2)	
		(i-2, i-1)	
TIM	β structure	(i, i + 1)	Triosephosphate isomerase /1o5x/
		(i + 1, i + 2)	
		(i + 5, i + 6)	

2000; Forwood et al. 2010), CRM1 (Monecke et al. 2012), ARM (Huber et al. 1997), tetratricopeptides (Allan et al. 2011) and other motifs induces conformational transitions. As ankyrin repeats (Lee et al. 2006) and clathrin (Jin et al. 2013), ARM/HEAT repeats show elastic properties important in mechanotransduction.

Consensus sequences, derived for four groups of ARM/HEAT (Andrade et al. 2001; Malik et al. 1997), were applied for the design of repeats with new properties.

ARM repeats are present in functionally unrelated proteins. Nuclear localization signal binding domain of importin α (Kobe 1999), β-catenin involved in cell adhesion and growth factor signaling (Huber et al. 1997), regulatory complex of the LKB1 tumor suppressor protein kinase mutated in Peutz-Jeghers cancer syndrome (Milburn et al. 2004), mTORC1 (Yang et al. 2017) and ubiquitin ligase (Tu et al. 2007) carry this structural motif. Such important proteins as huntingtin, elongation factor 3, protein phosphatase 2A, importin β, target of rapamycin, and exportin CRM1 contain HEAT repeats for which involvement in protein-protein interactions and vesicle trafficking is proposed (Andrade and Bork 1995).

As in all types of assemblies, amino acid combinations at helix-helix interfaces (Kurochkina 2007; Kurochkina 2008; Kurochkina and Choekyi 2011) and location of amino acids at helical edges (Kurochkina 2010) of HEAT/ARM repeat are important determinants of the fold and characterize positive and negative interhelical angles and overall shape of the helical assembly (Table 5.1). Distribution of contacts on the edges of helical surfaces and characteristic amino acid combinations of helix-helix interfaces in ankyrin and HEAT/ARM repeat assemblies are unique. For both left-handed and right-handed assemblies, involvement of edges of helical surfaces in the formation of helix-helix interfaces follow characteristic pattern. Amino acid combinations of parallel HEAT/ARM interfaces are homologous to those present at helix-helix interfaces with positive interhelical angles which helps to obtain good methods of fold recognition that distinguishes between the various types of repeats predict the structure of new proteins.

Ankyrin repeats and HEAT/ARM repeats comprise two rows of α-helices. In each repeat unit, two antiparallel α-helices, A and B, form interface with −149 to −162 interhelical angle. Stacked AB repeats form parallel interfaces (AA' and BB') with negative interhelical angles in ankyrins but positive interhelical angles in ARM/HEAT and TPR. In ankyrins (Fig 5.3A), A helices form inner (concave) row and B helices form outer (convex) row. In contrast, in repeats ARM and HEAT (Fig 5.3B) B helices form inner (concave) row whereas A helices form outer (convex) row, and these two repeat types are very similar in structure (Malik et al. 1997; Kobe 1999). Inner and outer rows of ankyrin repeats can be superimposed on those of ARM/HEAT repeats (Fig 5.3C) and we can see similarities and differences in the outline of the superhelix. Whereas ankyrin outline and overall HEAT/ARM outline show increment in the direction of the superhelical axes, ARM/HEAT repeats show alternating deviations from the major direction. Length of helix-helix interface is approximately between ½ of the turn and one complete turn of the helix.

For helix-helix interfaces, coiled coil nomenclature is applied to identify main residues involved in interface formation. In leucine zipper molecule, leucines at every seventh positions together with valines/asparagines form a heptad (3–4) repeat and their positions are designated by letters **a-g**. Leucines at positions **d**, valines/asparagines at positions **a**, and polar residues at positions **e** and **g** are main contributors to helix-helix interfaces (O'Shea et al. 1991; Krylov et al. 1994). This nomenclature, introduced for leucine zipper GCN4, was successfully used in the analysis of not only parallel coiled coil but of all types of interfaces (Gernert et al. 1995; Kurochkina 2007; Kurochkina 2008; Kurochkina and Choekyi 2011). Ankyrin repeat and HEAR/ARM repeat interfaces were assigned heptad positions. Positions **a** of interacting antiparallel helices A and B, named **a** and **a'**, show the outline of the inner and outer rows (Fig 5.3C).

Ankyrin repeats contain a consensus sequence and show similarity to each other (Fig 5.4A) as well as ARM/HEAT repeats do (Fig 5.4C). Since overall shape of the assembly and its curvature are very similar in the two proteins, this may be attributed to the similar pattern of contacts on the surfaces of the helices (Fig. 5.6B, C). In both proteins, contacts on the surface of each helix follow the same pattern. Inner row helices use three consecutive edges (i, i + 1), (i + 1, i + 2), and (i + 2, i + 3) to contact previous, antiparallel outer row, and following helices correspondingly in both proteins. In the outer row, edges (i, i + 1), (i, i + 2), and (i-2, i-1) are involved in contacts with previous, antiparallel inner row, and following helices. When core positions **a** and **d**, and flanking positions **e** and **g** were assigned to each interface, involvement of helical edges in the formation of helix-helix interfaces is clearly seen.

Assignment of positions **a** and **d** can be verified by the percent change of the accessible surface area calculated for each amino acid (see Appendix). In ankyrinR, for each α-helix in its solvent exposed state (A_i) and state of contact with one

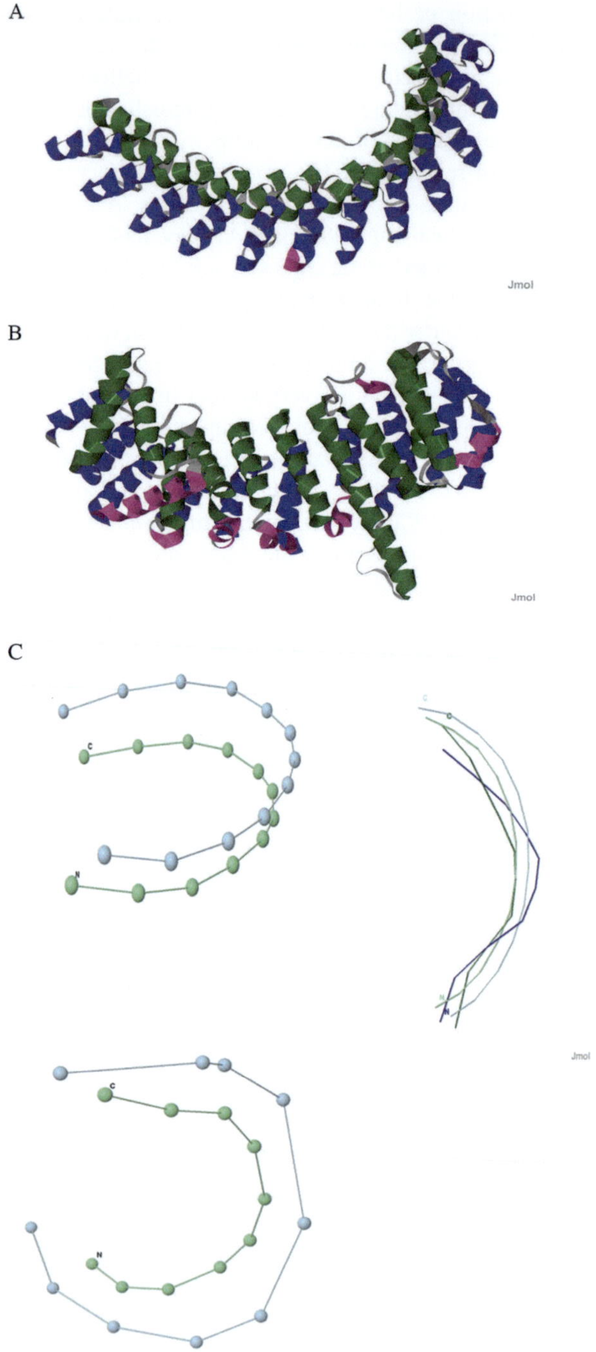

Fig. 5.3 Structure of (**A**) ankyrin repeats of *H. sapiens* ankyrinR (pdb code 1n11) and (**B**) HEAT repeats of *H. sapiens* importin β (pdb code 1gcj). Helices of the inner (green) and outer (blue) rows – ribbon. (**C**) Positions **a** from interacting antiparallel helices A and B (spheres at the Ca atoms), named **a** and **a**', show the outline of the inner (green) and outer (blue) rows of ankyrinR (left top, pdb code 1n11), ARM repeats of importin α (left down, pdb code 1bk5), and superimposed ankyrin repeats of ankyrinR (inner - lightgreen, outer – lightblue, pdb code 1n11) and ARM repeats of importin α (inner – green, outer – blue, pdb code 1iq1) (right)

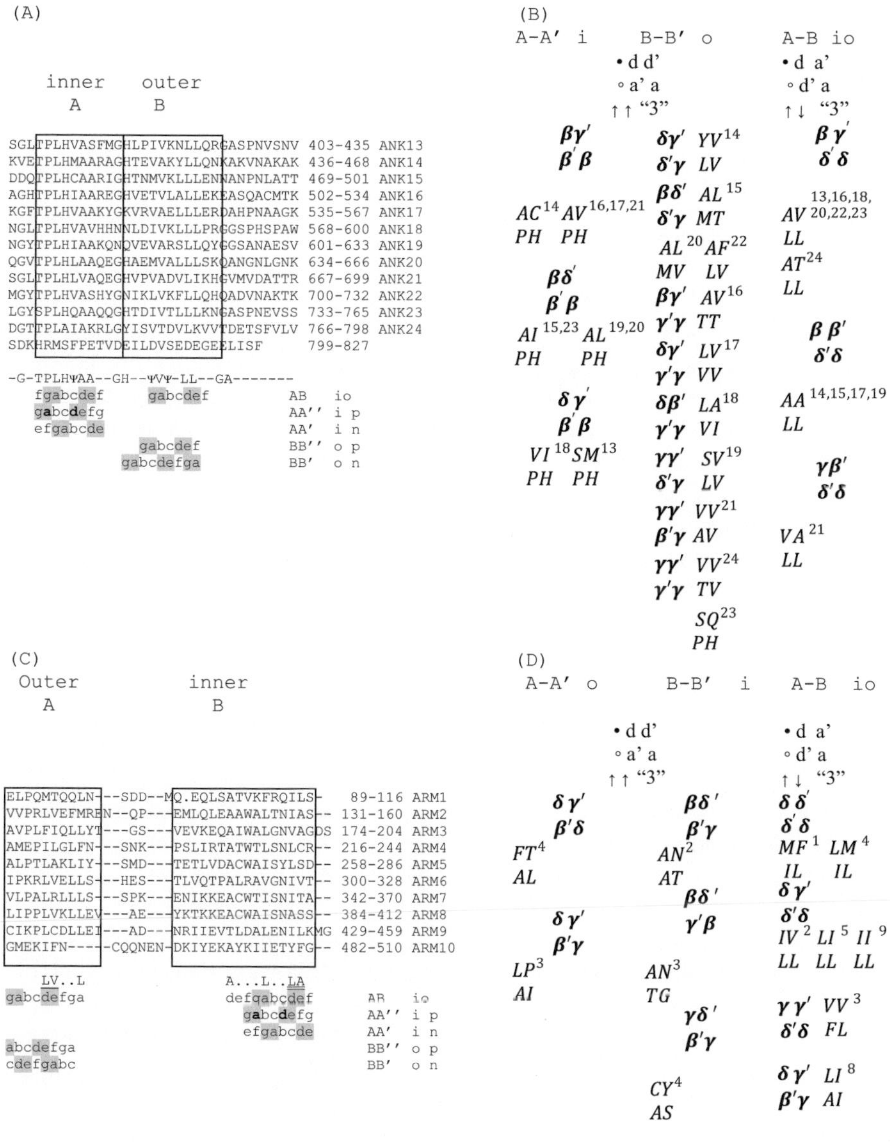

Fig. 5.4 Comparison of ankyrin repeats of ankyrinR (pdb code 1n11) and ARM repeats of importin α (pdb code 1bk5). Amino acid sequence alignment of repeats and their consensus sequence of ankyrinR (**A**) and importin β(**C**). Amino acid combinations at helix-helix interfaces of (**B**) ankyrinR and (**D**) importin α. Superscripts show repeat number; α-helices are boxed; ψ– hydrophobic residue; ↑↑ parallel and ↑↓ antiparallel orientation of helices

neighboring helix (A_c), accessible surface area change shows involvement in contacts (Fig. 5.5A). If accessible surface area is large, amino acid is mainly in contact with solvent. If accessible surface area is small, amino acid contacts neighboring helix and buries its accessible surface. Difference between solvent exposed state and folded state represents amount of surface area change upon folding. Nonzero value of this difference is an indicator that amino acid contributes to the formation of helix-helix interface.

Each ankyrin inner row helix (A) interacts with one outer row helix B (antiparallel interface) and two other inner row helices, preceding (A”) and following (A’) (parallel interfaces). Contacts of ankyrin inner row helix (amino acids 439–448) with three neighboring helices (Fig 5.6B) are given as an example of assignment of positions **a** and **d** and flanking positions **e** and **g**, main contributors to the interface (Fig. 5.5A). Similarly, contacts of the outer row helix (amino acids 449–460) show involvement of **a**, **d**, **e**, and **g** positions (Fig. 5.5B) in helix-helix interactions. Large amount of surface area buried at each helix-helix interface at positions **a** and **d** and smaller amount of area change at positions **e** and **g** are clearly seen. All three methods (surface area change, estimation of contacts, and visual inspection), used for the assignment of interface positions **a** and **d** and flanking positions **e** and **g** (Fig. 5.5), give consistent results.

The approach of assignment of positions **a**, **d**, **e**, and **g** was used for all inner and outer row helices of ankyrin repeats and HEAT/ARM repeats to identify involvement of edges of helix surface in the formation of helix-helix interfaces. We can see that for both types of repeats, ankyrin (Fig. 5.4A) and HEAT/ARM (Fig. 5.4C), pattern of contacts is the same. In the inner row helices of both proteins, three consecutive edges (i, i + 1), (i + 1, i + 2), and (i + 2, i + 3) carry **a**, **d**, **e**, and **g** positions involved in contacts with previous, antiparallel outer row, and following helices correspondingly. In the outer row, edges (i, i + 1), (i + 1, i + 2), and (i-2, i-1) are involved in contacts with previous, antiparallel inner row, and following helices (Fig. 5.4).

Amino acid combinations at each type of interface represent interactions of a given helix with three surrounding helices preceding, following, and antiparallel (Fig 5.4A–C).

At antiparallel interface of ankyrin repeats, the most frequently occurring combinations are $\frac{AV}{LL'}$ $\frac{AT}{LL'}$ $\frac{AS}{LL'}$ $\frac{AI}{LL'}$ and $\frac{AV}{LI}$ of the $\frac{\beta\gamma'}{\delta'\delta}$ type “3” group at positions $\frac{da'}{d'a}$ (Fig 5.4B). They contain leucine at position **a** and alanine at position **d** of the inner helix combined with valine at position **a’** and leucine at position **d’** of the outer helix. Alanine is one of the conserved residues of the ankyrin consensus sequence. However, ARM repeats achieve the same degree of curvature without presence of alanines at corresponding positions (Fig 5.4D). One possible reason for preference of alanines in ankyrin repeats is that they have negative angles of parallel helix-helix interfaces. Alanine residues are frequently found at parallel interfaces with negative interhelical angles in many proteins and form a group of TIM-barrel-like interfaces in contrast to leucine-zipper-like interfaces of positive interhelical angles

(A)

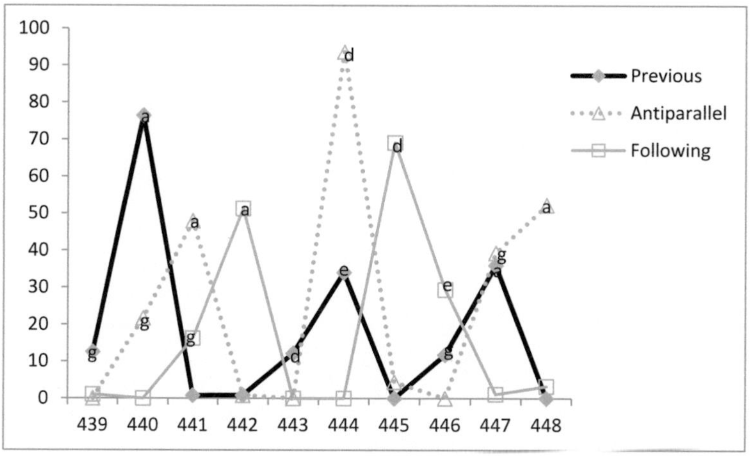

g a b c d e f g a b inner previous
g a b c d e f g a antiparallel
g a b c d e f g following

(B)

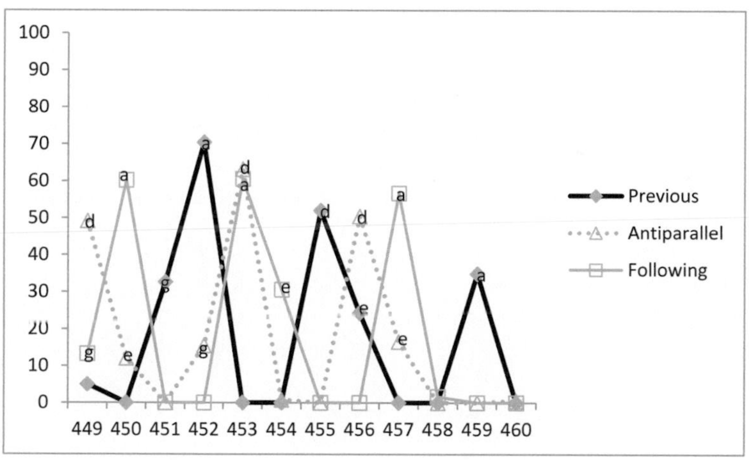

g a b c d e f g a outer previous
g a b c d e f g a antiparallel
g a b c d e f g a following

Fig. 5.5 Percent area change of amino acids of each helix observed in contacts with previous, antiparallel, and following helices in ankyrin repeats (**A**) inner row and (**B**) outer row helices. Assignment of positions a and d

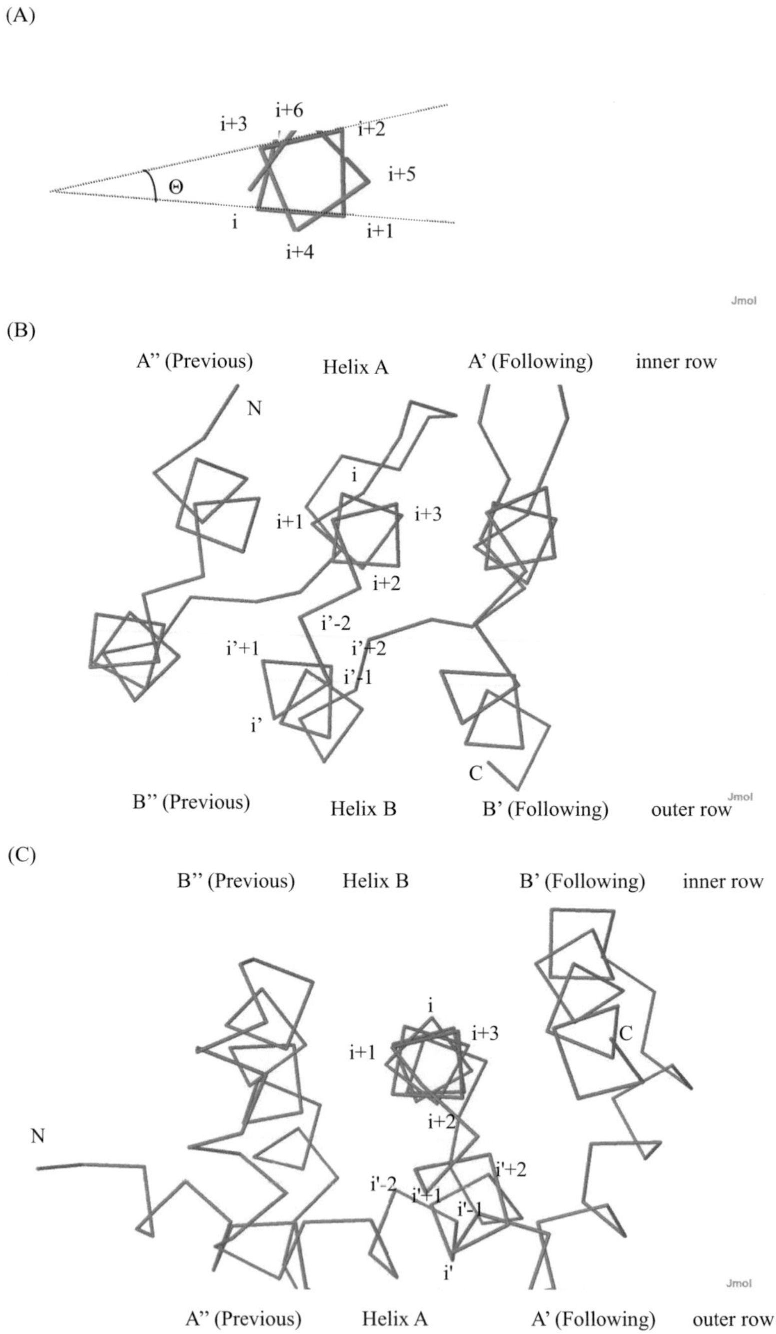

Fig. 5.6 Helix edges involved in helix-helix interfaces. (**A**) Helical wheel; (**B**) Ankyrin repeats - ankyrinR; (**C**) ARM repeats – importin α

(Kurochkina 2007; Kurochkina 2008). One of the most conserved groups of interfaces in ankyrins $\genfrac{}{}{0pt}{}{\gamma\beta'}{\beta'\beta}$ (repeats 2, 14, 16, 17, Fig 5.4B) is also a conserved group in TIM-barrel (Kurochkina 2008). Particularly clearly this preference can be seen in glutathione S-transferase (Kurochkina and Choekyi 2011). Two groups that are observed in parallel "3" type interfaces with negative interhelical angles of GST, $\genfrac{}{}{0pt}{}{\beta\gamma'}{\delta'\delta}$, $\genfrac{}{}{0pt}{}{\delta\gamma'}{\gamma'\gamma}$ and $\genfrac{}{}{0pt}{}{\delta\beta'}{\gamma'\gamma}$, are present in ankyrin repeats (Fig 5.4B, repeats 15, 17, 18, 20, and 22). Two groups $\genfrac{}{}{0pt}{}{\delta\gamma'}{\beta'\gamma}$ and $\genfrac{}{}{0pt}{}{\gamma\delta'}{\beta'\gamma}$ at parallel "3" type interfaces with positive interhelical angles of GST are repeated in ARM (Fig 5.4D, repeats 3 and 4).

Superhelical parameters curvature, twist, and lateral bending characterize relative orientation of the repeats in space (Madhurantakam et al. 2012; Lee et al. 2000; Forwood et al. 2010). Since consecutive edges (i, i + 1) and (i + 2, i + 3) of each helix (Table 5.1) are involved in contacts with preceding and following helices of the same row, angle between these two edges may define curvature. Indeed, this angle gives approximately 20 degrees value that is close to curvature angles observed in importin α (Madhurantakam et al. 2012). This angle is an intrinsic property of each α-helix (Fig 5.6A). The fact that there is a correlation between the angle between edges of the helix and curvature angle of the repeats assembly shows that distribution of contacts on the surface of the helix along the edges is important for the shape of the assembly. The same correlation was found for 4-helix bundle and TIM-barrel motifs (Kurochkina 2010).

Therefore, amino acid combinations characteristic for helix-helix interfaces with positive and negative interhelical angles together with location of these combinations at the helical edges provides rationale for the ankyrin and ARM/HEAT repeat assemblies and prediction of their tertiary structure. These two key elements give a basis for the development of fold recognition algorithm. Starting with amino acid sequence of the protein and predicted location of α-helices, algorithm generates all possible helix-helix interfaces and compares resulting amino acid combinations with characteristic amino acid combinations and their position relative to each other. Each protein is tested for both types of fold. In a set of 52 ankyrin repeats and ARM/HEAT repeats proteins, used to distinguish between the two types of fold, correct prediction of the fold is achieved.

Comparison of ankyrin and ARM/HEAT repeats highlights several important features. They can be superimposed; therefore, the shape of the superhelix is similar (Fig 5.3C). They contain approximately the same number of repeats per superhelical turn. Inner row helix is N-terminal to the outer row helix of the ankyrin repeat unit but inner row helix is C-terminal to its outer row helix of ARM/HEAT repeat. Antiparallel interfaces in both types of the repeats and parallel interfaces of ARM/HEAT repeats exhibit positive interhelical angles; however, parallel interfaces of ankyrins carry negative interhelical angles. The ankyrin outline shows constant increments in the direction of the superhelical axes whereas ARM/HEAT outline shows alternating deviations from the major direction. These similarities and differences originate in geometry and amino acid composition of helix-helix interfaces.

One complete turn of the superhelix comprises 24 repeats and therefore gives increment of 15 degrees per each repeat. Distribution of contacts on the surface of each helix exhibits the same pattern of edges (i, i + 1), (i + 1, i + 2) and (i + 2, i + 3) involved in interactions with three neighboring helices, preceding, antiparallel, and following, correspondingly. This pattern is present in both types of the repeats. In addition, angle between the two edges of the helix (i, i + 1) and (i + 2, i + 3) is approximately the same as the curvature angle of the repeats. These facts provide reasonable explanation of the overall shape similarity for these two types of assemblies in spite of differences in amino acid composition. Parameters of each type of the assembly twist, curvature, and lateral bending, deviate between repeats; the interrepeat angle variation shows no correlation with amino acid variability (Forwood et al. 2010). Conformational transitions of ankyrin and ARM/HEAT repeats affect pitch, radius, twist, curvature, and lateral bending of the superhelix as it was found for unbound (pdb code 1gcj), RanGTP-bound (pdb code 1ibr), and IBB-bound importin β (Lee et al. 2000; pdb codes 1qgk, 1qgr, and 2bku). However, the overall shape similarity supports the notion that curvature angle may depends on intrinsic properties of the α-helix and edges of the helical surface are the center points of helix-helix contacts. Another observation is that in antiparallel arrangement of α-helices of four-α-helix bundle or parallel arrangement of TIM-barrel helices, contacting planes of interacting helices are almost parallel but curvature of the assembly differs. There is a link between edges involved in contacts of interacting helices and angle of the assembly (Kurochkina 2010). All these observations show that curvature of the repeats may be a consequence of the angles between helical edges.

There exists a good agreement between amino acid combinations at helix-helix interfaces and interhelical angles. Combinations at parallel interfaces of ankyrin repeats are identical or homologous to combinations observed at interfaces of negative interhelical angles such as those present in TIM-barrel. Combinations at parallel interfaces of ARM/HEAT repeats are characteristic for interfaces with positive interhelical angles and are also found in leucine zipper (Kurochkina 2007; Kurochkina 2008). Some combinations are homologous to those of gluthathione S-transferase (Kurochkina and Choekyi 2011) and show consistency in interhelical angles (Fig. 5.4B–D).

Amino acid combinations at positions **a** and **d** of ankyrin repeats are more uniform and cluster into small number of groups. ARM/HEAT repeats show more variability. This variable composition of the core leads to alternating repeat increments. Design of ARM/HEAT repeats in which all middle repeating units of the assembly carry the same amino acids at positions **a** and **d** results in altered parameters of the superhelix and its ability to bind ligands (Madhurantakam et al. 2012). Repeat order was shown to be important for the assembly (Malik et al. 1997). Fold recognition based on amino acid combinations and distribution of amino acids on helical edges considers arrangement of groups in space and predicts those repeats that cannot be predicted with linear sequence analysis procedures.

TAL repeats In helical assemblies, toroids and barrels, helix-helix interactions play important role in determining the fold and ligand specificity. Pattern of interactions at helix-helix interfaces of assemblies and mutual orientation of the helical edges distinguish one type of the assembly from another and influence curvature of solenoids. Whereas ankyrin, ARM/HEAT, and PUM-HD repeats in interaction of the central α-helix with preceding and following helices utilize edges (i-1, i) and (i + 1, i + 2) in the inner row and edges (i-1, i) and (i-3, i-2) in the outer row, TAL repeats use edges (i + 5, i + 6) and (i + 3, i + 4) in the inner row and (i-3, i-2) and (i-1, i) in the outer row adding another example of enantiomer-selective binding. Arrangement of amino acids at helical edges of TAL repeats determines handedness of the helical assembly. Arrangement of contacts that DNA double helix makes with TAL repeats mimics the assembly pattern of the protein repeats. These results contribute to our understanding of specificity of protein-protein interactions and determinants of fold of protein molecules opening new possibilies for design of drugs and materials.

The transcription activator-like (TAL) effectors form a family of proteins that include virulence factors carrying important role in transcription of plant genes. They contain highly conserved N-terminal secretion/translocation region, C-terminal nuclear localization region, and central DNA binding domain. The central domain consists of tandem repeats; arrangement of the repeats, pairs of antiparallel α-helices, is similar to that of tetratricopeptide repeats (TPR) but TAL repeats are more similar to each other. One TAL repeat binds one DNA base pair by a hypervariable two-residue region located between two α-helices. Right-handed superhelix of the protein repeats fits right-handed double strands of DNA (Reviews: Scholze and Boch 2011; Bochtler 2012). TAL binding of double stranded DNA induces conformational transition, compression of the spiral. Transcription activator-like effector nucleases, engineered by fusion TAL effectors (TALE) and FokI nuclease, are used as genome disrupting agents against HIV and hepatitis B virus (Stone et al. 2013). TALE linked to nucleases, nickases, transposases, and recombinases provide a variety of molecules for genome editing. Genome visualization is another area of promising TALE applications (Miyanari 2014).

α-Helix, a major element of protein molecule, possesses unique properties and exhibits specific hydrogen bonding pattern (Pauling et al. 1951). Association of α-helices occurs as "knobs into holes" interaction of pairs of helices (Crick 1953). Complementary interacting surfaces, hydrogen bonding and van der Waals interactions are main factors that determine specificity and stability of fibrous and globular protein molecules (Chothia et al. 1981; Scheraga et al. 1982; Schulz and Schirmer 1982).

As covalent bonds, noncovalent bonds are also chiral (Kurochkina and Iadarola 2015a). Each type of four-α-helix bundle, α-barrel, or helical toroid exhibit specific mode of helix-helix interaction. Involvement of particular edges of helical surfaces in the formation of the interface and amino acid combinations at helix-helix interfaces are important characteristics of the fold (Kurochkina 2008; Kurochkina and Choekyi 2011) and determine chirality and curvature of the assembly (Kurochkina and Iadarola 2015a). Relationship between the helical edges and the shape of the assembly was previously demonstrated for 8α/8β TIM-barrel proteins, 4-α-helix subunits of tobacco mosaic virus (Kurochkina 2010), and inner and outer rows of

Table 5.2 Chiral Helical Assemblies. Arrangement of contacts on the helix surfaces

Protein	Repeats per turn, handedness	AA' angle BB' angle	AB angle	Inner helix	Outer helix
Ankyrin	24, left	Neg	Neg	A	B
ARM	24, right	Pos	Pos	B (α3)	A (α1)
HEAT	24, right	Pos	Pos	B	A
TPR HAT	8, right	Pos	Pos	A	B
SLR	8, right	Pos	Pos	A	B
PPR	8, right	Pos	Neg/Pos	A	B
LRR	24, right	Pos		β structure	3_{10}
PUM	8/halfturn, right	Pos	Pos	A (α2)	B (α3)
TIM-barrel	8	Neg			

Ankyrin, ARM/HEAT, PUM homology, and LRR repeats (Table 5.2). Inversion in the pattern of contacts leads to the change in handedness of the assembly: inner rows of ankyrin, ARM/HEAT, and PUM-HD repeats utilize edges (i-1, i) and (i + 1, i + 2) for the interaction of the given α-helix with preceding and following helices correspondingly; outer rows of these proteins and LRR repeats invert this pattern and utilize edges (i-1, i) and (i-3, i-2). As a result, binding of inner row to outer row is enantiomer selective (Kurochkina and Iadarola 2015a). Recognition between inner and outer rows of TAL repeats occurs in the same manner. TAL repeats involve edges (i + 5, i + 6) and (i + 3, i + 4) in the inner row and edges (i-3, i-2) and (i-1, i) in the outer row to interact with preceding and following helices. Although previous studies emphasized that DNA binding to TAL repeats involve mainly one DNA strand, it can be seen that second strand binds similarly to the first strand but with weaker contacts. Whereas the first DNA strand binds the inner row, second strand binds the outer row and DNA double helix reflects chirality of the assembly. Therefore, there is a link between protein conformation and DNA conformation.

As in ankyrin repeats, each of the two different contact patterns of TAL repeats corresponds to a unique helix arrangement and opposite direction of the assembly of the two rows. These results contribute to understanding of specificity of protein-protein interactions and recognition of ligands by protein molecules and can be used for the prediction of specific protein-protein interactions leading to the modeling and design of new drugs, nanostructures and nanomaterials and deciphering mechanisms of action of protein molecules.

Solenoids structure and pattern of interactions. Both Ankyrin and TAL repeats consist of two rows of α-helices, inner and outer (Fig. 5.7A, B). Helix A of the inner row is antiparallel to helix B of the outer row. AB pair is a repeat unit. In the inner row, helix A is parallel to the preceding helix, A", and following helix, A'. Similarly, each helix B of the outer row, except N-terminal and C-terminal, is surrounded by preceding B" and following B' helices. Helix-helix interface is centered around particular edges. Each edge is designated by the two adjacent amino acids that form a peptide bond. Ankyrin repeats were found to use edges (i-1, i) and (i + 1, i + 2) at the inner row parallel interfaces AA' and BB' (Fig 5.7C) and edges (i-1, i) and (i-3, i-2) at the outer row parallel interfaces AA" and BB" (Fig 5.7E). Several other types of

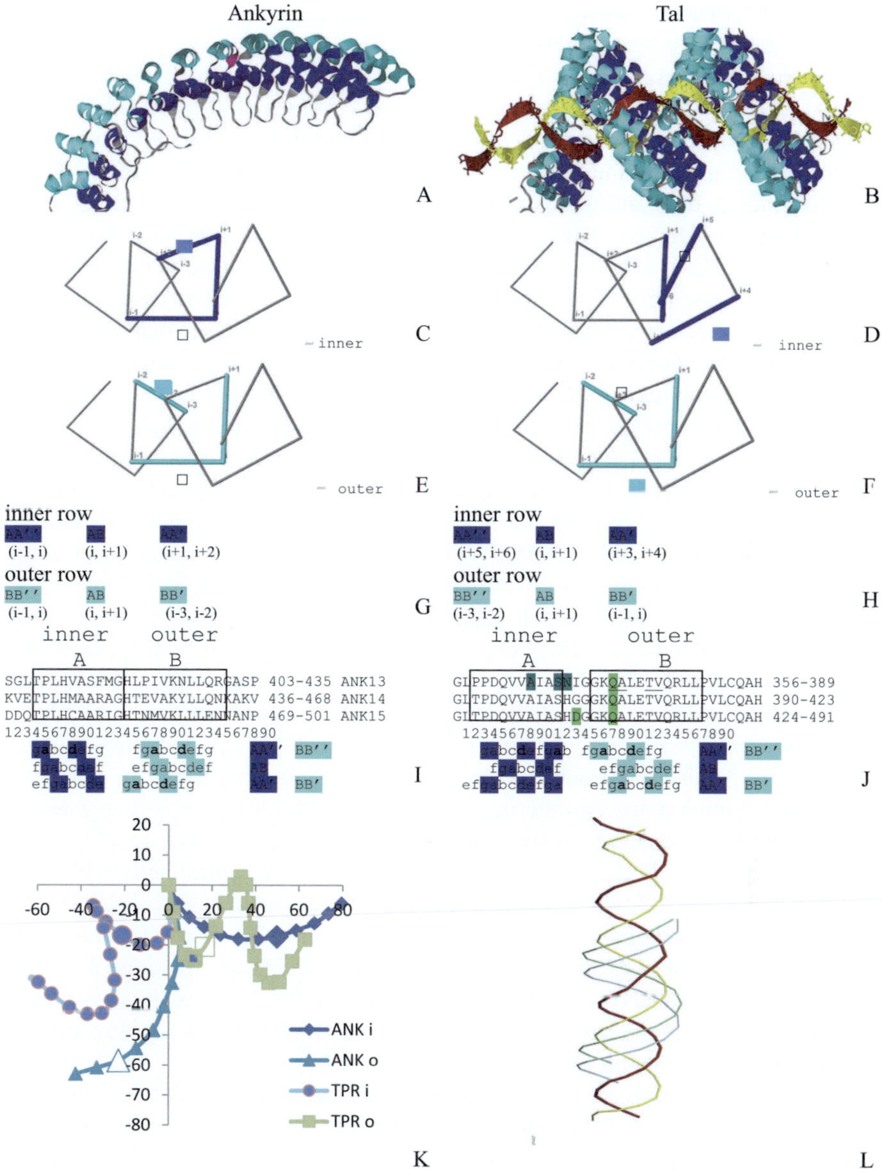

Fig. 5.7 Helical assemblies of ankyrin (left) and TAL (right) repeats. Three-dimensional structure of (**A**) ankyrin and (**B**) TAL. Inner row helices (blue), outer row helices (cyan), and DNA strands (brown, yellow). (**C D E F**) Helical wheel. Interactions of the central helix with preceding (□) and following ((■ ■) helices. Leading edges are designated by filled boxes. (**G H I J**) Assigned positions **a**-**g** at helix-helix interfaces in leucine zipper nomenclature of ankyrin (**GI**) and TAL (**HJ**). Amino acids interacting with DNA (green). (**K**) The line connecting positions **a** of the repeats in the XY plane of the coordinate system of (Kurochkina 2008). The lowest and highest points in the Z direction are shown by empty and filled enlarged markers correspondingly. (**L**) Superimposed TAL positions **a** of the inner (blue) and outer (cyan) rows on the DNA backbone double helix (brown, yellow)

repeats, ARM, PUM homology, and LRR, follow the same rule. When this approach of assigning central edges to helix-helix interfaces is applied to the TAL repeats, it can be seen that edges (i + 5, i + 6) and (i + 3, i + 4) in the inner row (Fig 5.7D) and edges (i-3, i-2) and (i-1, i) in the outer row are involved (Fig 5.7F). Helix-helix interface is formed by amino acids at positions **a** and **d** flanked by amino acids at positions **e** and **g** in leucine zipper nomenclature (Kohn et al. 1977) (Fig 5.7G–J). How positions **a**, **d**, **e**, and **g** at parallel AA', and AA" interfaces of the inner rows, parallel interfaces BB' and BB" of the outer rows, and antiparallel interfaces AB of ankyrin (Fig. 5.7G–I) and TAL (Fig. 5.7H–J) repeats are arranged is the key to the direction of the assembly. If positions **a** of the row helices of the two assembly types are compared in the same coordinate system (Fig 5.7K), we can see that direction of the assembly of α-helices in the outer row is opposite to that of the inner row. These two types of assemblies cannot be superimposed. The coordinate system applied for this purpose transforms (1) the α-carbon of amino acid at position **a** to the origin, and (2) the peptide group between residue at position **a** and residue at position **g** preceding **a** to XZ plane so that the vector from C_α at position **g** to C_β at position **a** is parallel to the X axis, and the negative end of the Y-axis points toward the interacting helix (Kurochkina 2008). With position **a** of the N-terminal helix at the origin and peptide group of the residues at positions **g** and **a** in XZ plane, all consecutive helices of the row will follow in the negative Y direction. We can see that inner and outer rows follow opposite X axis directions in both types of the repeats. Besides, each of the four rows in the two proteins adopts unique direction of the assembly. Two turns of the TAL assembly fits one turn of ankyrin assembly.

Structure of the repeat unit Ankyrin repeat unit consists of a pair of antiparallel α-helices and a β-hairpin whereas TAL repeat unit contains only a pair of α-helices. Repeats are homologous. Numbering of residues within a repeat is suggested as shown for three consecutive aligned repeats of ankyrin (Fig 5.7G) and Tal (Fig 5.7H) (Mosavi et al. 2002; de Lange et al. 2014). Contacts of the central helix with three surrounding helices follow a particular pattern that is repeated at each ankyrin unit. This pattern is unique characteristic of each type of assembly. Two residues in the loop between α-helices of the TAL repeat form DNA contacts. Residue 13 is the main determinant of DNA base binding whereas residue 12 is assigned a role in stabilizing the repeat; the remaining residues form a scaffold that brings recognition loop to dsDNA (Mak et al. 2012; de Lange et al. 2014).

Protein conformation and DNA conformation Binding of DNA B form to TAL repeats mainly occurs via TAL inner row α-helices and one DNA strand so that one TAL repeat binds one nucleotide base pair (Mak et al. 2012). Distances are in the range 3–3.5 Angstrom and TAL amino acids at positions **e** and **b** (or 13 and 17) are involved. Second DNA strand binds TAL outer row α-helices with weaker interaction: 4–4.5 Angstrom involving TAL positions **e**, **a**, and **b** (or 8, 11, and 12) (Fig 5.7K). DNA strands mimic protein curvature. There is good correlation between protein amino acids at positions **a** and DNA backbone atomic positions (Fig 5.7L).

Pocket geometry and chirality. TAL outer row contacts of the central helix with the preceding and following helices involve the same edges as ankyrin outer row contacts but order inversion (preceding i-1, i and following i-3, i-2 in ankyrin versus preceding i-3, i-2 and following i-1, i in TAL) results in the opposite direction of the assembly. Another point of difference is in positioning of the antiparallel interface relative to two parallel interfaces: in ankyrin preceding and following helices form contacts at the edges ±1 residue from antiparallel, whereas in TAL +3 and + 5. Order of the inner row edges is also inverted in TAL compared to ankyrin.

To summarize, a relationship between intrinsic properties of an α-helix and chirality and curvature of the helical assembly is demonstrated for TAL repeats proteins. Comparison of Patterns of contacts of both ankyrin and TAL assembly types demonstrate how inversion of contacts at the helix-helix interfaces of the inner and outer rows results in change of the direction of the solenoid. Contact sites of the helix with preceding and following helices in the same row go along the direction of the helix backbone and helix axis in ankyrins but opposite to this direction in TAL inner rows. In the outer rows, contact sites of the helix with preceding and following helices in the same row go along the direction of the helix backbone and helix axis in TAL but opposite to this direction in ankyrin.

Pocket geometry of an α-helix contributes to chirality and curvature of the helical assembly of TAL repeats: orientation of the helical edges determines direction of the assembly and influences curvature of the assembly.

Tetratricopeptide, TPR, is a 34 amino acids repeat of two α-helices (Main et al. 2003). First identified in CDC23 and nuc2+, TPR motifs were found in many proteins: protein phosphatase 5 (PP5) (Hong et al. 2016), cochaperone Hsp70 interacting protein (Hip) (Li et al. 2013), Hsp90 organizing protein (Hop), cyclophilin, p67phox, TOM complex (Perry et al. 2006), SycD (Buttner et al. 2008), APC/C complex (Zhang et al. 2010), and O linked GlcNAc transferase (Jinek et al. 2004). They are present in many organisms from plants to animals. Inner grove of TPR binds peptides (Hip/ Hsp70, Hop/Hsp90).

Designed TPRs carry a consensus sequence and are characterized by a set of parameters: angle between antiparallel helices A and B of one repeat unit (-149–-162), angle between parallel helices A and A’ of the two adjacent repeat units (11–32), radius and handedness (right) of the superhelix, twist angle formed by the two equivalent points in the adjacent repeats (45), pitch (53–76) (Main et al. 2003; Sawyer, N. et al. 2013). Although the length of the conserved two helix unit is 34 amino acids, it can be extended by insertions in the loop regions and lengthening of helices (Perry et al. 2006).

TPR proteins are involved in transcription regulation, protein translocation, cell cycle control, pre-mRNA splicing, GlcNAc modification (Jínek et al. 2004), and peptidyl prolyl isomerisation (D’Andrea and Regan 2003) and represent important pharmacological target (Holzer and Izzo 2014).

HAT is half a TPR. One repeat unit, a pair of α-helices, is a stable dimer. Twelve HAT repeats of RNA14 comprise aminoterminal (5 repeats) and carboxyterminal (7 repeats) subdomains of *K. lactis* cleavage factor IA (CFIA), protein implicated in 3’ processing of mRNA precursors and similar to murine CstF77 (Paulson and Tong 2012). Squamous cell carcinoma antigen recognized by T cell 3 (SART3) is a spliceosome recycling factor that binds USP15 (Ubiquitin-specific protease, USP) and brings it from cytosol to the nucleus. SART3 two subdomains, aminoterminal and carboxyterminal, compose twelve repeat HAT domain. Convex surface of the carboxyterminal HAT interacts with USP15 whereas concave surface mediate dimerization. Quaternary structure of SART3, RNA14, and CstF77 shows differences in dimerization modes (Zhang et al. 2016).

Pentatricopeptide (PPR) motif bind single stranded RNA (Hall et al. 2016). One repeat unit consists of two α-helices and C-terminal loop comprising in total 31–36 amino acid residues. Repeats are subdivided into P (35 residues), S (31 residues) and L (31 to 36 residues) types and P (with P type only) or PLS (with alternating P, L, or S types) classes. Eight repeat units form a turn of a superhelix and bind eight RNA bases to the concave row, one base per one unit in orientation that brings together 5' RNA end and protein C-terminus. Complexes such as that of PPR10 six PPRs with six nucleotide bases of 18 nt PSAJ element show modular organization (4 m59). Also, one or two extra bases can be incorporated into the RNA strand (Lu et al. 2009). Two amino acids of protein α-helix five residues apart recognize one RNA nucleotide and form specific protein/RNA contacts.

Protein only RNase P, PRORP1, processes precursor tRNA cleavage of 5′ leader sequence, function also carried out by ribozyme RNase P. It consists of three domains: PPR, central, and metallonuclease. PRORP1 PPR domain with five PPR repeats may interact with precursor tRNA. Removal of four PPR repeats results in impairment of substrate binding and enzymatic activity (Karasik et al. 2016). PPR domain in mitochondrial single subunit RNA polymerase is interaction site of AT rich loop binding to promoter DNA (Ringel et al. 2011). PPR repeat is a component of mitochondrial ribosome which synthesizes mitochondrial encoded membrane proteins (Gerber et al. 2015).

In plants, tandem repeats containing proteins including ankyrin, ARM/HEAT, TPR, PPR and other are essential components of cell homeostasis, stress and development pathways (Sharma et al. 2016). PLS type PPR are RNA editing factor interacting proteins (Yan et al. 2017). PPR proteins interact with MORF proteins as mediators of RNA editing, posttranslational cytidine to uridine modification in plastids and mitochondria of flowering plants (Yang et al. 2018). MORF binding induces conformational transitions in PLS PPR complexes and increase RNA binding (Yan et al. 2017). RNA induces dimerization of PPRs (Ke et al. 2013). PPRs are prototypes for the design of new scaffolds for RNA binding (Coquille et al. 2017).

SLR repeats bear name originating from *C. elegans sel1* gene product. These repeats are found in *Escherichia coli* C5321, a protective antigen present in uropathogenic strains CFT073 (Urosev et al. 2013) and mouse SEL1L involved in degradation of proteins from ER, *H. pylori* HcpB/C responsible for adaptation to hosts, yeast Hrd3, adaptor protein connecting membrane complexes, and *Klebsiella pneumonia* kpn. Repeat length of 36/38 amino acids and consensus sequence are similar to those of TPR (34 aa). HcpC contains interrepeat disulfide bonds.

PUM Pumilio/fem-3 mRNA binding factor (PUF) proteins bind single stranded mRNA in the 3' untranslated region via Pumilio (PUM) repeats (Hall et al. 2016). One repeat unit consists of three α-helices comprising 36 amino acid residues, α1, α2 (concave row), and α3 (convex row). Aminoterminal and carboxyterminal repeats are imperfect and contain one or two helices. Topology of PUM repeats is similar to that of ARM and LRR. Eight repeat units form a turn of a superhelix and bind eight or nine RNA bases (Lu et al. 2009) to the concave row in orientation that brings together 5' RNA end and protein C-terminus. Two amino acids of protein α-helix four residues apart recognize one RNA nucleotide and form specific protein/RNA contacts. The same length eight units protein sequence can recognize one base per one unit (Lu et al. 2009). Ideal "one nucleotide/one protein repeat" modular organization provides very convenient template for the design of new specific sequences (Wang et al. 2002) although multiple deviations from this pattern exist as can be seen in complexes of human Pumilio 1/Cyclin B reverse RNA, human Pumilio 1/PUF5 RNA, NRE1–14 RNA, NRE2–10 RNA, NRE1–19, erk2 NRR, and PUF6/5BE13 RNA. Bulge formed when one or two nucleotides are inserted in the middle of 8 nt sequence may provide binding site for various effector proteins.

Use of a programmable RNA binding domain PUF introduces new approaches to monitoring mRNA localization in the cytoplasm, local translation, phenotypical changes, and protection against degradation (Abil et al. 2017).

Recognition of RNA transcripts by PUF proteins is important for regulation of many biological functions such as maintenance of stem cell production, anterior/posterior patterning and organelle biogenesis during embryonic development. PUF proteins from *C. elegans, D. melanogaster, S. cereviciae* and recently *A. thaliana* proved their role in posttranscriptional regulation and consequently gene expression. *A. thaliana* 26 PUF proteins mediating organ polarity, branching, rhizogenesis, flowering show the mechanisms of development in plants (Jenkins et al. 2009; Wang et al. 2018).

Distribution of contacts on the edges of helical surfaces positions PUM together with ankyrin and HEAT/ARM repeat assemblies indicating similar structural alignment of helix helix units.

Leucine rich repeat (LRR) proteins regulate cell cell interactions. They were largely studied for their role in the development of multicellular organisms: formation of connections in neural circuits. LRR-containing proteins function as synaptic adhesion molecules during excitatory synapse formation, midline guidance (Slit). LRR-containing proteins are implicated in Tourette's syndrome (TS) and OCD (Slit and Trk-like family member 1), epilepsy (Lgi1), schizophrenia (*NRXN1, Ngr*), autism spectrum disorders (*LRRTM3* and *LRRN3)*, Alzheimer's disease (*LRRTM3)*, immune response (Toll like receptors, TLR) (de Wit et al. 2003).

LRR repeat unit is 20 to 30 amino acids and consists of a helix (outer row) and strand (inner row). Consensus sequence LxxLxLxxNxL shows high content of leucines (deWit et al. 2003). Crystallographic structures of bacterial invasion protein internalin (*Listeria monocytogenes* InlA) in complex with its human receptor Ecadherin (hEC1) (2omt), TLR9 complexes with inhibitory DNA and agonistic cytosine phosphate guanine (CpG) containing viral DNA (3wpc), TLR3 complex with double stranded DNA (3ciy), actin associated LRR protein C20 (1io0), tropomodulin LRR (actin binding site) complex to gelsolin (4pki) and other were determined. TLR3 consists of 23 protein repeats. Its extracellular domain binds dsRNA of 40 to 50 base pairs DNA and each interaction occurs with both RNA strands (3ciy). In Tropomodulin LRR (actin binding site) complex, gelsolin caps barbed end. Tropomodulin is the only protein known to cap the pointed end of tropomyosin (TM)-coated actin filaments.

Outer row of ankyrin, ARM/HEAT, PUM-HD and LRR repeats utilize edges (i-1, i) and (i-3, i-2) (Table 5.1).

Clathrin is main component of coated vesicles that deliver substances from one cellular compartment to another or from cell to cell (Yuan et al. 2015). Trafficking vesicles with packaged substances bud from one cellular compartment and fuse with the targeted cellular compartment to empty the content. Vesicle coats polymerize as polygonal cages, heteroheptameric complexes that assemble on membranes and bend them (Dodonova et al. 2015). Heterotetrameric adaptor proteins (AP) complexes provide a link between the clathrin coated vesicles and membrane bound cargo components. Large portion of AP complexes comprises HEAT repeats (Heldwein et al. 2004). Movement of clathrin coated vesicles inside the cell relies on the intracellular actin-based vesicular transport system (Khaitlina et al. 2014). Dynamin tetramers assemble as righthanded helical oligomers on clathrin coated vesicles mediating constriction and scission of the membrane. Mutations that disrupt tetramer result in Charcot Marie Tooth disease (Reuboold et al. 2015).

BIR repeat Baculovirus IAP repeat, BIR, is a zinc stabilized domain of approximately 80-amino acids which occurs as a tandem repeat of three units in IAP family proteins. Its arrangement of six subunits makes a circular assembly. Eight human IAP proteins neuronal apoptosis inhibitory protein (NAIP), cellular IAP (c-IAP1, c-IAP2), X-linked IAP (XIAP), melanomaIAP (ML-IAP), IAP-likeprotein 2 (ILP2), survivin and BIR repeat-containing ubiquitin-conjugating enzyme (BRUCE) carry out important functions in caspase inhibition and protection against apoptotic stimuli, regulation of cell division, cascade signaling, and transforming growth factor ß activation. Three cysteines and one histidine coordinate zinc (Shin et al. 2005). Mitochondrial Smac/DIABLO, in contrast, promotes apoptosis as its tetrapeptide binds the groove of the zinc-binding site in BIR domains (Wu ET Al. 2000). The NLR apoptosis inhibitory proteins (NAIPs) specifically recognize bacterial ligands. These cytosolic receptors utilize three structural domains, the pre-BIR, BIR1, and HD1 for anti-bacterial defenses (Yang et al. 2018).

BRC repeat is an evolutionarily conserved sequence in the breast cancer susceptibility protein BRCA2. The BRC repeat competes for binding by mimicking an oligomerization sequence in RAD51 and, therefore, prevents its assembly necessary for strand-pairing of the DNA recombination process (Pallegrini et al. 2002).

Arcs Groups of three or more protein helices assemble as helical rods and form arcs of rings and spirals. These arcs are present in all proteins with three or more interacting helices. Proteins of nucleotide binding Rossmann fold comprising central parallel β-sheet surrounded by two layers of helices, αβα, is one example. Representatives include aldehyde dehydrogenases NAD^+-binding proteins, tubulin and glycosyl transferases. The transport protein particle TRAPP, mediator of vesicular transport from ER to Golgi (Kummel et al. 2006; Ku¨mmel et al. 2008), response regulators (RR) of the bacterial signal transduction system (Narayanan et al. 2014; Toro-Roman et al. 2005), kinetochore Ska complex and KIF5C assemblies with microtubules (Abad et al. 2014; Ciferri et al. 2016; Singh R 2017), components of cell cycle regulation (Hutchins and Wray 2016; Zhavoronkov et al. 2016), pathways of regenerating liver PRL-1/SH3, RhoGAP, ERK1/2/RhoA (Bai et al. 2011) are only few examples of assemblies involving short helical arcs of various shapes (Fig. 5.8).

Protein/PDB entry	Helix orientation/ boundaries	Arc in protein tertiary structure
Glucokinase /1v4s / Hexokinase /1dgk/	↓ ↑ ↑ 345 331 410 355 339 416	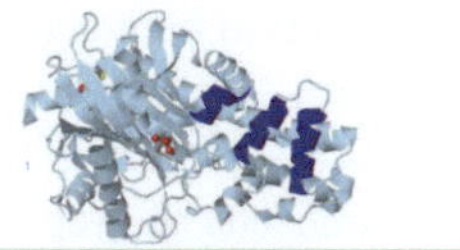
BET3-TPC6B of TRAPP /2cfh/	↑ ↓ ↑ 114 3 36 118 17 63 TPC6B BET3	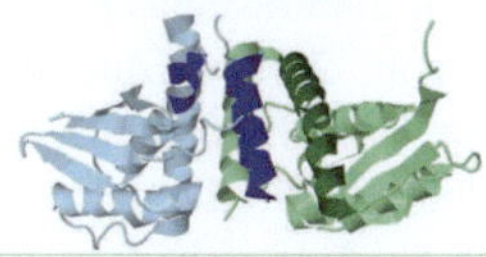
KDPE /1zh4 4kfc/ *E.coli* /1nxw/ *S. pnemoniae* /1p2f 3nns/ *T. maritima*	↑ ↑ ↑ ↑ ↑ 34 60 83 104' 10' 44 68 93 121' 23'	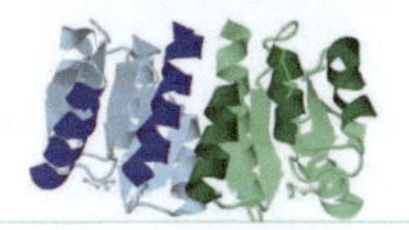
PRL-1 /3rz2/	↑ ↑ ↑ 78 110 125 95 121 135	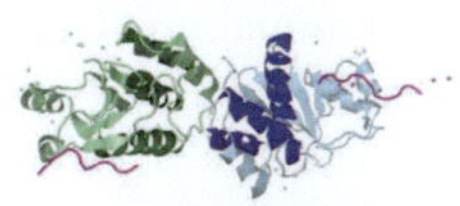
UDP-GlcNAc 2- epimerase /4zht/	↑ ↑ ↑ 257 225 339 268 244 351	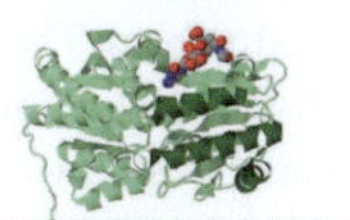
NS1 /4pp4/	↓ ↑ ↓ ↑ 53 80 141 166 71 105 161 180	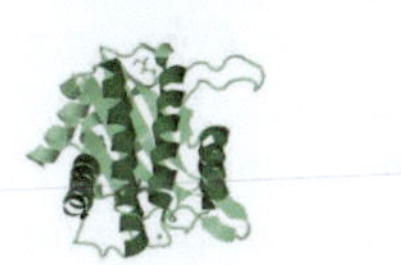
Autotaxin /2xr9/	↓ ↑ ↓ 705 682 688 718 686 697	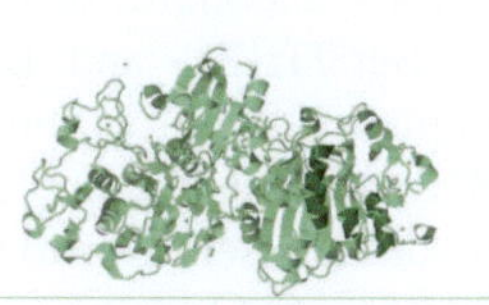
Tubulin / 5eyp/	↑ ↑ ↑ ↑ 109 144 182 382 129 161 198 401	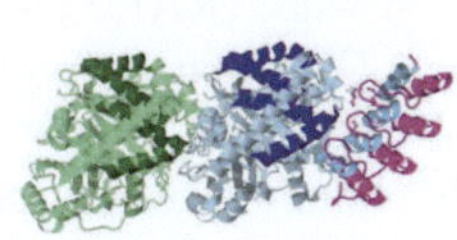

Fig. 5.8 Arcs

Different patterns of contact distribution on the helix surfaces correspond to distinct assembly shapes. Since shape of the arc is major component of specific interactions within the assembly and also affects ligand binding, it is important to explore the connection between contacts on the helical surfaces, shape of the arc, and binding properties.

UDP-GlcNAc 2-epimerase/ManNAc Kinase (GNE) enzyme, regulator of sialic acid synthesis, consists of 2-epimerase and kinase domains (Hinderlich et al. 2015). GNE glycosyl transferase has Rossmann fold (Chua et al. 2008 /2r60/; Schmidt et al. 2012; Chen et al. 2016). Glucokinase palm shape is formed by small and large domains (Kamata et al. 2004). Both UDP-GlcNAc 2-epimerase and ManNAc kinase contain arcs that exhibit the pattern of contacts as in the inner row helices of ankyrin repeats (Fig. 5.8). Arcs in human glucokinase (PDB entry 1v4s 3a0i) and hexokinase (PDB entry 1dgk) are located in the large domain and contain site of ADP binding. GNE enzyme and its homologs show a conserved pattern of interhelical contacts. Structure of the fragment containing arc is also similar to N-acetylmannosamine kinase group of helices located in the C-terminal lobe at the site of ADP binding (Martinez et al. 2012; Kurochkina et al. 2009). Only two corresponding helices of ManNAc kinase form a well defined interface. In GNE, this site carries several disease associated mutations (Huizing, M 2005; Kurochkina et al. 2009).

Receiver domain of the two component signal transduction system, transcription regulator KDPE, consists of two subunits (Narayanan et al. 2014) in which five-stranded parallel β-sheet is surrounded by five α-helices on each side. Structures of the receiver domain from *Thermotoga maritima* DrrB and DrrD, *Escherichia coli* PhoB, and *Bacillus subtilis* PhoP show similar organization (Bent et al. 2004). In *E. coli* complex, intersubunit interface is formed by I88 (position **a**), L91 (d), A110 (**a**), R113 (**d**), and V114 (**e**) of the helix-helix interfaces. Five helices, two from one subunit and three from another, form a row of two joined segments with two overlapping helices. Four helices are of the same concativity as ankyrin inner row whereas adjacent three helices as outer row (Toro-Roman et al. 2005).

TRAPP, the transport protein particle, mediates vesicular transport from ER to Golgi (Kummel et al. 2006; Kümmel et al. 2008). Crystal structures of yeast and mammalian TRAPP complexes show its association with multiple subunits such as Bet3/Tpc6b, Bet3/Trs31, and Bet3/Trs33. Association of multiple subunits in TRAPP complexes show conserved intersubunit interfaces. In Bet3/Tpc6b, Bet3/Trs31, Bet3/Trs33 subunit complexes, conserved pattern of contacts corresponds to the curvature of the assembly shown for the BET3-TPC6B /2cfh/ (Fig. 5.8). The row of three α-helices that has the same curvature as ankyrin inner row contains two helices from one subunit and one helix from another subunit. GTP-binding protein Ypt1 association with TRAPP complex promotes its activation, or transition from GDP- bound to GTP-bound form (Cai et al. 2008). GDP dissociation inhibitor subunit RabGDI regulates Ypt1/RabGDI complex (Rak et al. 2003 /1ukv 2bcg/).

MG^{+2}/BEF^{3-}bound receiver domain of two component signal transduction KDP potassium transport system has a response regulator KDPE which forms two fold rotationally symmetric dimer. It contains two subunits and its dimerization interface contains an arc of five helices (Narayanan et al. 2014). Symmetric dimerization is important for activation of response regulator and transcription. Similar mechanism is described for TorR (Toro-Roman et al. 2005).

Cell cycle, regulation kinases Cdk, their activators such as cyclins and p35, and inhibitors (Zhavoronkov et al. 2016) contain multiple substructures shaped as arcs.

Phosphatase of the regenerating liver (PRL-1) is a protein tyrosine phosphatase induced after liver damage. PRL-1 binds SH3 domain of RhoGAP and activates ERK1/2 and RhoA pathways enhancing cell proliferation (Bai 2011) which suggests its oncogenic role in cancer and metastasis progression. PRL-1 in complex with peptide reveals novel mode of interaction involving SH3 domain/ RhoGAP.

Minute virus of mice non-structural protein-1(NS1) N-terminal nuclease domain hosts a unique ZN^{2+} coordination in the active site that is required for ssDNA binding and cleavage. The structure demonstrates novel mode of replication origin recognition (Tewary et al. 2015). The nickase domain which consists of five stranded antiparallel beta sheet surrounded by four a helices on one side and three a helices on another side contains two arcs. NS1 guides virus replication in host organism.

Tubulin consists of α and β subunits that assemble in microtubules. These cytoskeleton filaments are dynamic bodies which polymerize and depolymerize in response to cellular signals. Crystallographic structures of αβ-tubulin heterodimer with stathmin-like coiled coil and GDP/GTP (Nawrotek et al. 2011) show curved and straight assemblies. Designed ankyrin repeats protein DARPin inhibits polymerization by binding to the β-tubulin (Pecqueur et al. 2012) and many mutants show improved affinity (Ahmad et al. 2016). Both tubulin subunits are proteins of Rossmann fold: a parallel β-sheet flanked by two layers of α-helices. Curvature and pattern of contacts of α-helical rows are the same as in ankyrin inner row (Fig. 5.8).

Binding of kinetochore Ska complex to microtubules involves Ska1-MTBD. Ndc80 complex is the site of kinetochore to microtubule attachment (Ciferri et al. 2016). It binds MT at the site of a- and b-tubulin interface (Abad et al. 2014). Rod of Ndc80 as long Ska coiled coil has tubulin on one end and kinetochore on the other end; together Ska1, Ska2, and Ska3 organize a W shaped structure (Jeyaprakash et al. 2014). The curvature of coiled helix correlates with curvature of ankyrin outer row.

Autotaxin contains cytosolic, transmembrane, somatomedin-B-like, catalytic and nuclease domains (Hausmann et al. 2011; 2xr9 3nkm). Mutations in both catalytic and nuclease domains are associated with GACI disorder, strongly affect interdomain and intradomain interfaces and lead to hypermineralization (Kato et al. 2012).

Many other protein assemblies contain arcs: suppressor of kinetochore protein 1 (Skp1) (Schulman 2000; Willhoft et al. 2017; Wong et al. 2017), PMS1 (Arana et al. 2010), PMS2 (Guarné et al. 2001), MLH1, (Wu et al. 2015;), proteins coded by DNA mismatch-repair gene MutL/MutS (Groothuizen et al. 2015); recognition complexes of mismatches (Natrajan et al. 2003), channels (Warren et al. 2003), human (MSH2, MSH3, MSH6, MLH1, PMS1, and PMS2) and E. coli mismatch repair genes (Gu et al. 1998).

Helical Repeats and SH Domains

Function of many proteins depends on protein-protein interactions with involvement of SH domains and helical repeats. Multiprotein complexes carrying these domains are essential components of many biological processes such as response to bacterial and viral infection, protein synthesis, sorting and targeting to various organelles, degradation, shuttling between cellular compartments, signaling and communication. These complexes consist of tens or even hundreds of proteins. SH domains and helical repeats bring together multiple modules of signal transduction complexes that regulate cellular pathways. Although helical repeats and SH domains differ in structure and exhibits high specificity toward bound atomic groups, they recognize similar sequence motifs. Their organization impacts the protein interactions with each other, nucleus acids, carbohydrates, and lipids and governs highly precise nature of structure and function of living organisms (Kurochkina and Iadarola 2015b).

SH2 and SH3 domains can be found in many adaptor proteins that lack enzymatic activity, such as Grb-2 and α-spectrin, and in proteins that possess enzymatic activity: protein tyrosine kinase (PTK), phospholipase C-γ1 (PLC-γ1), *ras* GTPase-activating protein (*ras* GAP), phosphatidyl-inositol 3-kinase (PI3-kinase) (Kefalas et al. 1995). Out of PTK six domains, N-terminal domain, site of lipid modifications important for PTK targeting to the plasma or intracellular membrane (SH4), unique domain (UD), catalytic domain (SH1), phosphotyrosine recognition domain (SH2), polyproline sequence specific (SH3) domain, and C-terminal tail with a regulatory tyrosine (Gmeiner and Horita 2001), SH2 and SH3 domains are the most important regulatory units. SH2 domains comprise a β-sheet flanked by one α-helix on each side (Filippakopoulos et al. 2009). SH3 domains consist of five to eight β-strands that form two antiparallel β-sheets or a barrel (Mayer 2001; Kurochkina and Guha 2014). Phosphorylation and mutations of tyrosine residues in SH2/SH3 domains indicate their involvement in assembly/disassembly of complexes (Kurochkina and Guha 2015).

Kinesin KIFC1 interacts with ADP ribosylation factor GTPase activating protein with SH3 domain, ankyrin repeat and PH domain 1 (ASAP1) (Lee et al. 2018).

Barrels

8α/8α barrel: ATPase molecules pump protons in plasma membranes and the acidic organelles. ATP synthases reside in the membranes of bacteria, mitochondria, and chloroplasts. They catalyze ATP synthesis and hydrolysis coupled with transmembrane proton or sodium ion transport. ATP synthase (F-ATPase) and vacuolar ATPase (V-ATPase) are proteins that use ATP hydrolysis to drive rotation of the central stalk and the membrane ring attached to it. F1 and V1 are cytosolic globular catalytic domains which hydrolyze ATP. They are attached to Fo /Vo domains, intrinsic membrane domains, by stalk domains. H^+ or Na^+ is pumped at the interface between the rotating ring and a static membrane portion of the molecule. Crystal structures of the Na + −bound/unbound K ring (Fig 5.9A) provide insight into the ion-transport mechanism of *E. hirae* V1Vo-ATPase (Mizutani et al. 2011).Ten subunits, each with four transmembrane helices, form the ring. In the *E. hirae*

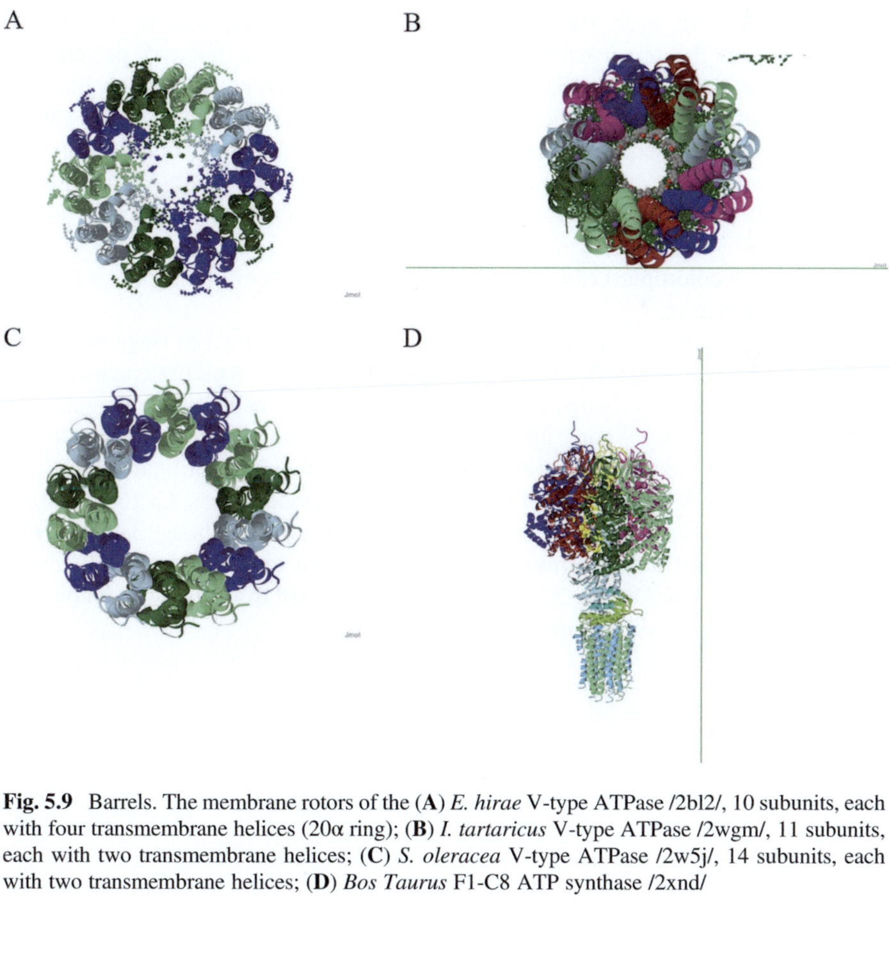

Fig. 5.9 Barrels. The membrane rotors of the (**A**) *E. hirae* V-type ATPase /2bl2/, 10 subunits, each with four transmembrane helices (20α ring); (**B**) *I. tartaricus* V-type ATPase /2wgm/, 11 subunits, each with two transmembrane helices; (**C**) *S. oleracea* V-type ATPase /2w5j/, 14 subunits, each with two transmembrane helices; (**D**) *Bos Taurus* F1-C8 ATP synthase /2xnd/

Na^+-V-ATPase, *N,N'*dicyclohexylcarbodiimide (DCCD) binds the conserved Glu^{139} (or yeast $cGlu^{59}$) inhibiting ATPase activity and proton translocation.

Alkaline pH of mitochondria or acidic pH of lysosome is critical for the organelle function.V type H^+ATPases pump protons and alter pH mainly in endocytic and secretory organelles regulating processes that depend on transmembrane electrical potential: action of enzymes, dissociation of ligands, and coupled membrane transport of substrates (Maxson and Grinstein 2014).

The membrane rotor of the *Ilyobacter tartaricus* Na^+-F-ATPase c-ring (Fig 5.9B) contains 11 monomers (Meier et al. 2009 /2wgm/). Comparison of the *Spinacia oleracea* chloroplast ATPase C14 rotor ring (Fig 5.9C) with *I. tartaricus* Na^+-F-ATPase C11 ring (Vollmar et al. 2009) shows that proton binding site at Glu61 differs. The concerved carboxylates involved in proton/sodium translocation are at approximately 10.8 A distance in both c rings.

The number of c subunits varies among organisms from 8 to 15 (Table 5.3). Stoichiometry of sodium ions per ATP is 3.3 for the yeast, 4.7 for the chloroplast, and 3.7 for the *I. tartaricus* ATPases (Stahlberg et al. 2001). The number of protons translocated for one complete ring rotation of 360 degrees is the same as the number of c subunits with each subunit carrying one carboxylate that is protonated/ deprotonated in the translocation process; total estimate is 3.3 to 5 protons per ATP (Watt et al. 2010) (Fig 5.9D). During the translocation process ion binds to carboxylate of the c ring and then unbinds it upon release. The ring rotation causes rotation of the attached γε stalk subunits and induces conformational change in catalytic β subunits comprising α3β3 complex with consequent ATP synthesis (Preiss et al. 2010). The yeast c10-ring is ~55 °A in diameter and 27 A° in the inner diameter.

The spinach chloroplast c14 ring has an external diameter of 52.30 A° and 40 A° in the inner diameter.

Connexin (Cx26) is a gap junction channel. It contains two 12α rings within six four-α-helix subunits (Maeda et al. 2009). Gap junctions intercellular channels such as connexin allow to exchange ions and small molecules between the neighboring cells. Upon injury, Ca^{2+} ions inhibit intercellular communication (Bennett et al. 2018). Calcium binds at the interface between two subunits coordinated by glutamate and inhibits K^+ translocation through the pore. Dodecameric Cx26 assembles from two hexameric hemichannels and contains four transmembrane helices, cytoplasmic N terminal anxC terminal domains and intracellular/extracellular loops. Connexin 43 (Cx43) (Delmar 2012) forms an intercalated disc structure. It communicates with complex of sodium channel and ankyrin G.

Table 5.3 Amino acid sequence of Na+-F-ATPase c-ring subunits. Residues involved in proton/sodium translocation

Source	PDB	Sequence	n
S. platensis	2wie	ESNLTTAASVIAAALAVGIGSIGPGLGQGQAAGQAVEGIARQPEEAEGKIRGTLLLSLAFM**E**ALTIYGLVVALVLLFANP FV	15
S. oleracea	2w5j	NPLIAAASVIAAGLAVGLASIGPGVGQGTAAGQAVEGIARQPEAEGKIRGTLLLSLAFM**E**ALTIYGLVVALALLFANP	14
B. Pseudoformis	2x2v	MAFLGAAIAAGLAAVAGAIAVAIIVKATIEGTTRQPELRGTLQTLMFIGVPLAEAVPIIAIVISLLILF	13
B.TA2.TA1	2qe7		13
P.Denitrificans	5dn6		12
E.Coli	1c17	MENLNMDLLYMAAAVMMGLAAIGAAIGIGILGGKFLEGAARQPDLIPLLRTQFFIVMGLV**D**AIPMIAVGLGLYVMFAVA	12
F.Nuc	3zk1	TAKTIVLGCSAVGAGLAMIAGLGPGIGEGYAAGKAVESVARQPEARGSIISTMILGQAVA**E**STGIYSLVIALILLYANPFL	11
I.Tar *P. modestum* *C. paradoxum*	2wgm / /	FAKTVVLAASAVGAGTAMIAGIGPGVGQGYAAGKAVESVARQPEAKGDIISTMVLGQAVA**E**STGIYSLVIALILLYANPFV	11
B.PS3	1wu0	MSLGVLAAAIAVGLGALGAGIGNGLIVSRTIEGIARQPELRPVLQTTMFIGVALV**E**ALPIIGVVFSFIYLGR	10
Y.Lip	5 fl7	QLVLAGKYIGAGLASIGLVGAGIGIAIVFAALINGVSRNPALKGQLFTYSILGFALS**E**ATGLFALMIAFLLLY	10
E.Coli	1a91	MENLNMDLLYMAAAVMMGLAAIGAAIGIGILGGKFLEGAARQPDLIPLLRTQFFIVMGLV**D**AIPMIAVGLGLYVMFAVA	10
Yeast	2xok	LVLAAKYIGAGISTIGLLGAGIGIAIVFAALINGVSRNPSIKDTVFPMAILGFALS**E**ATGLFCLMVSFLLLFQ	10
Enterococcus hirae	2bl2	MMDYLITQNGGMVFAVLAMATATIFSGIGSAKGVGMTGEAAAALTTSQPEKFGQALILQLLPGTQGLYGFVIAFLIFINL GSDMSVVQGLNFLGASLPIAFTGLFSGIAQGKVAAAGIQILAKKPEHATKGIIFAAMV**E**TYAILGFVISFLLVLNA	10
Bos taurus	2xnd		8

References

Abad MA, Medina B, Santamaria A, Zou J, Plasberg-Hill C, Madhumalar A et al (2014) Structural basis for microtubule recognition by the human kinetochore Ska complex. Nat Commun 5:2964. https://doi.org/10.1038/ncomms3964

Abil Z, Gumy LF, Zhao H, Hoogenraad CC (2017) Inducible control of mRNA transport using reprogrammable RNA-binding proteins. ACS Synth Biol 6(6):950–956. https://doi.org/10.1021/acssynbio.7b00025. Epub 2017 Mar 8. PubMed PMID: 28260376; PubMed Central PMCID: PMC5477001

Ahmad S, Pecqueur L, Dreier B, Hamdane D, Aumont-Nicaise M, Plückthun A et al (2016) Destabilizing an interacting motif strengthens the association of a designed ankyrin repeat protein with tubulin. Sci Rep 6:28922. https://doi.org/10.1038/srep28922

Alber T, Banner DW, Bloomer AC, Petsko GA, Phillips D, Rivers PS, Wilson IA (1981) On the three-dimensional structure and catalytic mechanism of triose phosphate isomerase. Philos Trans R Soc London 293:159–171

Allan RK, Ratajczak T (2011) Versatile TPR domains accommodate different modes of target protein recognition and function. Cell Stress and Chaperones *16*:353–367

Andrade MA, Bork P (1995) HEAT repeats in the Huntington's disease protein. Nat Genet *11*:115–116

Andrade MA, Perez-Iratxeta C, Ponting CP (2001a) Protein repeats: structures, functions, and evolutions. J Struct Biol 134:117–131

Andrade MA, Petosa C, O'Donoghue SI, Müller CW, Bork P (2001b) Comparison of ARM and HEAT protein repeats. J Mol Biol *309*:1–18

Arana ME, Holmes SF, Fortune JM, Moon AF, Pedersen LC, Kunkel TA (2010) Functional Residues on the Surface of the N-terminal domain of Yeast Pms1. DNA Repair 9(4):448–457. https://doi.org/10.1016/j.dnarep.2010.01.010

Bai Y, Luo Y, Liu S, Zhang L, Shen K, Dong Y et al (2011) PRL-1 protein promotes ERK1/2 and RhoA protein activation through a non-canonical interaction with the Src homology 3 domain of p115 Rho GTPase-activating protein. J Bioll Chem 286(49):42316–42324. https://doi.org/10.1074/jbc.M111.286302

Bent CJ, Isaacs NW, Mitchell TJ, Riboldi-Tunnicliffe A (2004) Crystal structure of the response regulator 02 receiver domain, the essential YycF two-component system of *Streptococcus pneumoniae* in both complexed and native states. J Bacteriol 186(9):2872–2879. https://doi.org/10.1128/JB.186.9.2872-2879.2004

Binz HK, Stumpp MT, Forrer P, Amstutz P, Pluckthun A (2003) Designing Repeat Proteins: Well-expressed, Soluble and Stable Proteins from Combinatorial Libraries of Consensus Ankyrin Repeat Proteins. J Mol Biol 332:489–503

Bochtler M (2012) Structural basis of the TAL effector–DNA interaction. Biol Chem 393:1055–1066

Chothia C, Levitt M, Richardson D (1981) Helix to helix packing in proteins. J Mol Biol 145:215–250

Chua TK, Bujnicki JM, Tan T-C, Huynh F, Patel BKC, Sivaraman J (2008) The structure of sucrose phosphate synthase from halothermothrix orenii reveals its mechanism of action and binding mode. Plant Cell 20:1059–1072

Clapham DE, Miller C (2011) A thermodynamic framework for understanding temperature-sensing by TRP channels. Proc Natl Acad Sci USA 108:19492–19497

Cordero-Morales JF, Gracheva EO, Julius D (2011) Cytoplasmic ankyrin repeats of transient receptor potential A1 (TRPA1) dictate sensitivity to thermal and chemical stimuli. Proc Natl Acad Sci 108:E1184–E1191

Crick F (1953) Acta Crystallogr 6:689

Cunha ES, Hatem CL, Barrick D (2016) Synergistic enhancement of cellulase pairs linked by consensus ankyrin repeats: determination of the roles of spacing, orientation and enzyme identity. Proteins 84(8):1043–1054. https://doi.org/10.1002/prot.25047

D'Andrea LD, Regan L (2003) TPR proteins: the versatile helix. Trends Biochem Sci 28(12):655–662. Review. PubMed PMID: 14659697

Dautant A, Velours J, Giraud MF (2010) Crystal structure of the Mg·ADP-inhibited state of the yeast F1c10-ATP synthase. J Biol Chem 285(38):29502–29510. https://doi.org/10.1074/jbc.M110.124529. Epub 2010 Jul 7. PubMed PMID: 20610387; PubMed Central PMCID: PMC2937982

Delmar M (2012) Connexin43 regulates sodium current; ankyrin-G modulates gap junctions: the intercalated disc exchanger. Cardiovasc Res 93(2):220–222. https://doi.org/10.1093/cvr/cvr343. Epub 2011 Dec 16. PubMed PMID: 22180603

Deng L, Guindon J, Vemuri VK, Thakur GA, White FA, Makriyannis A, Hohmann AG (2012) The maintenance of cisplatin- and paclitaxel-induced mechanical and cold allodynia is suppressed by cannabinoid CB2 receptor activation and independent of CXCR4 signaling in models of chemotherapy-induced peripheral neuropathy. Mol Pain 8:71–83

Dluzewski AR, Fryer PR, Griffiths S, Wilson RJ, Gratzer WB (1989) Red cell membrane protein distribution during malarial invasion. J Cell Sci 92(Pt4):691–699. PubMed PMID: 2532219

Dodonova SO, Diestelkoetter-Bachert P, von Appen A, Hagen WJ, Beck R, Beck M, Wieland F, Briggs JA (2015) Vesicular transport. A structure of the COPI coat and the role of coat proteins in membrane vesicle assembly. Science 349(6244):195–198. https://doi.org/10.1126/science.aab1121. PubMed PMID: 26160949

Edwards RA, Lee MS, Tsutakawa SE, Williams RS, Tainer JA, Glover JNM (2008) The BARD1 C-terminal domain structure and interactions with polyadenylation factor CstF-50. Biochemistry *47*:11446–11456

Ferreiro DU, Cervantes CF, Truhlar Stephanie ME, Cho SS, Wolynes PG, Komives EA (2007) Stabilizing IκBα by 'consensus' design. J Mol Biol 365:1201–1216

Filippakopoulos P, Müller S, Knapp S (2009) SH2 domains: modulators of nonreceptor tyrosine kinase activity. Curr Opin Struct Biol 19(6):643–649. https://doi.org/10.1016/j.sbi.2009.10.001. Epub 2009 Nov 18. Review. PubMed PMID: 19926274; PubMed Central PMCID: PMC2791838

Forwood JK, Lange A, Zachariae U, Marfori M, Preast C, Grubmüller H, Stewart M, Corbett AH, Kobe B (2010) Quantitative structural analysis of importin-β flexibility: paradigm for solenoid protein structures. Structure *18*:1171–1183

Gernert KM, Surles MC, Labean TH, Richardson JS, Richardson DC (1995) Alacoil: a very tight, antiparallel coiled-coil of helices. Protein Science 4:2252–2260

Gmeiner WH, Horita DA (2001) Implications of SH3 domain structure and dynamics for protein regulation and drug design. Cell Biochem Biophys 35(2):127–140. Review. PubMed PMID: 11892788

Groothuizen FS, Winkler I, Cristóvão M, Fish A, Winterwerp HH, Reumer A et al (2015) MutS/MutL crystal structure reveals that the MutS sliding clamp loads MutL onto DNA. eLife 4:e06744. https://doi.org/10.7554/eLife.06744

Groves MR, Hanlon N, Turowski P, Hemmings BA, Barford D (1999) The structure of the PROTEIN Phosphatase 2A PR65/A Subunit Reveals the Conformation of Its 15 Tandemly Repeated HEAT Motifs. Cell *96*:99–110

Gu L, Hong Y, McCulloch S, Watanabe H, Li GM (1998) ATP-dependent interaction of human mismatch repair proteins and dual role of PCNA in mismatch repair. Nucleic Acids Res 26(5):1173–1178. PubMed PMID: 9469823; PubMed Central PMCID: PMC147380

Guarné A, Junop MS, Yang W (2001) Structure and function of the N-terminal 40 kDa fragment of human PMS2: a monomeric GHL ATPase. The EMBO Journal 20(19):5521–5531. https://doi.org/10.1093/emboj/20.19.5521

Han D, Kim K, Oh J, Park J, Kim Y (2007) TPR domain of NrfG mediates complex formation between heme lyase and formate-dependent nitrite reductase in Escherichia coli O157:H7. Proteins *70*:900–914

Hausmann J, Kamtekar S, Christodoulou E, Day JE, Wu T, Fulkerson Z et al (2011) Structural basis for substrate discrimination and integrin binding by autotaxin. Nat Struct Mol Biol 18(2):198–204. https://doi.org/10.1038/nsmb.1980

Heldwein EE, Macia E, Wang J, Yin HL, Kirchhausen T, Harrison SC (2004) Crystal structure of the clathrin adaptor protein 1 core. Proc Natl Acad Sci U S A 101(39):14108–14113. Epub 2004 Sep 17. PubMed PMID: 15377783; PubMed Central PMCID: PMC521094

Hendrickson WA, Ward KB (1977) Pseudosymmetry in the structure of myohemerythrin. *J Biol Chem 252*:3012–3018

Hinderlich S, Weidemann W, Yardeni T, Horstkorte R, Huizing M (2015) UDP-GlcNAc 2-epimerase/ManNAc kinase (GNE), a master regulator of sialic acid synthesis. Topics in Current Chemistry 366:97–137. https://doi.org/10.1007/128_2013_464

Holm L, Sander C (1993) Structural alignment of globins, phycocyanins and colicin A. FEBS letters 315:301–306

Holzer S, Ban N, Klinge S (2013) Crystal structure of the yeast ribosomal protein rpS3 in complex with its chaperone Yar1. J Mol Biol 425(22):4154–4160. https://doi.org/10.1016/j.jmb.2013.08.022. Epub 2013 Sep 7. PubMed PMID: 24021814

Holzer P, Izzo AA (2014) The pharmacology of TPR channels. British J of Pharmacology 171:2469–2473

Huber AH, Nelson WJ, Weis WI (1997) Three-Dimensional Structure of the Armadillo Repeat Region of *b*-Catenin. Cell *90*:871–882

Jenkins HT, Baker-Wilding R, Edwards TA (2009) Structure and RNA binding of the mouse Pumilio-2 Puf domain. J Struct Biol. 167(3):271–276. https://doi.org/10.1016/j.jsb.2009.06.007. Epub 2009 Jun 18. PubMed PMID: 19540345

Jin AJ, Lafer EM, Peng JQ, Smith PD, Nossal R (2013) Unraveling protein–protein interactions in clathrin assemblies via atomic force spectroscopy. Methods *59*:316–327

Jínek M, Rehwinkel J, Lazarus BD, Izaurralde E, Hanover JA, Conti E (2004) The superhelical TPR-repeat domain of O-linked GlcNAc transferase exhibits structural similarities to importin alpha. Nat Struct Mol Biol 11(10):1001–1007. Epub 2004 Sep 12. PubMed PMID: 15361863

Kajander T, Cortajarena AL, Mochrieb S, Regan L (2007) Structure and stability of designed TPR protein superhelices: unusual crystal packing and implications for natural TPR proteins. Acta Cryst *D63*:800–811

Kajander T, Cortajarena AL, Regan L (2006) Consensus design as a tool for engineering repeat proteins. Methods Mol Biol *340*:151–170

Kajava AV (1998) Structural diversity of leucine-rich repeat proteins. J Mol Biol *277*:519–527

Karasik A, Shanmuganathan A, Howard MJ, Fierke CA, Koutmos M (2016) Nuclear protein-only ribonuclease P2 structure and biochemical characterization provide insight into the conserved properties of tRNA 5′ end processing enzymes. J Mol Biol 428(1):26–40. https://doi.org/10.1016/j.jmb.2015.11.025. Epub 2015 Dec 3. PubMed PMID: 26655022; PubMed Central PMCID: PMC4738078

Kaszas K, Keller JM, Coddou C, Mishra SK, Hoon MA, Stojilkovic S, Jacobson KA, Iadarola MJ (2012) Small molecule positive allosteric modulation of TRPV1 activation by Vanilloids and Acidic pH. *J. Pharm. Exp. Therapeutics 340*:152–160

Kato K, Nishimasu H, Okudaira S, Mihara E, Ishitani R, Takagi J et al (2012) Crystal structure of Enpp1, an extracellular glycoprotein involved in bone mineralization and insulin signaling. Proc Natl Acad Sciences U S A *109*(42):16876–16881. https://doi.org/10.1073/pnas.1208017109

Ke J, Chen RZ, Ban T, Zhou XE, Gu X, Tan MH, Chen C, Kang Y, Brunzelle JS, Zhu JK, Melcher K, Xu HE (2013) Structural basis for RNA recognition by a dimeric PPR-protein complex. Nat Struct Mol Biol 20(12):1377–1382. doi: 10.1038/nsmb.2710. Epub 2013 Nov 3. PubMed PMID: 24186060

Kefalas P, Brown TR, Brickell PM (1995) Signalling by the p60c-src family of protein-tyrosine kinases. Int J Biochem Cell Biol 27(6):551–563. Review. PubMed PMID: 7545532

Kim DH, Park M-J, Gwon GH, Silkov A, Xu Z-Y, Yang EC, Song S, Song K, Kim Y, Yoon HS, Honig B, Cho W, Cho Y, Hwang I (2014) Chloroplast targeting factor AKR2 evolved from an ankyrin repeat domain coincidentally binds two chloroplast lipids. Dev Cell 30(5):598–609

Kobe B (1999) Autoinhibition by an internal nuclear localization signal revealed by the crystal structure of mammalian importin. Nat Str Biol *6*:388–397

Kobe B, Deisenhofer J (1996) Mechanism of ribonuclease inhibition by ribonuclease inhibitor protein based on the crystal structure of its complex with ribonuclease A. J Mol Biol *264*:1028–1043

Kohl A, Binz HK, Forrer P, Stumpp MT, Plốckthun A, Grốtter MG (2002) Designed to be stable: crystal structure of a consensus ankyrin repeat protein. Proc Natl Acad Sci *100*:1700–1705

Kohn WD, Mant CT, Hodges RS (1977) Helical protein assembly motifs. J Biol Chem 272:2583–2586

Krylov D, Mikhailenko I, Vinson C (1994) A thermodynamic scale for leucine zipper stability and dimerization specificity: e and g interhelical interactions. *EMBO J. 13*:2849–2861

Krzywda S, Brzozowski AM, Higashitsuji H, Fujita J, Welchman R, Dawson S, Mayer RJ, Wilkinson AJ (2004) The Crystal Structure of Gankyrin, an Oncoprotein Found in Complexes with Cyclin-dependent Kinase 4, a 19 S Proteasomal ATPase Regulator, and the Tumor Suppressors Rb and p53. J Biol Chem *279*:1541–1545

Kummel D, Muller JJ, Roske Y, Henke N, Heinemann U (2006) Structure of the Bet3-Tpc6b core of trapp: two Tpc6 paralogs form trimeric complexes with Bet3 and Mum2. Ref J Mol Biol 361:22–32

Kümmel D, Oeckinghaus A, Wanga C, Krappmann D, Heinemann U (2008) Distinct isocomplexes of the TRAPP trafficking factor coexist inside human cells. FEBS Lett 582:3729–3733

Kurochkina N (2007) J Theor Biol 247:110–121

Kurochkina N (2008) J Theor Biol 255:188–198

Kurochkina N, Choekyi T (2011) Helix-helix interfaces and ligand binding. J Theor Biol 283:92–102

de Lange O, Binder A, Lahaye T (2014) From dead leaf, to new life: TAL effectors as tools for synthetic biology. Plant J 78:753–771

Lee G, Abdi K, Jiang Y, Michaely P, Bennett V, Marszalek PE (2006) Nanospring behaviour of ankyrin repeats. Nature 440:246–249

Lee SJ, Imamoto N, Sakai H, Nakagawa A, Kose S, Koike M, Yamamoto M, Kumasaka T, Yoneda Y, Tsukihara T (2000) The Adoption of a Twisted Structure of Importin-b is Essential for the Protein-protein Interaction Required for Nuclear Transport. *J Mol Biol 302*:251–264

Lepage PK, Lussier MP, Barajas-Martinez H, Bousquet SM, Blanchard AP, Francoeur N, Dumaine R, Boulay G (2006) Identification of two domains involved in the Assembly of Transient Receptor Potential Canonical Channels. J Biol Chem 2006(*281*):30356–30364

Li Y, Meng X, Xiang Y, Deng J (2010) Structure function studies of vaccinia virus host range protein K1 reveal a novel functional surface for Ankyrin repeat proteins. J Virol 84:3331

Li X, Song B, Chen X, Wang Z, Zeng M, Yu D, Hu D, Chen Z, Jin L, Yang S, Yang C, Chen B (2013) Crystal structure of a four-layer aggregate of engineered TMV CP implies the importance of terminal residues for oligomer assembly. PLoS One 8(11):e77717. https://doi.org/10.1371/journal.pone.0077717. eCollection 2013. PubMed PMID: 24223721; PubMed Central PMCID: PMC3817195

Lishko PV, Procko E, Jin X, Phelps CB, Gaudet R (2007) The ankyrin repeats of TRPV1 bind multiple ligands and modulate channel sensitivity. Neuron 54:905–918

Lu G, Dolgner SJ, Hall TM (2009) Understanding and engineering RNA sequence specificity of PUF proteins. Curr Opin Struct Biol 19(1):110–115. doi: https://doi.org/10.1016/j.sbi.2008.12.009. Epub 2009 Jan 29. PubMed PMID: 19186050; PubMed Central PMCID: PMC2748946

Madhurantakam C, Varadamsetty G, Grütter MG, Plückthun A, Mittl PRE (2012) Structure-based optimization of designed Armadillo-repeat proteins. Protein Sci *21*:1015–1028

Maeda S, Nakagawa S, Suga M, Yamashita E, Oshima A, Fujiyoshi Y, Tsukihara T (2009) Structure of the connexin 26 gap junction channel at 3.5 A resolution. Nature 458(7238):597–602. https://doi.org/10.1038/nature07869. PubMed PMID: 19340074

Main ER, Xiong Y, Cocco MJ, D'Andrea L, Regan L (2003) Design of stable alpha-helical arrays from an idealized TPR motif. Structure 11(5):497–508. PubMed PMID: 12737816

Mak AN-S, Bradley P, Cernadas RA, Bogdanove AJ, Stoddard BL (2012) The crystal structure of TAL effector PthXo1 bound to its DNA target. Science 335:716–719

Malik HS, Eickbush TH, Goldfarb DS (1997) Evolutionary specialization of the nuclear targeting apparatus. Proc Natl Acad Sci *94*:13738–13742

Martinez J, Nguyen LD, Hinderlich S, Zimmer R, Tauberger E, Reutter W et al (2012) Crystal structures of N-Acetylmannosamine Kinase provide insights into enzyme activity and inhibition. J Biol Chem 287(17):13656–13665. https://doi.org/10.1074/jbc.M111.318170

Maxson ME, Grinstein S (2014) The vacuolar-type H?-ATPase at a glance - more than a proton pump. J Cell Sci 127(Pt 23):4987–4993. https://doi.org/10.1242/jcs.158550. Review. PubMed PMID: 25453113

Mayer BJ (2001) SH3 domains: complexity in moderation. J Cell Science 114:1253–1263

Meier T, Krah A, Bond PJ, Pogoryelov D, Diederichs K, Faraldo-Gómez JD (2009) Complete ion-coordination structure in the rotor ring of Na+-dependent F-ATP synthases. J Mol Biol. 391 (2):498–507. doi:https://doi.org/10.1016/j.jmb.2009.05.082. Epub 2009 Jun 3. PubMed PMID: 19500592

Michaely P, Tomchick DR, Machius M, Anderson RGW (2002) Crystal structure of a 12 ANK repeat stack from human ankyrinR. EMBO J 21:6387–6396

Milburn CC, Boudeau J, Deak M, Alessi DR, van Aalten DMF (2004) Crystal structure of MO25α in complex with the C terminus of the pseudo kinase STE20-related adaptor. Nat Str & Mol Biol *11*:193–200

Mills RD, Mulhern TD, Cheng H-C, Culvenor JG (2012) Analysis of LRRK2 accessory repeat domains: prediction of repeat length, number and sites of Parkinson's disease mutations. Biochem Soc Trans 40:1086–1089

Miyanari Y (2014) TAL effector-mediated genome visualization (TGV). Methods 69:198–204

Mizutani K, Yamamoto M, Suzuki K, Yamato I, Kakinuma Y, Shirouzu M, Walker JE, Yokoyama S, Iwata S, Murata T (2011) Structure of the rotor ring modified with N,N′-dicyclohexylcarbodiimide of the Na+-transporting vacuolar ATPase. Proc Natl Acad Sci U S A 108(33):13474–13479. https://doi.org/10.1073/pnas.1103287108. Epub 2011 Aug 3. PubMed PMID: 21813759; PubMed Central PMCID: PMC3158168

Monecke T, Haselbach D, Vo B, Russek A, Neumann P, Thomson E, Hurt E, Zachariae U, Stark H, Grubmüller H, Dickmanns A, Ficner R (2012) Structural basis for cooperativity of CRM1 export complex formation. Proc Natl Acad Sci 110:960–965

Mosavi LK, Cammett TJ, Desrosiers DC, Peng Z (2004) The ankyrin repeat as molecular architecture for protein recognition. Protein Sci 13:1435–1448

Mosavi LK, Minor DL, Peng Z (2002) Consensus-derived structural determinants of the ankyrin repeat motif. Proc Natl Acad Sci 99:16031–16034

Nagatomo K, Ishii H, Yamamoto T, Nakajo K, Kubo Y (2010) The Met268Pro mutation of mouse TRPA1 changes the effect of caffeine from activation to suppression. Biophys J 99:3609–3618

Narayanan A, Kumar S, Evrard AN, Paul LN, Yernool DA (2014) Asymmetric hetero-domain interface stabilizes a response regulator-DNA complex. Nat Commun 5:3282. https://doi.org/10.1038/ncomms4282

Natrajan G, Lamers MH, Enzlin JH, Winterwerp HHK, Perrakis A, Sixma TK (2003) Structures of Escherichia coli DNA mismatch repair enzyme MutS in complex with different mismatches: a common recognition mode for diverse substrates. Nucleic Acids Res 31(16):4814–4821

Nawrotek A, Knossow M, Gigant B (2011) The determinants that govern microtubule assembly from the atomic structure of GTP-tubulin. J Mol Biol 412(1):35–42. https://doi.org/10.1016/j.jmb.2011.07.029

Nilius N, Prenen J, Owsianik G (2011) Irritating channels: the case of TRPA1. J Physiol 589 (7):1543–1549

O'Shea EK, Klemm JD, Kim PS, Alber T (1991) Crystal structure of GCN4 leucine zipper, a two-stranded parallel coiled coil. Science 254:539

Padmanabhan B, Adachi N, Kataoka K, Horikoshi M (2004) Crystal structure of the homolog of the oncoprotein gankyrin, an interactor of Rb and CDK4/6. J Biol Chem 279:1546–1552

Pan X, Xu S, Wu J, Duan Y, Zheng Z, Wang J, Song X, Zhou M (2018) Ankyrin-like protein AnkB interacts with CatB, affects catalase activity, and enhances resistance of Xanthomonas oryzae pv. oryzae and Xanthomonas oryzae pv. oryzicola to Phenazine-1-Carboxylic Acid. Appl Environ Microbiol. 84(4). pii: e02145–17. https://doi.org/10.1128/AEM.02145-17. Print 2018 Feb 15. PubMed PMID: 29180371; PubMed Central PMCID: PMC5795068

Pareek TK, Keller J, Kesavapany S, Agarwal N, Kuner R, Pant H, Iadarola MJ, Brady RO, Kulkarni AB (2006) Cyclin-dependent kinase 5 modulates nociceptive signaling through direct phosphorylation of transient receptor potential vanilloid 1. Proc Natl Acad Sci 104:660–665

Pauling L, Corey RB, Branson HR (1951) The structure of proteins: two hydrogen-bonded helical configurations of the polypeptide chain. Proc Natl Acad Sci USA 37:205

Paulson AR, Tong L (2012) Crystal structure of the Rna14-Rna15 complex. RNA Jun;18(6):1154–1162. https://doi.org/10.1261/rna.032524.112. Epub 2012 Apr 18. PubMed PMID: 22513198; PubMed Central PMCID: PMC3358638

Pecqueur L, Duellberg C, Dreier B, Jiang Q, Wang C, Plückthun A et al (2012) A designed ankyrin repeat protein selected to bind to tubulin caps the microtubule plus end. Proc Natl Acad Sci USA 109(30):12011–12016. https://doi.org/10.1073/pnas.1204129109

Perry AJ, Hulett JM, Likić VA, Lithgow T, Gooley PR (2006) Convergent evolution of receptors for protein import into mitochondria. Curr Biol 16:221–229

Phelps CB, Huang RJ, Lishko PV, Wang RR, Gaudet R (2008) Structural analyses of the Ankyrin Repeat Domain of TRPV6 and related TRPV ion channels. Biochemistry 47:2476–2484

Preiss L, Yildiz O, Hicks DB, Krulwich TA, Meier T (2010) A new type of proton coordination in an F(1)F(o)-ATP synthase rotor ring. PLoS Biol 8(8):e1000443. https://doi.org/10.1371/journal.pbio.1000443. PubMed PMID: 20689804; PubMed Central PMCID: PMC2914638.

Rak A, Pylypenko O, Durek T, Watzke A, Kushnir S, Brunsveld L, Waldmann H, Goody RS, Alexandrov K (2003) Structure of Rab GDP-dissociation inhibitor in complex with prenylated YPT1 GTPase. Science 302:646

Rao ST, Rossmann MG (1973) Comparison of super-secondary structure in proteins. J Mol Biol 76:241–256

Ringel R, Sologub M, Morozov YI, Litonin D, Cramer P, Temiakov D (2011) Structure of human mitochondrial RNA polymerase. Nature 478(7368):269–273. https://doi.org/10.1038/nature10435. PubMed PMID: 21947009

Rogowski A, Briggs JA, Mortimer JC, Tryfona T, Terrapon N, Lowe EC, Baslé A, Morland C, Day AM, Zheng H, Rogers TE, Thompson P, Hawkins AR, Yadav MP, Henrissat B, Martens EC, Dupree P, Gilbert HJ, Bolam DN (2015) Glycan complexity dictates microbial resource allocation in the large intestine. Nat Commun. 6:7481. https://doi.org/10.1038/ncomms8481. Erratum in: Nat Commun. 2016;7:10705. PubMed PMID: 26112186; PubMed Central PMCID: PMC4491172

Sanders SS, Mui KKN, Sutton LM, Hayden MR (2014) Identification of binding sites in Huntingtin for the Huntingtin Interacting Proteins HIP14 and HIP14L. PlosOne 28:e90669

Scheraga HA, Chou KC, Nemethy G (1982) In: Srinivasan R, Sarma RH (eds) Conformation in biology. Adenine Press, Gilderland

Schmidt H, Hansen G, Singh S, Hanuszkiewicz A, Lindner B, Fukase K et al (2012) Structural and mechanistic analysis of the membrane-embedded glycosyltransferase WaaA required for lipopolysaccharide synthesis. Proc Natl Acad Sci USA 109(16):6253–6258. https://doi.org/10.1073/pnas.1119894109

Scholze H, Boch J (2011) TAL effectors are remote controls for gene activation. Curr Opin Microbiol 14:47–53

Schulman (2000) 1fqv 1fs1
Schulz GE, Schirmer RH (1982) Principles of protein structure. Ed: Cantor, C. R.,
Chen S-C, Huang C-H, Lai S-J, Yang CS, Hsiao T-H, Lin C-H, Fu P-K, Ko T-P, Chen Y (2016) Mechanism and inhibition of human UDP-GlcNAc 2-epimerase, the key enzyme in sialic acid biosynthesis. Sci Rep 6:23274. https://doi.org/10.1038/srep23274
Shin H, Renatus M, Eckelman BP, Nunes VA, Sampaio CA, Salvesen GS (2005) The BIR domain of IAP-like protein 2 is conformationally unstable: implications for caspase inhibition. Biochem J 385(Pt 1):1–10. PubMed PMID: 15485395; PubMed Central PMCID: PMC1134667
Smith DF (2004) Tetratricopeptide repeat cochaperones in steroid receptor complexes. Cell Stress Chaperones 9:109–121
Stahlberg H, Müller DJ, Suda K, Fotiadis D, Engel A, Meier T, Matthey U, Dimroth P (2001) Bacterial Na(+)-ATP synthase has an undecameric rotor. EMBO Rep 2(3):229–233. PubMed PMID: 11266365; PubMed Central PMCID: PMC1083843
Stoll VS, Kimber MS, Pai EF (1996) Insights into substrate binding by D-2-ketoacid dehydrogenases from the structure of Lactobacillus pentosus D-lactate dehydrogenase. Structure 4(4):437–447. PubMed PMID: 8740366
Stone D, Kiem H-P, Jerome KR (2013) Targeted gene disruption to cure HIV. Curr Opin HIV AIDS 8:217–223
Tewary SK, Liang L, Lin Z, Lynn A, Cotmore SF, Tattersall P et al (2015) Structures of minute virus of mice replication initiator protein N-terminal domain: insights into DNA nicking and origin binding. Virology 476:61–71. https://doi.org/10.1016/j.virol.2014.11.022
Toro-Roman A, Wu T, Stock AM (2005) A common dimerization interface in bacterial response regulators KdpE and TorR. Protein Science?: A Publication of the Protein Society 14(12):3077–3088. https://doi.org/10.1110/ps.051722805
Tu D, Li W, Ye Y, Brunger AT (2007) Structure and function of the yeast U-box-containing ubiquitin ligase Ufd2p. Proc Natl Acad Sci *104*:15599–15606
Urosev D, Ferrer-Navarro M, Pastorello I, Cartocci E, Costenaro L, Zhulenkovs D, Maréchal JD, Leonchiks A, Reverter D, Serino L, Soriani M, Daura X (2013) Crystal structure of c5321: a protective antigen present in uropathogenic Escherichia coli strains displaying an SLR fold. BMC Struct Biol. 13:19. https://doi.org/10.1186/1472-6807-13-19. PubMed PMID: 24099525; PubMed Central PMCID: PMC3851747
Vollmar M, Schlieper D, Winn M, Büchner C, Groth G (2009) Structure of the c14 rotor ring of the proton translocating chloroplast ATP synthase. J Biol Chem 284(27):18228–18235. https://doi.org/10.1074/jbc.M109.006916. Epub 2009 May 7. PubMed PMID: 19423706; PubMed Central PMCID: PMC2709358
Yan J, Zhang Q, Guan Z, Wang Q, Li L, Ruan F, Lin R, Zou T, Yin P (2017) MORF9 increases the RNA-binding activity of PLS-type pentatricopeptide repeat protein in plastid RNA editing. Nat Plants 3:17037. https://doi.org/10.1038/nplants.2017.37. PubMed PMID: 28394309
Wang M, Ogé L, Perez-Garcia MD, Hamama L, Sakr S (2018) The PUF protein family: overview on PUF RNA targets, biological functions, and post transcriptional regulation. Int J Mol Sci 19 (2). pii: E410. https://doi.org/10.3390/ijms19020410. Review. PubMed PMID: 29385744; PubMed Central PMCID: PMC5855632
Wang X, McLachlan J, Zamore PD, Hall TM (2002) Modular recognition of RNA by a human pumilio-homology domain. Cell;110(4):501–512. PubMed PMID: 12202039
Wierenga RK, Noble MEM, Vriend G, Naughe S, Hol WGJ (1991) Refined 1.83 Angstrom structure of trypanosomal triosephospate isomerase, crystallized in presence of 2.4M- Ammonium sulphate. A comparison with the structure of the trypanosomal triosephosphate isomerase-glycerol-3-phosphate complex. J Mol Biol 220:995–1015
Wierenga RK (2001) The TIM-barrel fold: a versatile framework for efficient enzymes. FEBS Lett 492:193–198
Willhoft O, Kerr R, Patel D, Zhang W, Al-Jassar C, Daviter T et al (2017) The crystal structure of the Sgt1-Skp1 complex: the link between Hsp90 and both SCF E3 ubiquitin ligases and kinetochores. Sci Rep 7:41626. https://doi.org/10.1038/srep41626

Wong K, Perpich JD, Kozlov G, Cygler M, Abu Kwaik Y, Gehring K (2017) Structural mimicry by a bacterial F box effector hijacks the host ubiquitin-proteasome system. Structure 25(2):376–383. https://doi.org/10.1016/j.str.2016.12.015

Wu H, Zeng H, Lam R, Tempel W, Kerr ID, Min J (2015) Structure of the human MLH1 N-terminus: implications for predisposition to Lynch syndrome. Acta Crystallogr Sect F Struct Biol Commun 71(Pt 8):981–985. https://doi.org/10.1107/S2053230X15010183

Yang H, Jiang X, Li B, Yang HJ, Miller M, Yang A, Dhar A, Pavletich NP (2017) Mechanisms of mTORC1 activation by RHEB and inhibition by PRAS40. Nature 552(7685): 368–373. https://doi.org/10.1038/nature25023. Epub 2017 Dec 13. PubMed PMID: 29236692; PubMed Central PMCID: PMC5750076

Yuan T, Liu L, Zhang Y, Wei L, Zhao S, Zheng X, Huang X, Boulanger J, Gueudry C, Lu J, Xie L, Du W, Zong W, Yang L, Salamero J, Liu Y, Chen L (2015) Diacylglycerol guides the hopping of clathrin-coated pits along microtubules for exo-endocytosis coupling. Dev Cell 35(1):120-130. https://doi.org/10.1016/j.devcel.2015.09.004. Epub 2015 Oct 1. PubMed PMID: 26439397

Zhang Q, Harding R, Hou F, Dong A, Walker JR, Bteich J, Tong Y (2016) Structural basis of the recruitment of ubiquitin-specific protease USP15 by spliceosome recycling factor SART3. J Biol Chem 291(33):17283–17292. https://doi.org/10.1074/jbc.M116.740787. Epub 2016 Jun 2. PubMed PMID: 27255711; PubMed Central PMCID: PMC5016127

Zhang Z, Devarajan P, Dorfman AL, Morrow JS (1998) Structure of the ankyrin-binding domain of alpha-Na,K-ATPase. J Biol Chem;273(30):18681–18684. PubMed PMID: 9668035

Zhavoronkov A, Izumchenko E, Kanherkar RR, Teka M, Cantor C, Manaye K et al (2016) Pro-fibrotic pathway activation in trabecular meshwork and lamina cribrosa is the main driving force of glaucoma. Cell Cycle 15(12):1643–1652. https://doi.org/10.1080/15384101.2016.1170261

Zweifel ME, Leahy DJ, Hughson FM, Barrick D (2003) Structure and stability of the ankyrin domain of the *Drosophila* Notch receptor. Protein Sci 12:2622–2632

Chapter 6
Multiprotein Complexes

Abstract Cell processes involve assembly of many components and coordination of their function. Protein complexes assemble and disassemble in response to various signals. Complexes are necessary to sort and target cargoes to cellular organelles, move molecules along filaments, exchange and recycle materials between cellular compartments, handle degradation of toxic molecules, provide signaling and communication inside the cell and between cells. They consist of tens or even hundreds of proteins and their ligands which work in accord to support life functions. Formation of multiprotein complexes is regulated by posttranslational modifications and depends on protein interactions. This chapter describes molecular assemblies involved in vital cellular processes.

Keywords Complex · Assembly · Cellular organelle · Filament · Protein interactions

Each biological process requires precise multiprotein assembly of many components essential for its function. Protein complexes assemble and disassemble to respond to bacterial and viral infection, sort and target cargoes to cellular organelles, move molecules along filaments, exchange and recycle materials between cellular compartments, handle degradation of toxic molecules, provide signaling and communication inside the cell and between cells. Numerous complexes composed of tens or even hundreds of proteins and their ligands work in accord to support life functions. Assembly of multiprotein complexes in organized manner is regulated by posttranslational modifications and depends on specificic interactions. Protein specific motifs can be small peptides or domains with characteristic fold, such as β-barrel of SH3 or WW, coiled coil,8α/8β barrel of triose phosphate isomerase, all alpha 6α/6αbarrel of fucosyltransferase or all beta barrel of porin, helical toroids of ankyrin, Armadillo/Heat, clathrin, leucine rich, and pumilio homology repeats or βhelix. Proteins in various tissues undergo alternative splicing and are expressed in multiple isoforms which differ in domain composition. In spite of differences in structure, proteins can recognize similar sequence motifs defined by specific interactions of atomic groups. This chapter describes molecular assemblies involved in vital cellular processes.

N. Kurochkina, *Protein Structure and Modeling*,
https://doi.org/10.1007/978-981-13-6601-7_6

Signaling Complexes

Epidermal growth factor receptor, EGFR Cell surface receptors and adapter proteins form multiprotein complexes that activate signaling pathways. The two major routes of signal transduction are via cell surface receptor signaling, initiated with activation of the receptor in the cellular membrane, and intracellular receptor signaling with involvement of molecules in the cytoplasm, the nuclear membrane, and the nucleus. Intracellular adaptor proteins interact simultaneously with several other molecules bringing them together in a multiprotein complex and amplify signaling cascades. They relay growth factor signals for subsequent transcriptional and translational regulation important for cell fate determination (Kebache et al. 2007; Semela 2017; Dufour et al. 2018).

EGFR cell signaling involves ligand binding to extracellular domain (ECD), heterodimerization, kinase activation, and interactions of catalytic and noncatalytic domains and adaptor proteins (Kurochkina and Guha 2015). Structure of EGFR and its family members, very often mutated and overexpressed in cancer cells, is extensively studied (Mitchell et al. 2018). EGFR is mutated in glioblastoma (ECD, catalytic, carboxyterminal domains) and lung adenocarcinoma (L858R, deletion within exon 19, insertion in exon 20). Patients are treated with such inhibitors as gefitinib, erlotinib, afatinib (Park et al. 2015), and MIG6 (Maity et al. 2015). Specificity of EGFR is based on binding 'primed' substrate: two adjacent tyrosines are critical for its activity (Park et al. 2015). Particularly important is characterization of EGFR signaling mechanisms that involve regulation of receptor density, distribution of cellular compartments, endosomal sorting and postendocytic trafficking, dynamics and activity along multiple pathways (Di Fiore et al. 2002; Caldieri et al. 2018).

Patients may develop resistance to treatments with inhibitors targeting ATP binding site. New inhibitors emerge including allosteric inhibitors that target drug resistant mutants (EA1045, EAI001, neratinib, lapatinib), compounds that disrupt asymmetric dimer (cetuximab) and mutant selective irreversible inhibitors (osimertinib, rociletinib) (Jia et al. 2016).

Receptor tyrosine kinases (EGFR) and protein tyrosine kinases (Src, Fyn, Btk, Ltk) have common mechanism of regulation by pY/SH2 and polyproline/SH3. Receptor tyrosine kinase phosphorylation provides signals for assembly/disassembly via docking of adaptor proteins. RTK MERTK SH2 domain interacts with GRB2, PIK3R1, VAV3, and SRC in retinal pigment epithelium regulating phagocytosis essential for retina health (Shelby et al. 2013; 2015). PTK action is reversed by protein tyrosine phosphatase (PTP). Tyrosine phosphorylation/dephosphorylation by PTKs and PTPs suggests a signaling mechanism of cell homeostasis that involves tightly regulated cascades of events (Aschner et al. 2018).

PTKs and serine/threonine kinases (STKs) are similar in structure. PTKs divergence from STKs in the evolutionary process is associated with emergence of multicellular organisms and is viewed as accumulation of mutations in the hydrophobic core and active site which alter the protein stability, activity, dynamics, ligand binding properties and function (Mohanty et al. 2016; Kurochkina and Choekyi 2011).

SH2 and SH3 domains are important components of assembly and signaling in EGFR pathways. These and other protein interactions domains, such as helical toroids and beta-helices, are highly selective toward their ligands. Protein-protein and protein-ligand interactions to a large extent depend on specific interactions of amino acids at secondary structure interfaces that determine parameters characterizing angles, distances, chirality and shape of the assemblies (Kurochkina 2008; Kurochkina and Iadarola 2015a).

Glideosome is a multiprotein complex involved is pathogen invasion. Its major component is actomyosin (Boucher and Bosch 2015). Apicomplexans, such as *Plasmodium* spp., *Cryptosporidium*, and *Toxoplasma gondii*, invade multiple cell types. The invasion process includes expression by parasites of extracellular membrane protein receptors, adhesions and formation of specific pathogen–host complexes. Adhesins used to invade various cell types specifically recognize the receptors. The receptors provide link to an actomyosin motor and molecular pathways of the invasion. Actin filaments and motor protein myosin form an actomyosin complex (Pospich et al. 2017) which interacts with other components and generate force for gliding. Actin/myosin motor pulls adhesins and essential invasion proteins across the plasma membrane. Structural data (over 75 structures of parasite receptors and glideosome proteins in the Protein Data Bank) include adhesins, motor proteins, bridging proteins, inner membrane complex and cytoskeletal proteins and highlight key interactions necessary within the machinery complex.

Apoptosome Apoptosis and inflammation, two closely related processes, are mediated by interactions between domains, and this process is an important signal transduction event. Signaling proteins assemble in oligomers: apoptosome, death-inducing signaling complex, myddosome, piddosome, and many other.

High-resolution crystallographic structures for apoptosomes from *C. elegans* (CED-4), *D. melanogaster* (Dark) (Cheng et al. 2017) and *H. sapiens* (Apaf-1) (Zhou et al. 2015) show key interdomain interactions of the assemblies. The complexes have similar domain composition but differ in organization of protein interfaces, intradomain and interdomain interactions (Dorstyn et al. 2018). CED4 (Teng et al. 2010) is a tetramer of a dimer comprising eight CED4 molecules interacting vie CARD domains so that one tetrameric ring of CARD1a interacts with second tetrameric ring of CARD1b. Apaf1 is also octamer or heptamer. CED4, Apaf1 and Dark are AAA+ ATPases and share common domain organization: CARD, nucleotide binding α/β domain, two helical domains HD1 and HD2, and winged helix domain WHD. Apaf1 contains 15 beta propeller domains at the carboxy terminus.

Apoptosome consists of Apaf-1, caspase 9, and cytochrome c. Procaspase 9 is activated upon binding to Apaf-1 and cytochrome coassembled in apoptosome. Structure of the apoptosome-procaspase-9 complex at 9.5Å (Yuan et al. 2010) resolution contains the central hub constructed like other AAA+ protein rings. Regulatory regions that contain Apaf-1 β-propeller domains and embraced cytochrome c domains occupy the outer ring of the apoptosome. Apaf-1 CARD and caspase 9 CARD are assembled in a disk structure above the central hub. Regions of Apaf-1 CARD are disordered. Apaf-1 was shown to interact with procaspase 9 via CARD-CARD domains interface (Weber et al. 2001). Solution structure of Apaf-1 CARD containing six-helix bundle (Vaughn 1999) and proposed model of interaction between Apaf-1 CARD and Caspase-9 (Zhou et al. 1999) and high resolution crystal structure (Li et al. 2017) show similarities of these complexes with FADD DED/ FAS DD (Fig. 6.1B), PIDD DD / RAIDD DD and RAIDD CARD caspase 2 CARD (Fig. 6.1C), and MYDD (Fig. 6.1D). Crystallographic structure of the residues 1–591 of APAf-1 includes CARD, nucleotide binding domain (NBD), helix domain 1 (HD1), winged helix domain (WHD) and helix domain 2 (HD2) (Riedl et al. 2005). Structure of the complex between Apaf-1 CARD and procaspase 9 CARD revealed that interaction site of these two proteins mainly involves positively charged helices H1a/H1b and H4 of procaspase 9 and negatively charged helices of Apaf-1 CARD (Qin et al. 1999). Superposition of one protein relative to another can be achived by rotation of approximately 90°around their axis. The two proteins exhibit amino acid and structural similarities. They both are six-helix antiparallel bundles.

Apoptosome assembly occurs in several oligomerization states: tetramer of *C. elegans* (Yan et al. 2005), heptamer of mammalian and octamer of drosophila (Yu et al. 2005) complexes.

Myddosome MYD88 (Myeloid differentiation primary response protein), IRAK4 (Interleukin-1 receptor-associated kinase 4), and IRAK2 (Interleukin-1 receptor-associated kinase 2), signaling mediators of the TLR/IL1-R pathway, assemble as a left-handed helical oligomer (Fig. 6.1D) (Lin et al. 2010) called myddosome. Each monomer is a six-helix bundle (Lasker et al. 2005). Structure of a six-helix bundle with a Greek key topology is found in intracellular domains of receptor proteins involved in apoptotic signaling, interleukin I receptor, ankyrin. Similarly, protein kinases Pelle and Tube form a myddosome. There are three types of interfaces between monomers. One type of interface involves helices H1/H4 from one monomer and helices H2/H3 of another monomer.

Two MYD88 molecules in myddosome MYD88, IRAK4, and IRAK2, monomers of the assembled left-handed helical oligomer (pdb code 3mop), can be superimposed on the structure of the complex between Apaf-1 CARD and procaspase 9 CARD.

Apaf-1 CARD and procaspase 9 CARD may also be assembled as left-handed helical oligomer in the disk located above the apoptosome platform. They have similar arrangement of interacting helices from both monomers as well as similar angles between subunits.

There is no significant sequence similarity between myddosome and apoptosome proteins but structurally they align very well (Fig. 6.1).

Being associated with cancer, autoimmune and neurodegenerative diseases, apoptosome is an attractive target for the design of new therapies. Many apoptotic signaling pathways initiate formation of the apoptosome, assembly of the complex in a shape of disk which consists of seven Apaf1 and seven cytochrome c molecules

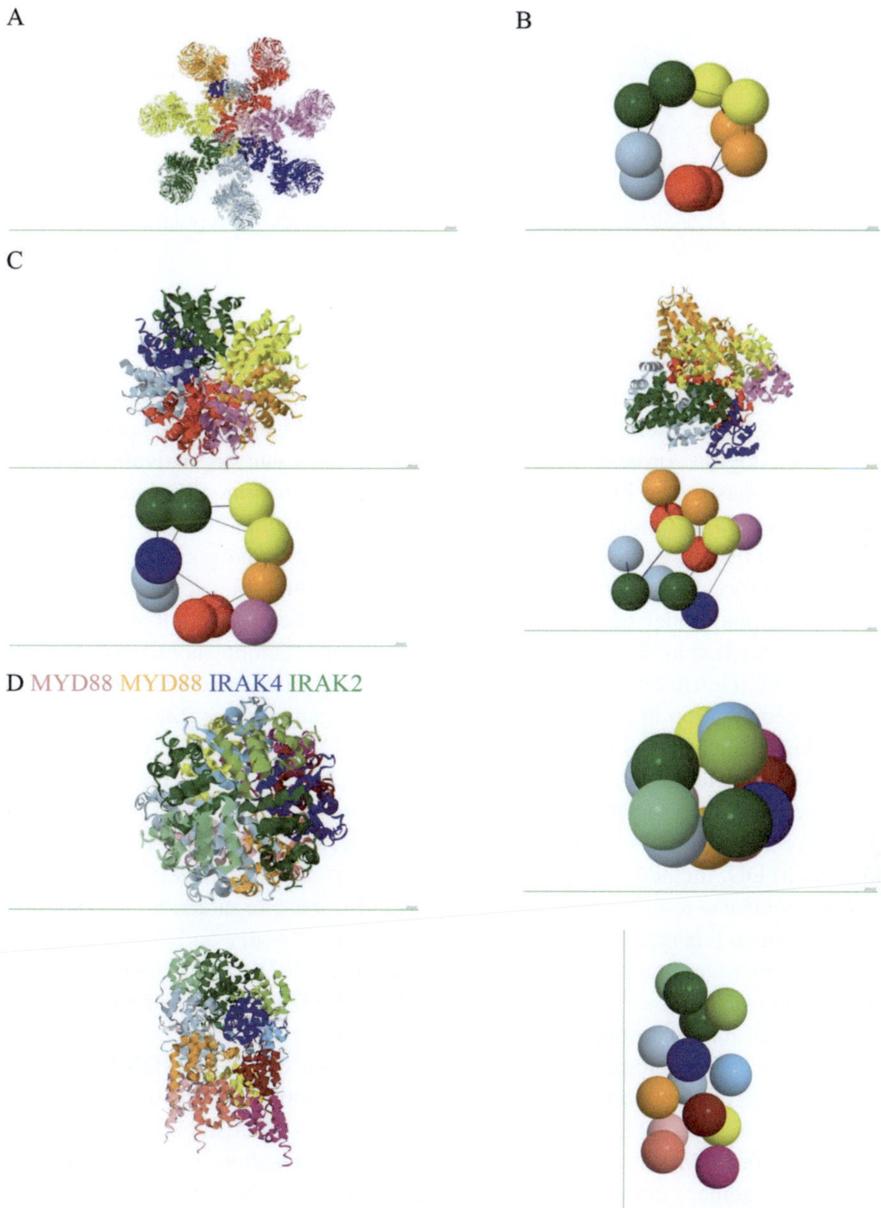

Fig. 6.1 Signaling complexes (**A**) Apoptosome (**B**) FAS FADD /3oq9/ (**C**) PIDDsome /2of5/ (**D**) MYDDosome /3mop/

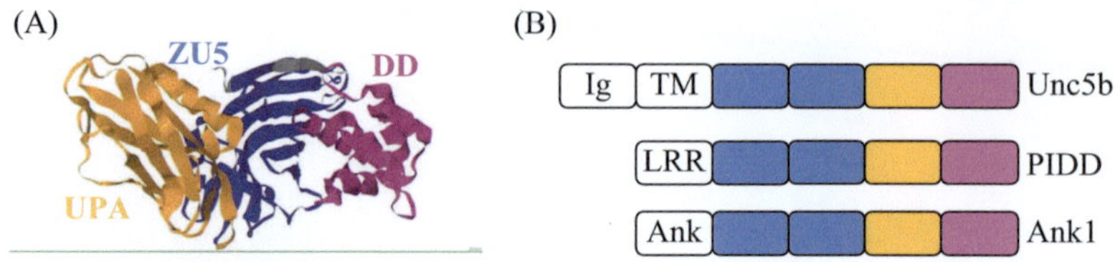

Fig. 6.2 UNK5b ZU5/UPA/DD supramodule (Wang et al. 2009 3g5b) (**A**) crystal structure and (**B**) a unit of UNK5b, PIDD1, and Ankyrin 1

binding procaspase 9 and activating a cascade of procaspase 3 and procaspase 7. Apaf1 and procaspase 9 interact by their CARD domains.

Piddosome comprises PIDD–RAIDD–caspase-2 complex. Its scaffold and adaptor proteins, p53-inducible protein 1 with a DD (PIDD1) and RIP-associated Ich-1/CED homologous protein with DD (RAIDD), associate to activate caspase2 although caspase2 can be active without complex formation. PIDD and RAIDD assemble via DD domains interactions. Ionizing radiation induces PIDDosome formation in response to DNA damage followed by mitotic progression, and inhibition of this process for example by BUBR1 has important implications for mediation of apoptosis and understanding links between the mitotic checkpoint and apoptotic processes (Shah et al. 2016). Discovered as mediator of apoptosis in DNA damage response, PIDDosome is assigned new roles in maintenance of centrosome function and polyploidization in cell differentiation during organogenesis (Sladky 2017).

Piddosome assembles via PIDD DD / RAIDD DD and RAIDD CARD / caspase 2 CARD interactions. In crystallographic structure of the complex (Park et al. 2007 2of5), seven RAIDD DDs and five PIDD DDs compose three layers: two RAIDD DDs, five RAIDD DDs, and five PIDD DDs (Fig. 6.1C). PIDD ten domains, in addition to DD, include seven LRR repeats, and two ZU5 domains (Fig. 6.2B).

Inflammasome multiprotein complex forms in response to infection. Recognition of pathogen and danger associated molecular patterns, PAMPs and DAMPs, results in activation of inflammatory caspases (Isaka et al. 2016). As CARD domains, pyrindomains (PYD) are assembly domains. PYD/PYD interactions regulate polymerization of inflammasome. The two families of inflammasomes include (1) ALR with Absent in Melanoma 2 (AIM2) as a receptor for cytosolic dsDNA which is important in recognition of viruses; (2) NLR with nucleotide binding domain (NBD) and leucine rich repeat (LRR) containing receptors. AIM2/dsDNA interaction or NLR activation initiates PYD/PYD interactions and formation of helical filaments bringing together caspase domains which activate via dimerization after transautocleavage. NLRP3 inflammasome for example recognizes activation signals of extracellular ATP, uric acid crystals, and pathogens (Fig. 6.3) (Vajjhala et al. 2014). Mutated inflammasomes are linked to Alzheimer's disease, psoriasis, type II diabetes, inflammatory bowel disease and other.

Assembly domains Death domain (DD) and pyrindomain (PYD) are main assembly domains of helical filaments. DD domain superfamily is subdivided into the DD, death effector (DED), caspase recruitment (CARD), and pyrin (PYD) domain

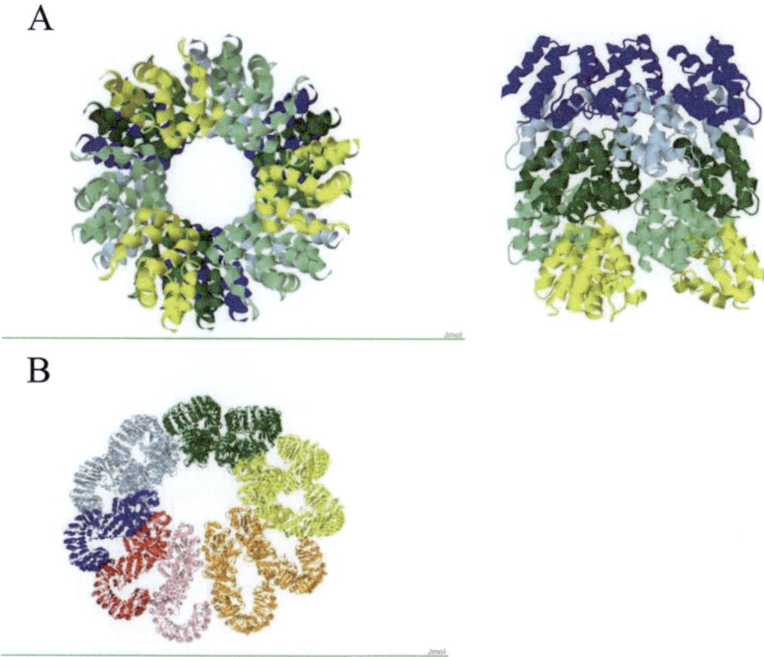

Fig. 6.3 Inflammasome (**A**) ALR /3j63/ and (**B**) NLR/3jbl/

subfamilies. Domains interact in assembly of oligomeric platforms and binding downstream effectors (caspases). Proximity induced autoactivation (dimerization, transphosphorylation, proteolysis) is one of the suggested mechanisms. Caspases activate via dimerization.

DD is also a part of ZU5/UPA/DD supramodule (Fig. 6.2), a conserved three domain combination of Ankyrin B, PIDD, and UNK5b proteins. This UNK5b ZU5/UPA/DD supramodule is a stable monomer in solution (Wang et al. 2009). DDs are important for assembly (Fig. 6.4). Intramolecular DD/ZU5 interaction is regulatory as it prevents UNK5b DD from interaction with other DDs. Also, this interaction is important for axonal repulsive responses of UNK5 in Drosophila (Keleman and Dickson 2001). Nervous system development involves action of many protein complexes. Receptors of the Deleted in Colorectal cancer (DCC) and UNC5 families together with secreted molecules such as Netrin guide neuronal growth cones through an action of axonal attraction and repulsion (Keleman and Dickson 2001). Specific domains of protein isoforms expressed in various tissues of the organism are important signaling modules. Receptors of the DCC and UNC5 families are transmembrane receptors with cytoplasmic regions containing ZU5, DD, and other domains as sites of intracellular assembly, regulation, and signaling. Netrin binding to the extracellular part of these receptors mediates interactions in the cytoplasmic portion of the complex. DD interactions contribute to the assembly of the signaling complex. DD is a part of ZU5/UPA/DD supramodule, a conserved

```
                              α1                     α2                        α3                    α4
MYd88   MLPLA------ALNMRVRRRLSLFLNVRTQVAADWTALAEEM-------DFEYLEIRQLETQA-----DPTGRLLDAWQG
IRAK4   MGPITPSTYVRCLNVGLIRK--LSDFID--PQEGWKKLAVAIKKPSGDDRYNQFHIRRFEALLQT-GKSPTSELLFDW-G
IRAK2   ------ACYIYQLPSWVLDD--LCRNMDALSEWDWMEFASYV-------ITDLTQLRKIKSMEWVQGVSITRELLWWW-G
PELLE                 LLPLPVRAQ-LCAHLDALD--VWQQLATAVKL-----YPD--QVEQISSQKQ-RGRSASNEFLNIW-G
TUBE                  IIPKGMDVQ-ACSGAGCLN--FPAEIKKGFKY-----TAQDVFQIDEAANRLPPDQSKSQMMIDEW-K
UNC5B                 IPLSIRQKICNSLDAPNSRGNDWRLLAQKL------SMDRYLNY--FATKA-----SPTGVILDLWEA

                   α5          α6
MYd88       RPGASVGRLLELLTKLGRDDVLLELG---PSIEEDCQKYIAAALEHHHHHH /3mop
IRAK4       TTNCTVGDLVDLLIQNEFFAPASLLL---PD---------AVPLEHHHHHH /3mop
IRAK2       MRQATVQQLVDLLCRLELYRAAQIILNWKPAPEIRCP----IPAFPDSVKP /3mop
PELLE   GQ----YNHTVQTLFALFKKLKLHNAMRLIKDY                   /1ik7
TUBE    TSGKLNERPTVGVLLQLLVQAELFSAADFVALDFLNESTPARPVDGPGALISLELL /1d2z
UNC5B      RQQ-GASVGRLLELLTKLGRDDVLLELGPSIEEDC                /3g5b
```

Fig. 6.4 Sequence alignment of DD domains

three domain combination of Ankyrin B, PIDD, and UNK5b proteins. The UNK5b ZU5/UPA/DD supramodule (Wang et al. 2009) (Fig. 6.2) is important for assembly and regulation. Intramolecular DD/ZU5 interaction prevents UNK5b DD from interaction with other DDs. This interaction is important for axonal repulsive responses of UNK5 in Drosophila. In *Xenopus,* cytoplasmic P3 domains interaction and multimerization of DCC receptors stimulates attraction, whereas P1 DCC and DB UNC5 interaction induces repulsion.

Apoptosome, Piddosome, Myddosome and other complexes associate signaling platforms via (FAS DD –FADD DED caspase8), (PIDD DD / RAIDD DD and RAIDD CARD caspase 2 CARD), (Pelle DD /Tube DD, Drosophila Toll signaling), (Apaf-1 CARD procaspase 9 CARD, caspase 9 activation), and Pyrin domains. They exhibit a variety of assemblies and modes of interactions.

Signaling complexes assemble via specific interactions of proteins with multiple ligands. Membrane associated guanylate kinases (MAGUK) cluster and anchor glutamate receptors and other proteins at synapses. The MAGUK family of proteins includes PSD95, PSD93, SAP102, SAP97, essential components of postsynaptic density, PSD. Stabilization of SAP102 at the PSD depends on SH3/GK domain. Actin, a core skeletal component in spines, interacts with multiple proteins of PSD (Zheng et al. 2010). MAGUKs indirectly bind to actin via complexes SAP97 – MyosinVI – actin, PSD-95 – SPAR – actin, PSD-95 – GKAP – Shank – cortactin – actin, and PSD-95 – NMDAR – actinin – actin (Petralia et al. 2012).

Ion channels function in large complexes involving signaling, cytoskeletal, scaffolding, and other components (Bankston et al. 2012).

SH2 and SH3 domains are important components of assembly of signaling complexes. These and other protein interactions domains, such as helical toroids and beta-helices, are highly selective toward their ligands. Protein-protein and protein-ligand interactions to a large extent depend on specific interactions of amino acids at secondary structure interfaces that determine parameters characterizing angles, distances, chirality and shape of the assemblies (Kurochkina 2008; Kurochkina and Iadarola 2015a).

Phagosome Phagocytic cells (macrophages, dendritic cells, neutophils) defend organism against infection. Phagocytosis process involves recognition of invading pathogens and processing of microbe filled vesicles (Hoffmann et al. 2018). Phagocytosis occurs with extensive restructuring of membranes and cytoskeleton. Coordinated action of phagocytosis and degradation through phagolysosome is essential for elimination of pathogens and maintenance of cell homeostasis (Niedergang and Grinstein 2018). Molecules such as vacuolar proton ATPase (VATPase) form complexes that regulate acidic environment necessary for protein degradation (Maxson and Grinstein 2014). Proteomics analysis of Mtb infection identified upregulated NADPH subunits (p67phox, p47phox) and macrophage migratory inhibitory factor. 61 host proteins, mainly those involved in blood coagulation, apoptosis, and oxidative phosphorylation, are overexpressed. Degradative capacity of phagolysosome is mediated by lipids, Ca2+, and channels such as Phosphoinositide gated lysosomal Ca2+ channel (TRPLM1) activated to promote phagosome lysosome fusion (Dayam et al. 2015).

Nuclear pore complex Eukaryotic Nuclear pore complex (NPC) consists of approximately 30 nucleoporins (Nups) with total mass of almost 125 MDa residing in the nuclear envelope and providing selectivity barrier for proteins and RNA shuttling between nuclear and cytoplasmic compartments (Ruba and Yang 2016).

Signaling NF-κB/IκB signaling regulates transcription and is a major component in development, cell growth and apoptosis 1. In many diseases such as cancer, arthritis, asthma, diabetes, AIDS and viral infections 2, NF-κB/IκB signaling is disregulated. Inhibitor of kappa-B (IκB) proteins 1 regulates NF-κB subcellular localization and transcriptional activity. IκB family members contain six ankyrin repeats which bind to the carboxy terminal Rel homology regions of NF-κB (Jacobs and Harrison 1998).

Postsynaptic Density

Ankyrin repeats of postsynaptic density: structure and function Ankyrin repeats domains are recognized for their role in assembly of multiprotein complexes and regulation of cellular pathways critical for cell division, proliferation and growth, development, migration and cytoskeletal modifications. Structural studies of ankyrin repeats containing proteins and their complexes by crystallography, NMR, electron microscopy, and other methods show unique organization and diverse modes of interactions. Designed ankyrin repeats find many applications in drug development and protein crystallization. Peptides, ATP and lipids that bind ankyrin domains are promising drug candidates. Studies of structure, function, and applications of ankyrin repeats containing proteins emphasize their involvement in synaptic transmission.

Postsynaptic density, PSD, is viewed as structural interconnected framework where signals of several transduction pathways involving membrane of excitatory synapses and nuclear receptors meet. PSD at the neuronal synapses assembles large number of proteins (Sheng and Hoogenraad 2007; Kurochkina and Iadarola 2015a; Tomasetti et al. 2017). Many mRNAs are associated with PSD and are proposed to carry signals between synapse and nucleus mediating transcription and translation (Shirai and Suzuki 2017). PSD comprises proteins of diverse functionality including channels, kinases, phosphatases, GTPases, cytoskeletal proteins, and scaffolding proteins such as SHANK, Homer, PSD95 and membrane-associated guanilate kinases (MAGUKs). PSD95/MAGUKs are main components of PSD which perform their function of regulating synaptic plasticity together with numerous binding proteins (Won et al. 2017). Scaffold proteins organize glutamate receptor molecules complex formation and traffic and provide connection to calcium storage; therefore, clustering, rearrangements at synapses and signal transduction of NMDA receptors strongly depend on these PSD components (Tomasetti et al. 2017). PSD has unique architecture (Baron et al. 2006; Hayashi et al. 2009). It is a dynamic assembly and super-resolution imaging reveals details of its structural organization (Sainlos et al. 2011; Constals et al. 2015; Hafner et al. 2015). Mobility of the receptors clustered at PSD and their lateral diffusion influence synaptic transmission (Heine et al. 2008). Diffusion and stationary phases alternate with increasing stationary phase as neurons finalize their developmental stages (Borgdorff and Choquet 2002). Diffusion plays important role in the formation of postsynaptic structures (Choquet and Triller 2003). Thermal fluctuations affect fluidity of membranes, retention of integral membrane proteins in plasma membrane, neurovascular homeostasis, and threshold of lymphocyte activation and cell signaling. Cytoskeletal and membrane cortex proteins including actin, spectrin, vimentin, and ankyrin, which mediate lateral diffusion behavior of plasma membrane residing lipids and proteins, and change clustering and diffuse patterns of distribution (Devaux et al. 2010; Zynda et al. 2015; Zamani et al. 2016; Ghazawi et al. 2016).

One of the major PSD residing ankyrin repeats proteins is SHANK. Other ankyrin repeats proteins localized in PSD include AIDA (Dosemeci et al. 2015), Ank3 (Nanavati et al. 2011), ARMS/Kidins220, (Arévalo et al. 2010; Cortés et al. 2007), TANC (Suzuki et al. 2005), CASCIN (Tabuchi et al. 2002), Synamon (Yao et al. 1999), and PTP-BAS (Maekawa et al. 1999).

Recent advancements in structure, function, and applications of ankyrin repeats containing proteins emphasize their involvement in synaptic transmission.

Ankyrin repeats structure Nervous system contains numerous scaffold proteins that assemble multiprotein complexes and regulate signal transduction. Scaffold proteins consist of multiple domains, each exhibiting unique structure and specificity toward ligands. Intramolecular and intermolecular interactions of protein domains are important for proper assembly of the components. PSD scaffold proteins with ankyrin repeats (Fig. 6.5) show diverse domain composition and many modes of regulation.

Ankyrin repeats comprise pairs of antiparallel α-helices stacked parallel to each other (Mosavi et al. 2004; Li et al. 2006). The overall shape of the molecule represents a toroid, spiral of helices. Each repeat unit contains two α-helices, A and B, flanked by a short β-structure. Repeats show amino acid similarities represented by a consensus sequence (Fig. 6.6). All helices A form inner row (concave) whereas all helices B form outer row (convex) (Fig. 6.7). A unique pattern of contacts on the surface of helices A differs from that of helices B which results in

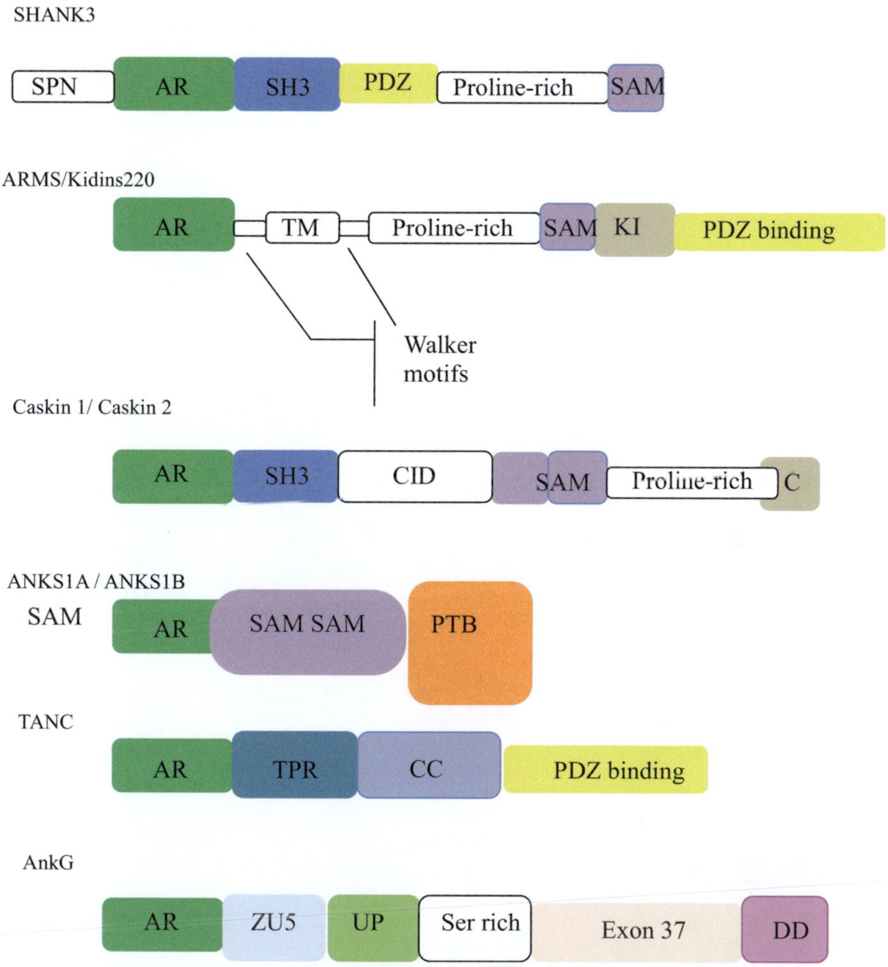

Fig. 6.5 Domain organization of PSD ankyrin repeats proteins

two types of assemblies that cannot be superimposed but can form a specific antiparallel interface (Kurochkina and Iadarola 2015a). Therefore, amino acid sequence of each helix determines whether it assembles its neighboring helices as inner row or outer row.

Similarly to protein domains, each ankyrin repeat is a microdomain. Conservation of amino acids results in similar core structure of the repeat unit, whereas dissimilarities are roots of specificity in ligand recognition.

Shank proteins Shanks are adaptor proteins and major scaffold components of postsynaptic density (Sarowar and Grabrucker 2016; Chen et al. 2014). Mutations in all three isoforms of SHANK have links to autistic spectrum disorders (ASD) (Mameza et al. 2013). Three existing isoforms SHANK1, 2, and 3 share sequence similarities and domain organization (Yao et al. 1999; MacGillavry et al. 2016).

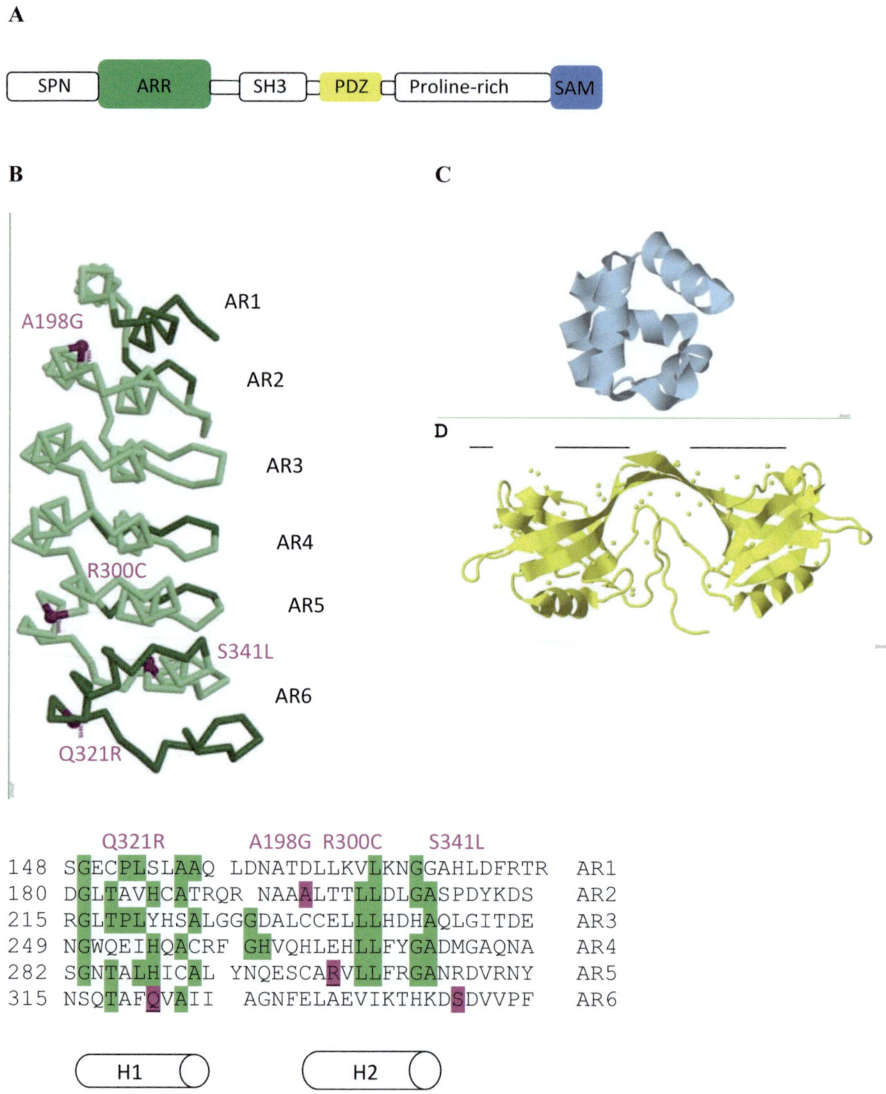

Fig. 6.6 Structural model of SHANK3. Crystal structures of (**A**) SPN and ARR (pdb designation 5g4x) with superimposed mutations (magenta); (**B**) SAM domain (pdb designation 2f3n); (**C**) PDZ domain (pdb designation 1q3p)

Shank3 contains Shank/ProSAP N-terminal (SPN), ankyrin repeats (ARR), SH3 and PDZ domains, a proline-rich region and a sterile alpha motif (SAM) (Fig. 6.6). Each domain is a site of specific interactions: ARR (Fig. 6.6A) binds SPN, α-fodrin and sharpin; SAM (Fig. 6.6B) is involved in oligomerization; and PDZ (Fig. 6.6C) interacts with GKAP and G-protein-coupled receptors. Other Shank-interacting proteins are actin-binding protein-1 (Abp1), Abelson interacting protein-1 (Abi-1), Arp2/3, Wave, neuroligin, and Rich2 (PDZ).

The SHANK3 ARR (amino acids 86–418) consists of six ankyrin repeats, AR1 to AR6 (Fig. 6.6A), and is flanked by SPN and SH3 domains. ARR intramolecular interaction with SPN carries regulatory function. Mutations in both regions, ARR and SPN, affect ligand binding to ankyrin repeats domain (Mameza et al. 2013). Crystal structure of the N-terminal domain of wild type SHANK3 also highlights interactions of SPN with RAS and important role of SHANK3 in G-protein signaling. As counteraction to activation of integrin by RAP1, SHANK1 and SHANK3 inhibit this activation (Lilja et al. 2017).

Mutations in SHANK3 ARR are associated with autism (Mameza et al. 2013). Since C-terminal spectrin motif of α-fodrin and C-terminus of sharpin, component of NFκB and integrin-dependent signaling pathways, are interaction sites of SHANK3 ARR, mutations in SHANK3 ARR (A198G, R300C, Q321R, S341 L) and SPN (R12C, L68P) associated with autism were studied with respect to these ligands. Expression of R12C, R300C, and Q321R mutants result in changes of spine morphology by possibly affecting actin polymerization-dependent processes. Mutations L68P and Q321R slightly increase whereas R300C slightly decreases α-fodrin binding to Shank3. All three mutants show sharpin increased binding. Mutations in ARR weaken actin polymerization and decrease number of actin filaments and spine density (Durand et al. 2012).

Three of the four ARR mutations are located on the convex surface and one mutation on the concave surface (Fig. 6.6A). Similarly, mutations found in individuals exhibiting Charcot-Marie-Tooth disease type 2C are distributed on the surface of TRPV4 (Sullivan et al. 2015).

Mutations in SHANK3 found in ASD patients show that different domains are involved in different synaptic and behavioral phenotypes. SHANK3 proline-rich domain in mouse models is associated with excitatory synaptic transmission (Bariselli and Bellone 2016). Shank proteins function as connectors between cell surface receptors, numerous scaffold proteins and actin (Tao-Cheng et al. 2015). By interacting with cortactin SH3 domain via C-terminal proline-rich region, they modulate actin regulation of synapse morphology. Shank mediates cortactin localization and mobility in spines; in the absence of Shanks cortactin moves quicker. Also, synapse morphology depends on Shank-cortactin interactions (MacGillavry 2016). Knocking out of all three isoforms disrupts spine morphology and diminishes stability of the actin cytoskeleton in spines. Shank localization in PSD decreases with lowered Ca2+ levels and increases upon depolarization with high K+ or NMDA (Tao-Cheng et al. 2015). Assembly of PSD complex NLGN-PSD-95-SHANK and interactions with G-protein coupled receptor *GPR85* were shown to be critical in ASD and schizophrenia patients possibly due to abnormalities in regulation of excitatory and inhibitory receptors (Fujita-Jimbo et al. 2015). SHANK3 mutations contribute to impaired Zn transport and provide possible link to Phelan McDermid Syndrome and autism spectrum disorders. The correlation between expression levels of Zn uptake transporters ZIP2/ZIP4 and SHANK3, as well as interactions between ZIP4 and SHANK3 suggest a link and involvement of SHANK3 in assembly of transcription factors as multiprotein complexes (Pfaender et al. 2017).

Ankyrin repeat-rich membrane spanning (ARMS/Kidins220) protein ARMS protein, also known as a kinase D-interacting substrate of 220 kDa (Kidins 220), is one of the signaling molecules of neurotrophin and ephrin pathways. Its involvement in MAP kinase and Trk signaling suggests a role as an oncogene. Also, via p21-cyclinD1/CDK4 it contributes to cell cycle progression (Jung et al. 2014). ARMS mediates basal synaptic strength in hippocampal neurons via influence on AMPAR function, mainly GluA1 subunit phosphorylation and localization (Cortés et al. 2007; Arévalo et al. 2010). ARMS also binds NMDAR (Gamir-Morralla et al. 2015), receptor critical for synaptic signaling and learning (Li et al. 2001). It is involved in basal synaptic transmission, neurodegeneration (Scharfman et al. 2013; Kranz et al. 2016), schizophrenia (Malaspina et al. 2016), and axonal transport (Schmieg et al. 2014). ARMS and synembryn-B expression levels regulate development and differentiation through NGF-mediated secretion (López-Benito et al. 2016). ARMS contains an N-terminal ankyrin repeats domain with 11 repeats, four transmembrane segments, a C-terminal PDZ binding motif, a kinesin-1 interacting motif (KIM), a sterile alpha motif (SAM) domain, and a proline rich domain (Neubrand et al. 2012) (Fig. 6.5). Ankyrin repeats concave surface (Fig. 6.7A) binds Rho–guanine nucleotide exchange factor (Neubrand et al. 2010). Interactions of polyproline sequences with SH3 domain of Crk links ARMS to Trk receptor tyrosine kinases signaling pathway (Arévalo et al. 2006). Being one of P-loop

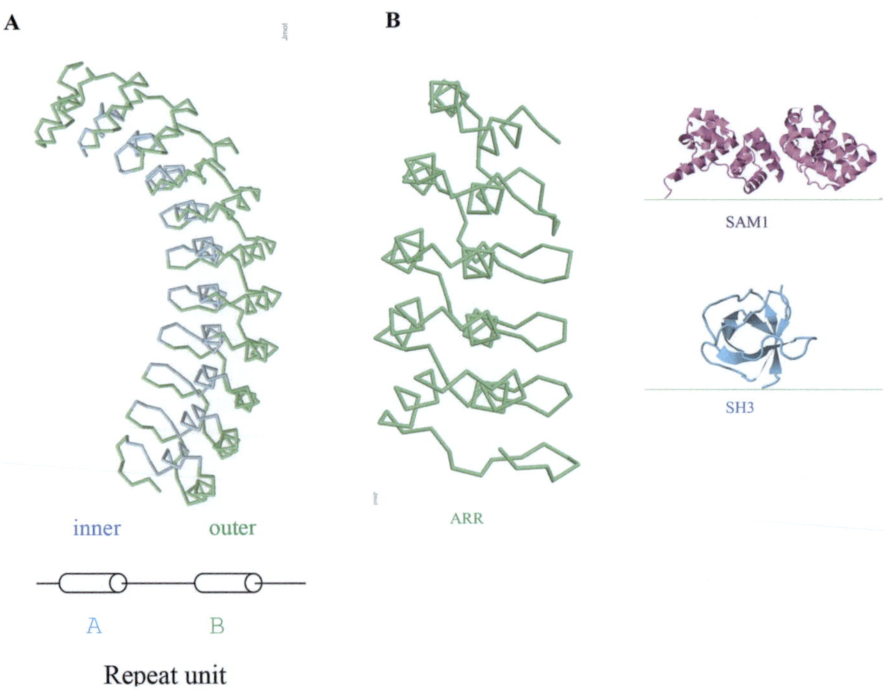

Fig. 6.7 Structures of (**A**) ARMS/Kidins220 ARR (theoretical model); (**B**) Caskin1 ARR (theoretical model), tandem SAM and SH3

nucleotide phosphatases, ARMS is proposed to regulate assembly of protein complexes at the inner membrane (Jung et al. 2014) and coordinate action of signaling cascades (Singh et al. 2015).

CASKIN Caskin 1 and Caskin 2 proteins (Pesti et al. 2012) comprise six N-terminal ankyrin repeats, SH3 domain, two SAM domains, a proline-rich sequence and a C-terminal domain (Fig. 6.5). This domain organization is similar to that of Shank although sequence similarity is not significant. Theoretical model of ARR and crystal structures of tandem SAM and SH3 domains provide details of its assembly (Fig. 6.7B). Calmodulin kinase domain of synaptic adaptor protein CASK binds Caskin which competes with adaptor protein MINT. Interestingly, Caskin is mainly found in vertebrates brain whereas CASK in invertebrates (Tabuchi et al. 2002). MINT1 and 2 are components of synaptic vesicle exocytosis and regulate Munc18/neurexins interactions (Biederer and Südhof 2000). One of important pathways associated with MINT is Notch signaling pathway implicated in cell fate determination, neurogenesis, somitogenesis, vasculogenesis, immune development and cancer. Notch1 activates β1 integrins and GTP-binding protein R-RAS in ankyrin repeats dependent manner which stimulates cellular adhesion to fibronectin (Hodkinson et al. 2007).

*Ankyrin repeat and sterile alpha motif domain-containing proteins 1A (ANKS1A) and 1B (ANKS1B AIDA-1)*ANKS1A and ANKS1B consist of six ankyrin repeats, two SAM, and one PTB domains (Fig. 6.5).

In the ER, where ANKS1A is localized upon phosphorylation on Ser647 and Ser663, ankyrin repeats of ANKS1A bind to tyrosine kinase domain of EphA2 and PTB binds to Sec23 of COPII vesicles, thus regulating loading of EphA2 into vesicles (Park 2016). Absence of ankyrin repeats domain impairs protein function in anterograde transport from the ER (Lee et al. 2016b). ANKS1A is implicated in susceptibility to systemic lupus erythematosus (Molineros et al. 2017), breast cancer (Park 2016), schizophrenia (De Rubeis et al. 2014) and ASD (Buxbaum et al. 2012).

ANKS1B localization in PSD excitatory neurons is activity dependent (Dosemeci et al. 2015; 2016). ANKS1B plays important role in learning and memory as a component of LTP circuit (Faurobert and Albiges-Rizo 2010). It interacts with many proteins such as Alzheimer's disease associated amyloid-beta protein precursor (Ghersi et al. 2004) and cerebral cavernous malformation protein-1 (CCM1/KRIT1) (Herberich et al. 2015). ANKS1B binds to the intracellular domain of amyloid-beta protein and provides communication between synapses and the nucleus (Jacob et al. 2010). Similarly to other ankyrin repeats proteins, it organizes localization of receptors: delivers NMDARs with GluN2B subunit to synapses (Tindi et al. 2015).

ANKS1A and ANKS1B are similar in amino acid sequence, domain organization and tertiary structure. Both proteins contain helical SAM domains. First SAM domain of ANKS1A interacts with SAM domain of EphA2 (Mercurio et al. 2012). Two tandem SAM domains of ANKS1B form an interface that rearranges when protein interacts with its receptor. Domain rearrangements regulate ANKS1A/ANKS1B in their receptor complexes and translocation to nucleus.

Ankyrin G (Ank3) Product of *Ank3* gene, ankyrin G, belongs to a group of fifteen proteins that are linked to five out of six major neuropsychiatric disorders, attention deficit hyperactivity disorder (ADHD), ASD, bipolar disorder (BD), major depressive disorder (MDD), and schizophrenia (SCZ) (Lotan et al. 2014; Nanavati et al. 2011). Ankyrin G regulates clustering of membrane sodium channels at axonal initial segment (AIS) and nodes Ranvier, sites of initiation and propagation of the action potentialin the mammalian nervous system. βIV spectrin links Ankyrin G to the actin cytoskeleton (Bouzidi et al. 2002; Yoshimura et al. 2016). Its levels are regulated by p65/NF-κB pathway and its role in development of axons and neuronal polarity was demonstrated (Jenkins et al. 2015; König et al. 2016). Downregulation of Ankyrin G and consequent reduction of E-cadherin levels at the cell membrane result in increased β-catenin levels in the nucleus and its activity in transcription; therefore, by participating in Wnt signaling, Ankyrin G plays important role in regulation of progenitor cell proliferation and brain development (Durak et al. 2015). Ankyrin G consists of membrane binding ankyrin repeats domain, a spectrin binding domain, a C-terminal region followed by very large exon domain and a death domain (Jenkins et al. 2015) (Fig. 6.5). Ankyrin G mutants R47A/R50A do not bind Na_v or KCNQ channels (Xu and Cooper 2015).

As structural component of extrasynaptic membrane, Ankyrin G stabilizes GABAergic interneuron synapses interacting with GABARAP and other proteins of endocytosis regulatory machinery (Oka et al. 2013; Tseng et al. 2015).

High homology between AnkR/B/G ankyrin repeats domains (80%) and even higher conservation at the inner peptide binding surface and finger loops suggest presence of common ligands, whereas low similarity in other domains indicates their distinct biological roles. Binding of AnkR C-terminal autoinhibitory segment and a peptide from sodium channel Nav1.2 to AnkB was studied by crystallography (pdb codes 4rlv and 4rly; Wang et al. 2014). Multiple binding sites are observed. Autoinhibitory segment binding site partially overlaps with Nav1.2 and neurofascin sites. Similarly, interaction of Ankyrin G with Nav1.2 involves multiple sites.

Cell structural components maintain cell function. Cellular processes involve signaling pathways of the transport machinery that deliver cargoes along retrograde and anterograde routes. In all cell types, numerous signaling complexes coordinate cell and secretory machinery functions important for balance and homeostasis (Kuijpers et al. 2016). Ankyrin G is glycosylated. O-linked N-acetylglucosamine modification of Ankyrin G isoforms occurs at nodes of Ranvier. O-GlcNAc-glycosylated serine-rich domain of AnkG accumulates at nodes of Ranvier of myelynated axon. Phosphorylation and glycosylation sites of AnkG overlap (Zhang and Bennett 1996).

Cochlea development progresses with appearance of auditory nerve spikes in auditory nerve fibers (ANF) and depends on assembly of voltage gated sodium and potassium channels by ankyrin G which determines channels subcellular positioning (Kim et al. 2016). ANF response to sound generates patterns of spikes at afferent ribbon synapses triggered by sensory inner hair cells.

TANC TANC1 and TANC2 reside in excitatory neurons where they interact with PSD-95 and regulate function of dendritic spines, spatial memory and embryonic development (Han et al. 2010). This protein contains ten ankyrin repeats, three tetratricopeptide repeats (TPRs), a coiled-coil domain, and a C-terminal PDZ (PSD-95/Dlg/ZO-1) domain-binding motif (Fig. 6.5). Binding to PDZ of PSD-95 localizes TANC in excitatory neurons and is important for its function. Ankyrin repeats, TPRs, and a coiled-coil domain interact with fodrin (Suzuki et al. 2005). TANC1 role in regulation of myoblast fusion has important implications in development of new treatments for rhabdomyosarcoma, a malignancy of muscle myoblasts (Avirneni-Vadlamudi et al. 2012). Psychomotor retardation is another disease that has link to TANC1 (Granot-Hershkovitz et al. 2011). TANC2 is one of genes candidates for food quality improvement via modifications of fat content in muscles (Zheng et al. 2016).

PTP-BAS is a protein tyrosine phosphatase, PTP, bound to a membrane. It consists of band-4.1 homology region, five PDZ domains, and PTP domain. Some splice variants contain FERM domain (Abaan and Toretsky 2008). Interaction of IκBα three N-terminal ankyrin repeats (Fig. 6.8A) with PTP-BAS first PDZ domain (Fig. 6.8B) (Lee et al. 2016a) reveals new mode of NF-κB regulation by PTP-BAS: suppression of NF-κB activation. If IκBα/PTP-BAS binding is inhibited by introducing a dominant-negative mutant form of PTP-BAS lacking phosphatase activity, tyrosine phosphorylation of IκBα is observed; therefore, IκBα may be a PTP-BAS substrate *in vivo* (Maekawa et al. 1999). PTP-BAS plays important roles in apoptosis, tumor progression and tumor suppression (Abaan and Toretsky 2008), adipose tissue growth and differentiation and insulin signaling (Glondu-Lassis et al. 2009).

Significance and potential for drug design Ability of ankyrin repeats to assemble protein complexes and mediate interactions has important consequences for many cellular processes. Each domain is the site of specific assemblies and similarly to

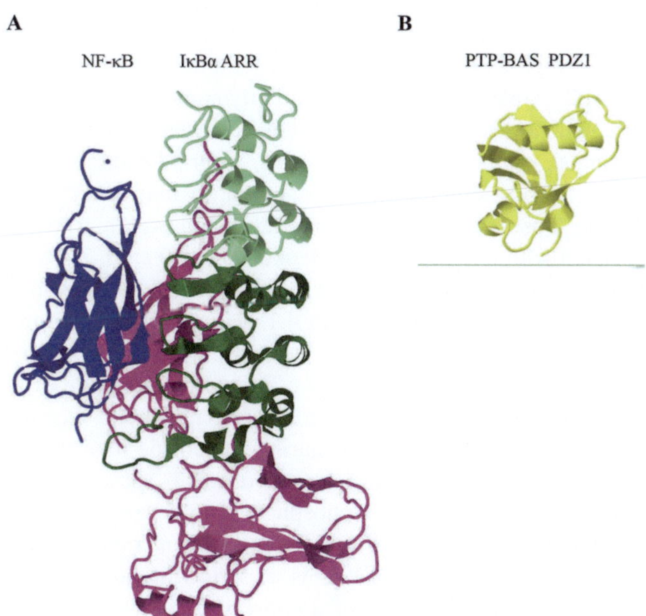

Fig. 6.8 Regulation of PTP-BAS by IκBα three N-terminal ankyrin repeats. (**A**) IκBα ARR and NFκB complex (pdb designation 1nfi 1ikn). (**B**) PDZ domain (pdb designation 5glj)

other domains, such as SH3 (Kurochkina and Guha 2014; Kurochkina 2015), the regulatory site. Tissue specific distribution of isoforms is critical in the design and evaluation of treatment approaches (Yardeni et al. 2013; Liu et al. 2016). Mutations at the ankyrin repeats domains affect structure, stability, specificity of interactions, and posttranslational modifications and, therefore, result in impaired function of cellular pathways. Design of new ankyrin based therapies by targeting of antagonistic opsins to specialized cellular compartments with potential to treat retina (Greenberg et al. 2011), regulating expression profiles and density of PSD proteins in psychiatric patients (Nanavati et al. 2011), applying Ank3/Na channel variants of cardiac proarrhythmogenic syndrome patients (Willis et al. 2015) or acting on SHANK3 pathway of ASD affected individuals (Costales and Kolevzon 2015) are emerging promising directions. Future studies are needed to assess how ARR regions participate in intra and intermolecular interaction, how mutations contribute to abnormalities, and what cellular pathways are modified. Progress in structure determination and structure-based design of ankyrin repeats of PSD proteins and their complexes contributes to the success of new approaches and sets important goals.

Enzymes and Multiprotein Complexes

Enzymes and nonenzymatic factors comprise multiprotein complexes such as multi-tRNA synthetase complex (MSC) that plays roles in protein synthesis and assembly. *H. sapiens* MSC components, methionyl-tRNA synthetase, glutaminyl-prolyl-tRNA synthetase, AIMP2, AIMP3, are organized in a dynamic complex (Cho et al. 2015 5bmu). AIMPs nonenzymatic factors are involved in immune response, hormonal regulation of glucose metabolism, angiogenesis, wound healing, response to DNA damage. AIMP3 tertiary structure is very similar to that of GST. Mutations of AIMP3, including those located at helix interfaces and proximal to ligand binding sites, are associated with chronic myeloid leukemia (Kim et al. 2008).

Hemoglobin, oxygen carrier, covalently binds to membrane and forms complex with spectrin, ankyrin, and band3, protein 4.1 being redox active and capable of binding ligands (Welbourn et al. 2017). Proteins that interact with hemoglobin include redox regulatory proteins (peridoxin 2, Cu/Zn superoxide dismutase, catalase, aldehyde dehydrogenase 1, flavin reductase), metabolic enzymes (carbonic anhydrase 1, selenium binding protein 1, purine nucleoside phosphorylase, nucleoside diphosphate kinase), membrane proteins (α and β spectrin, ankyrin, band3. protein 4.1, actin, glyceraldehydes 3 phosphate dehydrogenase) which highlight hemoglobin important role in homeostasis (Basu and Chakrabarti 2015).

Glycogen synthase kinase 3, GSK3, is a serine/threonine kinase. Its main isoforms GSK3α and GSK3β consist of a conserved calalytic domain flanked by the less conserved aminoterminal and carboxyterminal domains which are important for its activity. Regulation of GSK3 activity involves prerequisite phosphorylation of its substrates (priming), GSK3 phosphorylation p38 MAPK (GSK3β), and interactions with other proteins. Both isoforms act similarly toward β catenin. Tau phosphorylation at Ser396 by GSK3α but not GSK3β requires prerequisite priming. The phosphorylation of the aminoterminal GSK3α (Ser21) or GSK3β (Ser9) is important for insulin but not Wnt signaling. Phosphorylation results in the aminoterminus acting as a pseudosubstrate which binds to the active site and inhibits the kinase. Carboxyterminus is more important for the kinase activity. Its deletion results in misfolded and deficient kinase. Various GSK3 based therapies for treatment of Alzheimer's are proposed to prevent Ser396 Tau phosphorylation which results in pathological change and microtubule destabilization. Development of isoform specific GSK3 inhibitors (Buescher et al. 2010) is one of the approaches. Abnormal GSK3 activity is also involved in cancer and diabetes.

Helical repeats such as ankyrin, Pum, and PPR regulate function of many enzymes. Allosteric regulation occurs if structural changes in the binding site are induced by the interactions at another distant site of the enzyme molecule. Enzyme regulation by scaffold proteins ankyrin repeats (Perez Riba et al. 2018), TPR (Levine et al. 2018), LRR (Liu et al. 2018), and PUM (Miles et al. 2016) shows that many enzymes work in complexes and are subject to allosteric regulation. Helical repeats transduce allosteric effects through the sequence of repeated units, arrays. Tandem helical repeats possess folding cooperativity. PPR repeat is a component of mitochondrial ribosome which synthesizes mitochondrial encoded membrane proteins (Greber et al. 2015). A large group of proteins in mitochondria, PRORP, processes precursor tRNA cleavage of 5' leader sequence. This reaction is also carried out by ribozyme RNase P. PRORP consists of three domains: PPR, central, and metallonuclease. PRORP1 and PRORP2 PPR domain with five PPR repeats is essential for substrate binding and enzymatic activity and may interact with precursor tRNA (Fig. 6.9B). Removal of four PPR repeats results in impairment of enzyme activity (Karasik et al. 2016). Pum domain is a module of a pseudoenzyme KREBP4 in which RNAse III domain is noncatalytic and mediates assembly of editosome (McDermott et al. 2017).

Mitochondrial single subunit RNA polymerase PPR domain is interaction site of AT rich loop binding to promoter DNA (Ringel et al. 2011). Signal recognition

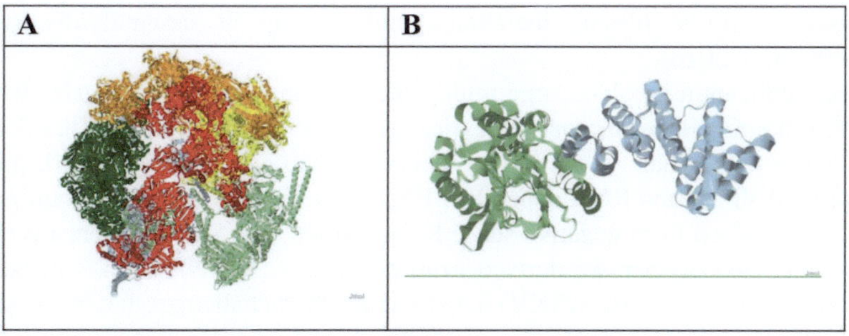

Fig. 6.9 Enzyme multiprotein complexes (**A**) *Saccharomyces cerevisiae* cerulenin-inhibited fungal fatty acid synthase type I multienzyme complex (PDB entry 2vkz); (**B**) Nuclear *A. thaliana* proteinaceous RNASE P 2 (PDB entry 5idz); catalytic (green) and PPR (blue) domains

particle (SRP) in chloroplasts forms a complex with light harvesting chlorophyll proteins (LHCPs). Its ankyrin repeats domain binds LHCP L18 domain and also regulates LHCP dimerization (Jonas-Straube et al. 2011). Proteins such as fatty acid synthase (FAS) type I function as multienzyme complexes. In eukaryotes and some later prokaryotes FAS I system consists of multifunctional enzymes (Fig. 6.9A), whereas in bacteria, FAS II system comprises separate enzymes catalyzing individual steps of the reaction cycle (Johansson et al. 2008).

The target of rapamycin (TOR) kinase functions in multiprotein complexes TORC1 and TORC2. These complexes have important roles in the regulation of metabolism, growth, and proliferation. Immunosuppressive properties of rapamycin suggest Tor1 and Tor2 kinases as the targets for the treatment of cancer, metabolic syndrome. Proteins exhibit typical kinase fold and conserved catalytic residues (Gaubitz et al. 2016; Laribee et al. 2018). PI3K-AKT-mTOR1 signaling plays important role in T-cell acute lymphoblastic leukemia via regulation by Notch1 (Hales et al. 2018). mTOR pathway is altered during the early stage of Alzheimer's Disease (Uddin et al. 2018). Communication between PI3K /AKT/mTOR and O-GlcNAcylation signaling pathway has important implications in cancer, type-2 diabetes, and cardiovascular, and neurodegenerative diseases.

Design of enzymes that assemble in multiprotein complexes proved efficient. Chemically modified cyclic peptide hormone somatostatin selectively targets human A549 lung cancer cells expressing SST-2 receptors. The multidomain protein complex, SST3-Avi-C3, retargets *Clostridium botulinum* C3 enzyme into non-small cell lung A549 cancer cells exhibiting very efficient tumor inhibition and increasing tumor sensitivity to chemotherapy (Kuan et al. 2018).

Wnt signaling regulates cellular pathways important for embryonic development, fate determination, stem cell homeostasis and cancers. Molecular components of the signaling, the tumor suppressors APC and Axin, organize the core of the multiprotein destruction complex and mark the Wnt-effector beta-catenin for phosphorylation, ubiquitination and degradation. Complex stability is critical for proper function. Loss of APC and Axin (negative regulators) results in proliferation and contributes to colon cancer. Wnt signals reverse this process. The ability of Wnt signaling to turn off the negative regulatory destruction complex machinery depends on the ratio of the levels of Axin and APC2. Wnt affects the destruction complex by recruiting it to the plasma membrane and altering its assembly/disassembly (Schaefer et al. 2018).

Virus replication involves large multiprotein complexes that synthesize RNA in EBOV, influenza B, respiratory syncytical virus (RSV), hepatitis C, Zika (ZIKV), and other viruses (Chap. 7). Active protein complexes comprise the large L protein, viral protein VP35, and RNA dependent RNA polymerases. Studies of multiprotein complexes derived from negative-sense RNA viruses RSV5 and influenza B (FluB) and monomeric polymerases derived from positive-sense RNA viruses hepatitis C virus (HCV) and Zika virus (ZIKV) help to understand challenges that expression of the negative-sense RNA viruses experiences. Negative-sense RNA viruses (influenza viruses, Measles virus, Mumps virus, RSV, and Ebola) are important human pathogens and viral RNA-dependent RNA polymerases, essential for replication of RNA viruses, represent important drug targets (Tchesnokov et al. 2018).

Organelles

Lysosomes residing in the cytoplasm and attached to membrane are major sites of protein degradation. Large group of lysosome related organelles in melanocytes (melanosomes), lymphocytes (lytic granules), platelets (delta granules), lung type 2 epithelial cells (lamellar bodies) form by the same pathway and maintain acidic environment (Huizing et al. 2008). Lysosomes and related organelles share common protein components in the form of large membrane associated complexes. Lysosomes related organelle complex 1, 2, and 3 (Proteins BLOC1, BLOC2 and BLOC3) regulate protein sorting from membrane microcompartments of the early endosomes to lysosomes (John Peter et al. 2013).

In eukaryots, lysosomal self-degrading and self-renewal pathway, macroautophagy, is conserved. Vesicle fusion events in eukaryotic cells are mediated by SNARE complex which consists of Syntaxin 17, Snap29 (two domains) and Vamp7. Other proteins also contribute: in Drosophila, autophagosome-lysosome fusion involves Ykt6 and HOP (Takáts et al. 2018).

Major event in autophagy is fusion of autophagosomes via eukaryotic membranes with involvement of such factors as Rab family GTPases, tethering complexes and membrane anchored SNARE containing two alpha helical coiled coil motifs that forms a heterotetrameric complex (Takáts et al. 2018).

Mitochondria origin is endosymbiotic and it has its own genome. Mitochondria are surrounded by two membranes, inner and outer, and also contain a cristae membrane. Mitochondrial contact site and cristae organizing system (MICOS) is a large multiprotein complex that resides in the inner mitochondrial membrane. It is essential for the formation and maintenance of variable in size and shape cristae structure, invaginations toward matrix and sites of entrance in the inner membrane (Kozjak-Pavlovic et al. 2017).

Organelle contact sites MICOS (mitochondrial contact site), ERMES (ER-mitochondria encounter structure) (Fig. 6.10), vCLAMP (vacuole and

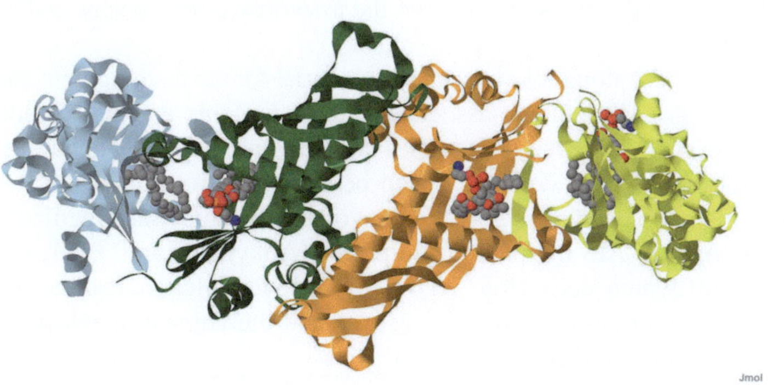

Fig. 6.10 Mitochondrial distribution and morphology protein 12, a component of the ERMES complex

mitochondria patch) and NVJ (nuclear-vacuole junction) are sites where various membranes join and bring together contributing factors (Tamura et al. 2018). Cristae membranes regulate the diffusion of ADP/ATP and metabolites across the inner membrane which is dynamically subcompartmentalized.

F1FO-ATP synthase, MICOS complex and its numerous components (MIC10/MINOS1, MIC13/Qil1, MIC19/CHCHD3, MIC25/CHCHD6, MIC26/APOO, MIC27/APOOL, and CHCHD10) regulate cristae junction formation (Eydt et al. 2017). Abnormalities in cristae membrane structures have been associated with deregulation of apoptosis, neurological pathologies, Down's syndrome, Parkinson's disease, schizophrenia, and cancer.

Nuclear mTOR, integrator of androgen signaling pathway, activates metabolic gene networks and reprogrammes mitochondrial metabolism (Audet Walsh et al. 2017).

Microtubules Motor proteins at AIS are proposed to be responsible for selective sorting of cargoes into axons and dendrites. According to the model based on actin/myosin, movement away from axon toward dendrites is favored and actin prevents movement of somatodendritic vesicles into axonal compartments (Lewis et al. 2009). Another model suggests that motor proteins associated with microtubules (MT) are axonal targeting (MT plus-end-directed kinesin movements) and somatodendritic targeting (MT minus-end-directed dynein driven retrograde axonal transport). The unidirectional movement depends on the motor proteins and MT organization (Kapitein and Hoogenraad 2011).

Dyneins are motor proteins, 0.7–1.8 MDa in size, moving cargo along microtubules in one specific direction arbitrarily named plus end to minus end. A cargo can be an organelle, such as mitochondrion, endosome or nucleus, a virus, or protein/RNA complex. Mitosis and beating of cilia and flagella involve various isoforms of dynein. Major components include a motor with six AAA+ domains and four extensions and stalk interacting with microtubules. Large (AAAL) subdomain interacts with small (AAAS) subdomain of the counterclockwise neighbor (Schmidt and Carter 2016; Soldati-Favre 2008).

Interesting questions arise as to whether the directionality of motion, sequence of ATP hydrolysis, and stereospecificity of the assembly units, helices and sheets, are linked together.

Cargo movement along microtubules is essential for the development and homeostasis of neurons. Microtubule-associated motor proteins are key components of polarized transport. Selectivity of vesicles targeting into axon and dendrites relies on sorting mechanisms. Can dendritic cargo be excluded from the axon? One of the suggested modes of action is based on involvement of dynein regulator NDEL1 which is brought to AIS by the scaffold protein Ankyrin-G and. LIS1, mediator of NDEL1-based dynein activation. NDEL1 or LIS1 depletion results in the local direction reversal of somatodendritic vesicles and the entry of dendritic cargo into the proximal axon (Kuijpers et al. 2016).

References

Abaan OD, Toretsky JA (2008) PTPL1: a large phosphatase with a split personality. Cancer Metastasis Rev 27:205–214

Arévalo JC, Pereira DB, Yano H, Teng KK, Chao MV (2006) Identification of a switch in neurotrophin signaling by selective tyrosine phosphorylation. J Biol Chem 281:1001–1007

Arévalo JC, Wu SH, Takahashi TZH, Yu T, Yano H et al (2010) The ARMS/Kidins220 scaffold protein modulates synaptic transmission. Mol Cell Neurosci 45:92–100

Avirneni-Vadlamudi U, Galindo KA, Endicott TR, Paulson V, Cameron S, Galindo RL (2012) *Drosophila* and mammalian models uncover a role for the myoblast fusion gene *TANC1* in rhabdomyosarcoma. J Clin Invest 122:403–407

Bankston JR, Camp SS, DiMaio F, Lewis AS, Chetkovich DM, Zagotta WN (2012) Structure and stoichiometry of an accessory subunit TRIP8b interaction with hyperpolarization-activated cyclic nucleotide-gated channels. Proc Natl Acad Sci USA 109(20):7899–7904. https://doi.org/10.1073/pnas.1201997109. Epub 2012 May 1. PubMed PMID: 22550182; PubMed Central PMCID: PMC3356637

Bariselli S, Bellone C (2016) VTA DA neuron excitatory synapses in Shank3 ?ex4-9 mouse line. Synapse. https://doi.org/10.1002/syn.21955

Baron MK, Boeckers TM, Vaida B, Faham S, Gingery M, Sawaya MR, Salyer D, Gundelfinger ED, Bowie JU (2006) An architectural framework that may lie at the core of the postsynaptic density. Science 311:531–535

Basu A, Chakrabarti A (2015) Hemoglobin interacting proteins and implications of spectrin hemoglobin interaction. J Proteomics 128:469–475. https://doi.org/10.1016/j.jprot.2015.06.014. Epub 2015 Jun 30. PubMed PMID: 26141508

Biederer T, Südhof TC (2000) Mints as adaptors. Direct binding to neurexins and recruitment of munc18. J Biol Chem 275(51):39803–39806

Borgdorff AJ, Choquet D (2002) Regulation of AMPA receptor lateral movements. Nature 417:649–653

Boucher LE, Bosch J (2015) The apicomplexan glideosome and adhesins -- structures and function. J Struct Biol 190(2):93–114. https://doi.org/10.1016/j.jsb.2015.02.008

Bouzidi M, Tricaud N, Giraud P, Kordeli E, Caillol G, Deleuze C, Couraud F, Alcaraz G (2002) Interaction of the Nav1.2a subunit of the voltage-dependent sodium channel with nodal ankyrinG. In vitro mapping of the interacting domains and association in synaptosomes. J Biol Chem 277:28996–28900

Buxbaum JD, Daly MJ, Devlin B, Lehner T, Roeder K, State MW, The Autism Sequencing Consortium (2012) The autism sequencing consortium: large scale, high throughput sequencing in autism spectrum disorders. Neuron 76:1052–1056

Caldieri G, Malabarba MG, Di Fiore PP, Sigismund S (2018) EGFR trafficking in physiology and cancer. Prog Mol Subcell Biol 57:235–272. https://doi.org/10.1007/978-3-319-96704-2_9. PubMed PMID: 30097778

Cheng TC, Akey IV, Yuan S, Yu Z, Ludtke SJ, Akey CW (2017) A near-atomic structure of the dark apoptosome provides insight into assembly and activation. Structure 25(1):40–52. https://doi.org/10.1016/j.str.2016.11.002. Epub 2016 Dec 1. PubMed PMID: 27916517; PubMed Central PMCID: PMC5214966

Cho HY, Maeng SJ, Cho HJ, Choi YS, Chung JM, Lee S, Kim HK, Kim JH, Eom CY, Kim YG, Guo M, Jung HS, Kang BS, Kim S (2015) Assembly of multi-tRNA Synthetase complex via heterotetrameric glutathione transferase-homology domains. J Biol Chem 290 (49):29313–29328. https://doi.org/10.1074/jbc.M115.690867. Epub 2015 Oct 15. PubMed PMID: 26472928; PubMed Central PMCID: PMC4705937

Choquet D, Triller A (2003) The role of receptor diffusion in the organization of the postsynaptic membrane. Nat Rev Neurosci 4:251–265

Constals A, Penn AC, Compans B, Toulmé E, Phillipat A, Marais S, Retailleau N, Hafner AS, Coussen F, Hosy E, Choquet D (2015) Glutamate-induced AMPA receptor desensitization increases their mobility and modulates short-term plasticity through unbinding from Stargazin. Neuron 85:787–803

Cortés RY, Arévalo JC, Magby JP, Chao MV, Plummer MR (2007) Developmental and activity-dependent regulation of ARMS/Kidins220 in cultured rat hippocampal neurons. Dev Neurobiol 67:1687–1698

Costales JL, Kolevzon A (2015) Phelan–McDermid Syndrome and SHANK3: Implications for Treatment. Neurotherapeutics 12:620–630

Dayam RM, Saric A, Shilliday RE, Botelho RJ (2015) The Phosphoinositide-Gated Lysosomal Ca (2+) channel, TRPML1, Is required for Phagosome maturation. Traffic 16(9):1010–1026. https://doi.org/10.1111/tra.12303. Epub 2015 Jun 18. PubMed PMID: 26010303

De Rubeis S, He X, Goldberg AP, Poultney CS, Samocha K, Cicek AE et al (2014) Synaptic, transcriptional, and chromatin genes disrupted in autism. Nature 515:209–215

Dorstyn L, Akey CW, Kumar S (2018) New insights into apoptosome structure and function. Cell Death Differ 25(7):1194–1208. https://doi.org/10.1038/s41418-017-0025-z. Epub 2018 May 15. Review. PubMed PMID: 29765111; PubMed Central PMCID: PMC6030056

Dosemeci A, Toy D, Reese TS, Tao-Cheng J-H (2015) AIDA-1 moves out of the postsynaptic density core under excitatory conditions. PLoS ONE 10:e0137216

Dosemeci A, Weinberg RJ, Reese TS, Tao-Cheng J-H (2016) The Postsynaptic Density: there is more than meets the eye. Front Synaptic Neurosci 8:23

Dufour A, Bellac CL, Eckhard U, Solis N, Klein T, Kappelhoff R, Fortelny N, Jobin P, Rozmus J, Mark J, Pavlidis P, Dive V, Barbour SJ, Overall CM (2018) C-terminal truncation of IFN-? inhibits proinflammatory macrophage responses and is deficient in autoimmune disease. Nat Commun 9(1):2416. https://doi.org/10.1038/s41467-018-04717-4. PubMed PMID: 29925830; PubMed Central PMCID: PMC6010466

Durak O, de Anda FC, Singh KK, Leussis MP, Petryshen TL, Sklar P, Tsai L-H (2015) Ankyrin-G regulates neurogenesis and Wnt signaling by altering the subcellular localization of β-catenin. Mol Psychiatry 20:388–397

Durand CM, Perroy J, Loll F, Perrais D, Fagni L, Bourgeron T et al (2012) *SHANK3* mutations identified in autism lead to modification of dendritic spine morphology via an actin-dependent mechanism. Mol Psychiatry 17:71–84

Eydt K, Davies KM, Behrendt C, Wittig I, Reichert AS (2017) Cristae architecture is determined by an interplay of the MICOS complex and the F(1)F(O) ATP synthase via Mic27 and Mic10. Microb Cell 4(8):259–272. https://doi.org/10.15698/mic2017.08.585. PubMed PMID: 28845423; PubMed Central PMCID: PMC5568431

Faurobert E, Albiges-Rizo C (2010) Recent insights into cerebral cavernous malformations: a complex jigsaw puzzle under construction. The Febs Journal 277:1084–1096

Fujita-Jimbo E, Tanabe Y, Yu Z, Kojima K, Mori M, Li H et al (2015) The association of GPR85 with PSD-95-neuroligin complex and autism spectrum disorder: a molecular analysis. Molecular Autism 6: 17 *Proc*. Natl Acad Sci *100*:13821–13826

Gamir-Morralla A, López-Menéndez C, Ayuso-Dolado S, Tejeda GS, Montaner J, Rosell A et al (2015) Development of a neuroprotective peptide that preserves survival pathways by preventing Kidins220/ARMS calpain processing induced by excitotoxicity. Cell Death & Disease 6:e1939. https://doi.org/10.1038/cddis.2015.307

Gaubitz C, Prouteau M, Kusmider B, Loewith R (2016) TORC2 structure and function. Trends Biochem Sci 41(6):532–545: https://doi.org/10.1016/j.tibs.2016.04.001. Epub 2016 May 5. Review. PubMed PMID: 27161823

Ghazawi FM, Faller EM, Parmar P, El-Salfiti A, MacPherson PA (2016) Suppressor of cytokine signaling (SOCS) proteins are induced by IL7 and target surface CD127 protein for degradation in human CD8 T cells. Cell Immunol 306-307:41–52. https://doi.org/10.1016/j.cellimm.2016.07.002

Ghersi E, Noviello C, D'Adamio L (2004) Amyloid-beta protein precursor (AbetaPP) intracellular domain-associated protein-1 proteins bind to AbetaPP and modulate its processing in an isoform-specific manner. J Biol Chem 279:49105–49112

Glondu-Lassis M, Dromard M, Chavey C, Puech C, Fajas L, Hendriks W, Freiss G (2009) Downregulation of protein tyrosine phosphatase PTP-BL represses adipogenesis. Int J Biochem Cell Biol 41:2173–2180

Kapitein LC, Hoogenraad CC (2011) Which way to go? Cytoskeletal organization and polarized transport in neurons. Mol Cell Neurosci 46(1):9–20. https://doi.org/10.1016/j.mcn.2010.08.015. Epub 2010 Sep 9

Granot-Hershkovitz E, Raas-Rothschild A, Frumkin A, Granot D, Silverstein S, Abeliovich D (2011) Complex chromosomal rearrangement in a girl with psychomotor-retardation and a de novo inversion: inv(2)(p15;q24.2). Am J Med Genet A 155A:1825–1832

Greber BJ, Bieri P, Leibundgut M, Leitner A, Aebersold R, Boehringer D, Ban N (2015) Ribosome. The complete structure of the 55S mammalian mitochondrial ribosome. Science 348 (6232):303–308. doi: https://doi.org/10.1126/science.aaa3872. Epub 2015 Apr 2. PubMed PMID: 25837512

Greenberg KP, Pham A, Werblin FS (2011) Differential targeting of optical neuromodulators to ganglion cell soma and dendrites allows dynamic control of center-surround antagonism. Neuron 69:713–720

Hafner AS, Penn AC, Grillo-Bosch D, Retailleau N, Poujol C, Philippat A, Coussen F, Sainlos M, Opazo P, Choquet D (2015) Lengthening of the Stargazin Cytoplasmic Tail Increases Synaptic Transmission by Promoting Interaction to Deeper Domains of PSD-95. Neuron 86:475–489

Han S, Nam J, Li Y, Kim S, Cho SH, Cho YS, Choi SY, Choi J, Han K, Kim Y, Na M, Kim H, Bae YC, Choi SY, Kim E (2010) Regulation of dendritic spines, spatial memory, and embryonic development by the TANC family of PSD-95-interacting proteins. J Neurosci 30:15102–15112

Hayashi MK, Tang C, Verpelli C, Narayanan R, Stearns MH, Xu R-M et al (2009) The postsynaptic density proteins Homer and Shank form a polymeric network structure. Cell 137:159–171

Heine M, Groc L, Frischknecht R, Béïque J-C, Lounis B, Rumbaugh G et al (2008) Surface Mobility of Postsynaptic AMPARs Tunes Synaptic Transmission. Science 320:201–205

Herberich SE, Klose R, Moll I, Yang W-J, Wüstehube-Lausch J, Fischer A (2015) ANKS1B interacts with the cerebral cavernous malformation Protein-1 and controls endothelial permeability but Not sprouting angiogenesis. PLoS ONE 10:e0145304

Hodkinson PS, Elliott PA, Lad Y, McHugh BJ, MacKinnon AC, Haslett C, Sethi T (2007) Mammalian NOTCH-1 activates beta1 integrins via the small GTPase R-Ras. J Biol Chem 282:28991–29001

Hoffmann E, Machelart A, Song OR, Brodin P (2018) Proteomics of mycobacterium infection: moving towards a better understanding of pathogen-driven immunomodulation. Front Immunol 9:86. https://doi.org/10.3389/fimmu.2018.00086. eCollection 2018. Review. PubMed PMID: 29441067; PubMed Central PMCID: PMC5797607

Huizing M, Helip-Wooley A, Westbroek W, Gunay-Aygun M, Gahl WA (2008) Disorders of lysosome-related organelle biogenesis: clinical and molecular genetics. Annu Rev Genomics Hum Genet. 9:359–386. https://doi.org/10.1146/annurev.genom.9.081307.164303. Review. PubMed PMID: 18544035; PubMed Central PMCID: PMC2755194

Isaka Y, Takabatake Y, Takahashi A, Saitoh T, Yoshimori T (2016) Hyperuricemia-induced inflammasome and kidney diseases. Nephrol Dial Transplant 31(6):890–896. https://doi.org/10.1093/ndt/gfv024. Epub 2015 Mar 31. Review. PubMed PMID: 25829326

Jacob AL, Jordan BA, Weinberg R (2010) The organization of amyloid-β protein precursor intracellular domain-associated protein-1 in the rat forebrain. J Comp Neurol 518:3221–3236

Jacobs MD, Harrison SC (1998) Structure of an IkappaBalpha/NF-kappaB complex. Cell 95 (6):749–758. PubMed PMID: 9865693

Jenkins PM, Kim N, Jones SL, Tseng WC, Svitkina TM, Yin HH, Bennett (2015) Giant ankyrin-G: A critical innovation in vertebrate evolution of fast and integrated neuronal signaling. Proc Natl Acad Sci USA 112:957–964

Jia Y, Yun CH, Park E, Ercan D, Manuia M, Juarez J, Xu C, Rhee K, Chen T, Zhang H, Palakurthi S, Jang J, Lelais G, DiDonato M, Bursulaya B, Michellys PY, Epple R, Marsilje TH, McNeill M, Lu W, Harris J, Bender S, Wong KK, Jänne PA, Eck MJ (2016) Overcoming

EGFR(T790M) and EGFR(C797S) resistance with mutant-selective allosteric inhibitors. Nature 534(7605):129–132. doi: https://doi.org/10.1038/nature17960. Epub 2016 May 25. PubMed PMID: 27251290; PubMed Central PMCID: PMC4929832

John Peter AT, Lachmann J, Rana M, Bunge M, Cabrera M, Ungermann C (2013) The BLOC-1 complex promotes endosomal maturation by recruiting the Rab5 GTPase-activating protein Msb3. J Cell Biol 201(1):97-111. doi: https://doi.org/10.1083/jcb.201210038. PubMed PMID: 23547030; PubMed Central PMCID: PMC3613695

Johansson P, Wiltschi B, Kumari P, Kessler B, Vonrhein C, Vonck J, Oesterhelt D, Grininger M (2008) Inhibition of the fungal fatty acid synthase type I multienzyme complex. Proc Natl Acad Sci U S A. 105(35):12803–12808. https://doi.org/10.1073/pnas.0805827105. Epub 2008 Aug 25. PubMed PMID: 18725634; PubMed Central PMCID: PMC2529065

Jung H, Shin J-H, Park Y-S, Chang M-S (2014) Ankyrin repeat-rich membrane spanning (ARMS)/ Kidins220 scaffold protein regulates neuroblastoma cell proliferation through p21. Mol Cells 37:881–887

Karasik A, Shanmuganathan A, Howard MJ, Fierke CA, Koutmos M (2016) Nuclear protein-only ribonuclease P2 structure and biochemical characterization provide insight into the conserved properties of tRNA 5′ end processing enzymes. J Mol Biol 428(1):26–40. https://doi.org/10.1016/j.jmb.2015.11.025. Epub 2015 Dec 3. PubMed PMID: 26655022; PubMed Central PMCID: PMC4738078

Kebache S, Ash J, Annis MG, Hagan J, Huber M, Hassard J, Stewart CL, Whiteway M, Nantel A (2007) Grb10 and active Raf-1 kinase promote Bad-dependent cell survival. J Biol Chem 282 (30):21873–21883. Epub 2007 May 29. PubMed PMID: 17535812

Keleman K, Dickson BJ (2001) Short- and long-range repulsion by the Drosophila Unc5 netrin receptor. Neuron. 32(4):605–617. PubMed PMID: 11719202

Kim KJ, Park MC, Choi SJ, Oh YS, Choi EC, Cho HJ, Kim MH, Kim SH, Kim DW, Kim S, Kang BS (2008) Determination of three-dimensional structure and residues of the novel tumor suppressor AIMP3/p18 required for the interaction with ATM. J Biol Chem 283 (20):14032–14040. https://doi.org/10.1074/jbc.M800859200. Epub 2008 Mar 14. PubMed PMID: 18343821

Kranz TM, Berns A, Shields J, Rothman K, Walsh-Messinger J, Goetz RR et al (2016) Phenotypically distinct subtypes of psychosis accompany novel or rare variants in four different signaling genes. EBioMedicine 6:206–214

Kuan SL, Fischer S, Hafner S, Wang T, Syrovets T, Liu W, Tokura Y, Ng DYW, Riegger A, Förtsch C, Jäger D, Barth TFE, Simmet T, Barth H, Weil T (2018) Boosting antitumor drug efficacy with chemically engineered multidomain proteins. Adv Sci (Weinh). 5(8):1701036. doi: 10.1002/advs.201701036. eCollection 2018 Aug. PubMed PMID: 30128225; PubMed Central PMCID: PMC6097141

Kuijpers M, van de Willige D, Freal A, Chazeau A, Franker MA, Hofenk J, Rodrigues RJ, Kapitein LC, Akhmanova A, Jaarsma D, Hoogenraad CC (2016) Dynein regulator NDEL1 controls polarized cargo transport at the axon initial segment. Neuron 89(3):461–471. https://doi.org/10.1016/j.neuron.2016.01.022

Kurochkina N (2008) J Theor Biol 255:188–198

Kurochkina N, Choekyi T (2011) Helix-helix interfaces and ligand binding. J Theor Biol 283:92–102

Lasker MV, Gajjar MM, Nair SK (2005) Molecular Structure of the IL-1R-Associated Kinase-4 Death Domain and Its Implications for TLR signaling. J Immun 175:4175–4179

Lee SO, Lee MK, Ku B, Bae KH, Lee SC, Lim HM, Kim SJ, Chi SW (2016a) High-resolution crystal structure of the PDZ1 domain of human protein tyrosine phosphatase PTP-Bas. Biochem Biophys Res Commun 478:1205–1210

Lee H, Noh H, Mun J, Gu C, Sever S, Park S (2016b) Anks1a regulates COPII-mediated anterograde transport of receptor tyrosine kinases critical for tumorigenesis. Nat Commun 7:12799

Levine ZG, Fan C, Melicher MS, Orman M, Benjamin T, Walker S (2018) O-GlcNAc transferase recognizes protein substrates using an asparagine ladder in the tetratricopeptide repeat (TPR) Superhelix. J Am Chem Soc 140(10):3510–3513. https://doi.org/10.1021/jacs.7b13546. Epub 2018 Mar 5. PubMed PMID: 29485866; PubMed Central PMCID: PMC5937710

Lewis TL, Mao T, Svoboda K, Arnold DB (2009) Myosin-dependent targeting of transmembrane proteins to neuronal dendrites. Nat Neurosci *12*(5):568–576. https://doi.org/10.1038/nn.2318

Li B-S, Sun M-K, Zhang L, Takahashi S, Ma W, Vinade L et al (2001) Regulation of NMDA receptors by cyclin-dependent kinase-5. Proc Natl Acad Sci USA 98:12742–12747. https://doi.org/10.1073/pnas.211428098

Li Y, Zhou M, Hu Q, Bai XC, Huang W, Scheres SH, Shi Y (2017) Mechanistic insights into caspase-9 activation by the structure of the apoptosome holoenzyme. Proc Natl Acad Sci U S A 114(7):1542–1547. https://doi.org/10.1073/pnas.1620626114. Epub 2017 Jan 31. PubMed PMID: 28143931; PubMed Central PMCID: PMC5320974

Lilja J, Zacharchenko T, Georgiadou M, Jacquemet G, Franceschi N, Peuhu E, Hamidi H, Pouwels J, Martens V, Nia FH, Beifuss M, Boeckers T, Kreienkamp HJ, Barsukov IL, Ivaska J (2017) SHANK proteins limit integrin activation by directly interacting with Rap1 and R-Ras. Nat Cell Biol 19:292–305

Lin S-C, Lo Y-C, Wu H (2010) Helical assembly in the MyD88–IRAK4–IRAK2 complex in TLR/IL-1R signaling. Nature 465:885–890

Liu C, Peng X, Hu C, Li C, Li Q, Xu X (2016) Developmental profiling of ASD-related *shank3* transcripts and their differential regulation by valproic acid in zebrafish. Dev Genes Evol 226:389–400. https://doi.org/10.1007/s00427-016-0561-4

Liu JJ, Lu L, Hu FQ, Yuan H, Xu Q, Qin YF, Gong JH (2018) Methylene blue attenuates renal ischemia-reperfusion injury by negative regulation of NLRP3 signaling pathway. Eur Rev Med Pharmacol Sci 22(9):2847–2853. https://doi.org/10.26355/eurrev_201805_14986. PubMed PMID: 29771438

López-Benito S, Lillo C, Hernández-Hernández A, Chao MV, Arévalo JC (2016) ARMS/Kidins220 and synembryn-B levels regulate NGF-mediated secretion. J Cell Sci 129:1866–1877

Lotan A, Fenckova M, Bralten J, Alttoa A, Dixson L, Williams RW, van der Voet M (2014) Neuroinformatic analyses of common and distinct genetic components associated with major neuropsychiatric disorders. Front Neurosci 8:331

MacGillavry HD, Kerr JM, Kassner J, Frost NA, Blanpied TA (2016) Shank–cortactin interactions control actin dynamics to maintain flexibility of neuronal spines and synapses. Eur J Neurosci 43:179–193

Maekawa K, Imagawa N, Naito A, Harada S, Yoshie O, Takagi S (1999) Association of protein-tyrosine phosphatase PTP-BAS with the transcription-factor-inhibitory protein IkappaBalpha through interaction between the PDZ1 domain and ankyrin repeats. Biochem J 337 (Pt 2):179–184

Maity TK, Venugopalan A, Linnoila I, Cultraro CM, Giannakou A, Nemati R, Zhang X, Webster JD, Ritt D, Ghosal S, Hoschuetzky H, Simpson RM, Biswas R, Politi K, Morrison DK, Varmus HE, Guha U (2015) Loss of MIG6 accelerates initiation and progression of mutant epidermal growth factor receptor-driven lung adenocarcinoma. Cancer Discov 5(5):534–549. https://doi.org/10.1158/2159-8290.CD-14-0750. Epub 2015 Mar 3. PubMed PMID: 25735773; PubMed Central PMCID: PMC4560174

Malaspina D, Kranz TM, Heguy A, Harroch S, Mazgaj R, Rothman K et al (2016) Prefrontal neuronal integrity predicts symptoms and cognition in schizophrenia and is sensitive to genetic heterogeneity. Schizophr Res 172:94–100

Mameza MG, Dvoretskova E, Bamann M, Hönck H-H, Güler T, Boeckers TM et al (2013) SHANK3 gene mutations associated with autism facilitate ligand binding to the Shank3 ankyrin repeat region. J Biol Chem 288:26697–26708

Maxson ME, Grinstein S (2014) The vacuolar-type H?-ATPase at a glance - more than a proton pump. J Cell Sci 127(Pt 23):4987–4993. https://doi.org/10.1242/jcs.158550. Review. PubMed PMID: 25453113

Mercurio FA, Marasco D, Pirone L, Pedone EM, Pellecchia M, Leone M (2012) Solution structure of the first sam domain of odin and binding studies with the epha2 receptor. Biochemistry 51:2136–2145

Mitchell RA, Luwor RB, Burgess AW (2018) The Epidermal Growth Factor Receptor: Structure-Function Informing the Design of Anticancer Therapeutics. Exp Cell Res pii: S0014-4827 (18):30635–30639. https://doi.org/10.1016/j.yexcr.2018.08.009. [Epub ahead of print] Review. PubMed PMID: 30098332

Mohanty S, Oruganty K, Kwon A, Byrne DP, Ferries S, Ruan Z, Hanold LE, Katiyar S, Kennedy EJ, Eyers PA, Kannan N (2016) Hydrophobic core variations provide a structural framework for Tyrosine Kinase evolution and functional specialization. PLoS Genet 12(2):e1005885. https://doi.org/10.1371/journal.pgen.1005885. eCollection 2016 Feb. Erratum in: PLoS Genet. 2016 Aug;12(8):e1006265. PubMed PMID: 26925779; PubMed Central PMCID: PMC4771162

Molineros JE, Yang W, Zhou XJ, Sun C, Okada Y, Zhang H, Chua K. H, Lau YL, Kochi Y, Suzuki A, Yamamoto K, Ma J, Bang SY, Lee HS, Kim K, Bae SC, Zhang H, Shen N, Looger LL, Nath SK (2017) Confirmation of five novel susceptibility loci for Systemic Lupus Erythematosus (SLE) and integrated network analysis of 82 SLE susceptibility loci. Hum Mol Genet pii: ddx026

Mosavi LK, Cammett TJ, Desrosiers DC, Peng Z (2004) The ankyrin repeat as molecular architecture for protein recognition. Protein Sci 13:1435–1448

Nanavati D, Austin DR, Catapano LA, Luckenbaugh DA, Dosemeci A, Manji HK et al (2011) The effects of chronic treatment with mood stabilizers on the rat hippocampal postsynaptic density proteome. J Neurochem 119:617–629

Neubrand VE, Thomas C, Schmidt S, Debant A, Schiavo G (2010) Kidins220/ARMS regulates Rac1-dependent neurite outgrowth by direct interaction with the RhoGEF Trio. J Cell Sci 123:2111–2123

Neubrand VE, Cesca F, Benfenati F, Schiavo G (2012) Kidins220/ARMS as a functional mediator of multiple receptor signalling pathways. J Cell Sci 125:1845–1854

Niedergang F, Grinstein S 2018 How to build a phagosome: new concepts for an old process. Curr Opin Cell Biol 50:57–63. https://doi.org/10.1016/j.ceb.2018.01.009. Epub 2018 Feb 20. Review. PubMed PMID: 29471269

Oka Y, Butnaru M, von Buchholtz L, Ryba NJP, Zuker CS (2013) High salt recruits aversive taste pathways. Nature 494:472–475

Park S (2016) Defective Anks1a disrupts the export of receptor tyrosine kinases from the endoplasmic reticulum. BMB Rep 49:651–652

Park E, Kim N, Ficarro SB, Zhang Y, Lee BI, Cho A, Kim K, Park AKJ, Park WY, Murray B, Meyerson M, Beroukhim R, Marto JA, Cho J, Eck MJ (2015) Structure and mechanism of activity-based inhibition of the EGF receptor by Mig6. Nat Struct Mol Biol 22(9):703–711. https://doi.org/10.1038/nsmb

Park HH, Logette E, Raunser S, Cuenin S, Walz T, Tschopp J, Wu H (2007) Death domain assembly mechanism revealed by crystal structure of the oligomeric PIDDosome core complex. Cell 128:533–546

Pesti S, Balázs A, Udupa R, Szabó B, Fekete A, Bőgel G, Buday L (2012) Complex formation of EphB1/Nck/Caskin1 leads to tyrosine phosphorylation and structural changes of the Caskin1 SH3 domain. Cell Commun Signal 10:36. https://doi.org/10.1186/1478-811X-10-36

Pfaender S, Sauer AK, Hagmeyer S, Mangus K, Linta L, Liebau S et al (2017) Zinc deficiency and low enterocyte zinc transporter expression in human patients with autism related mutations in SHANK3. Sci Rep 7:45190

Pospich S, Kumpula EP, von der Ecken J, Vahokoski J, Kursula I, Raunser S (2017) Near-atomic structure of jasplakinolide-stabilized malaria parasite F-actin reveals the structural basis of filament instability. Proc Natl Acad Sci USA 114(40):10636–10641. https://doi.org/10.1073/

pnas.1707506114. Epub 2017 Sep 18. PubMed PMID: 28923924; PubMed Central PMCID: PMC5635891

Qin H, Srinivasula SM, Wu G, Fernandez-Alnemri T, Alnemri ES, Shi Y (1999) Structural basis of procaspase-9 recruitment by the apoptotic protease-activating factor 1. Nature 399:549

Riedl SJ, Li W, Chao Y, Schwarzenbacher R, Shi Y (2005) Structure of the apoptotic protease-activating factor bound to ADP. Nature 434:926

Ringel R, Sologub M, Morozov YI, Litonin D, Cramer P, Temiakov D (2011) Structure of human mitochondrial RNA polymerase. Nature 478(7368):269–273. https://doi.org/10.1038/nature10435. PubMed PMID: 21947009

Ruba A, Yang W (2016) O-GlcNAc-ylation in the nuclear pore complex. Cell Mol Bioeng 9 (2):227–233. https://doi.org/10.1007/s12195-016-0440-0. Epub 2016 Apr 26. PubMed PMID: 28638491; PubMed Central PMCID: PMC5475274

Sainlos M, Tigaret C, Poujol C, Olivier NB, Bard L, Breillat C, Thiolon K, Choquet D, Imperiali B (2011) Biomimetic divalent ligands for the acute disruption of synaptic AMPAR stabilization. Nat Chem Biol 7:81–91

Sarowar T, Grabrucker AM (2016) Actin-dependent alterations of dendritic spine morphology in shankopathies. Neural Plast 8051861

Schaefer KN, Bonello TT, Zhang S, Williams CE, Roberts DM, McKay DJ, Peifer M (2018) Supramolecular assembly of the beta-catenin destruction complex and the effect of Wnt signaling on its localization, molecular size, and activity in vivo. PLoS Genet 14(4): e1007339. https://doi.org/10.1371/journal.pgen.1007339. eCollection 2018 Apr. PubMed PMID: 29641560; PubMed Central PMCID: PMC5912785

Schmidt H, Carter AP (2016) Review: structure and mechanism of the dynein motor ATPase. Biopolymers 105(8):557–567. https://doi.org/10.1002/bip.22856

Schmieg N, Menendez G, Schiavo G, Terenzio M (2014) Signalling endosomes in axonal transport: travel updates on the molecular highway. Semin Cell Dev Biol 27:32–43

Semela D (2017) Hepatitis C – Diagnostik und Therapie. Ther Umsch 74(3):101–108. https://doi.org/10.1024/0040-5930/a000892. German. PubMed PMID: 28777053

Shelby SJ, Colwill K, Dhe-Paganon S, Pawson T, Thompson DA (2013) MERTK interactions with SH2-domain proteins in the retinal pigment epithelium. PLoS One 8(2):e53964. https://doi.org/10.1371/journal.pone.0053964. Epub 2013 Feb 4. PubMed PMID: 23390493; PubMed Central PMCID: PMC3563642

Sheng M, Hoogenraad CC (2007) The postsynaptic architecture of excitatory synapses: a more quantitative view. Annu Rev Biochem 76:823–847

Singh VB, Wooten AK, Jackson JW, Maggirwar SB, Kiebala M (2015) Investigating the role of ankyrin-rich membrane spanning protein in Human Immunodeficiency Virus Type-1 Tat-induced microglia activation. J Neurovirol 21:186–198

Soldati-Favre D (2008) Molecular dissection of host cell invasion by the apicomplexans: the glideosome. Parasite 15:197–205

Sullivan JM, Zimanyi CM, Aisenberg W, Bears B, Chen D-H, Day JW et al (2015) Novel mutations highlight the key role of the ankyrin repeat domain in TRPV4-mediated neuropathy. Neurology: Genetics 1:e29

Suzuki T, Li W, Zhang JP, Tian QB, Sakagami H, Usuda N, Kondo H, Fujii T, Endo S (2005) A novel scaffold protein, TANC, possibly a rat homolog of Drosophila rolling pebbles (rols), forms a multiprotein complex with various postsynaptic density proteins. Eur J Neurosci 21:339–350

Tabuchi K, Biederer T, Butz S, Sudhof TC (2002) CASK participates in alternative tripartite complexes in which Mint 1 competes for binding with caskin 1, a novel CASK-binding protein. J Neurosci 22:4264–4273

Takáts S, Glatz G, Szenci G, Boda A, Horváth GV, Hegedus K, Kovács AL, Juhász G (2018) Non-canonical role of the SNARE protein Ykt6 in autophagosome-lysosome fusion. PLoS Genet. 14(4): e1007359. https://doi.org/10.1371/journal.pgen.1007359. eCollection 2018 Apr. PubMed PMID: 29694367; PubMed Central PMCID: PMC5937789

Tamura Y, Kawano S, Endo T (2018) Organelle contact zones as sites for lipid transfer. J Biochem https://doi.org/10.1093/jb/mvy088. [Epub ahead of print] PubMed PMID: 30371789

Tao-Cheng J-H, Yang Y, Reese TS, Dosemeci A (2015) Differential Distribution of Shank and GKAP at the Postsynaptic Density. PLoS ONE 10:e0118750

Tchesnokov EP, Raeisimakiani P, Ngure M, Marchant D, Götte M (2018) Recombinant RNA-dependent RNA polymerase complex of Ebola Virus. Sci Rep 8(1):3970. https://doi.org/10.1038/s41598-018-22328-3. PubMed PMID: 29507309; PubMed Central PMCID: PMC5838098

Tindi JO, Chávez AE, Cvejic S, Calvo-Ochoa E, Castillo PE, Jordan BA (2015) *ANKS1B* gene product AIDA-1 controls hippocampal synaptic transmission by regulating GluN2B subunit localization. J Neurosci 35:8986–8996. https://doi.org/10.1523/JNEUROSCI.4029-14.2015

Tomasetti C, Iasevoli F, Buonaguro EF, De Berardis D, Fornaro M, Fiengo ALC et al (2017) Treating the synapse in major psychiatric disorders: the role of postsynaptic density network in dopamine-glutamate interplay and psychopharmacologic drugs molecular actions. Int J Mol Sci 18:135

Tseng WC, Jenkins PM, Tanaka M, Mooney R, Bennett V (2015) Giant ankyrin-G stabilizes somatodendritic GABAergic synapses through opposing endocytosis of $GABA_A$ receptors. Proc Natl Acad Sci USA 112:1214–1219

Uddin MS, Mamun AA, Labu ZK, Hidalgo-Lanussa O, Barreto GE, Ashraf GM (2018) Autophagic dysfunction in Alzheimer's disease: cellular and molecular mechanistic approaches to halt Alzheimer's pathogenesis. J Cell Physiol https://doi.org/10.1002/jcp.27588. [Epub ahead of print] Review. PubMed PMID: 30362531

Wang C, Wei Z, Chen K, Ye F, Yu C, Bennett V, Zhang M (2014) Structural basis of diverse membrane target recognitions by ankyrins. eLife 3:e04353

Welbourn EM, Wilson MT, Yusof A, Metodiev MV, Cooper CE (2017) The mechanism of formation, structure and physiological relevance of covalent hemoglobin attachment to the erythrocyte membrane. Free Radic Biol Med 103:95–106. https://doi.org/10.1016/j.freeradbiomed.2016.12.024. Epub 2016 Dec 20. PubMed PMID: 28007575; PubMed Central PMCID: PMC5282401

Willis BC, Ponce-Balbuena D, Jalife J (2015) Protein assemblies of sodium and inward rectifier potassium channels control cardiac excitability and arrhythmogenesis. Am J Physiol Heart and Circ Physiol 308:H1463–H1473

Won S, Levy JM, Nicoll RA, Roche KW (2017) MAGUKs: multifaceted synaptic organizers. Curr Opin Neurobiol 43:94–101

Xu M, Cooper EC (2015) An ankyrin-G N-terminal gate and protein kinase CK2 dually regulate binding of voltage-gated Sodium and KCNQ2/3 Potassium channels. J Biol Chem 290:16619–16632

Yan N, Chai J, Lee ES, Gu L, Liu Q, He J, Wu JW, Kokel D, Li H, Hao Q, Xue D, Shi Y (2005) Structure of the CED-4-CED-9 complex provides insights into programmed cell death in Caenorhabditis elegans. Nature 437(7060):831–837. PubMed PMID: 16208361

Yao I, Hata Y, Hirao K, Deguchi M, Ide N, Takeuchi M, Takai Y (1999) Synamon, a novel neuronal protein interacting with synapse-associated protein 90/postsynaptic density-95-associated protein. J Biol Chem 274:27463–27466

Yoshimura T, Stevens SR, Leterrier C, Stankewich MC, Rasband MN (2016) Developmental changes in expression of βIV spectrin splice variants at axon initial segments and nodes of Ranvier. Front Cell Neurosci 10:304

Yu X, Acehan D, Ménétret JF, Booth CR, Ludtke SJ, Riedl SJ, Shi Y, Wang X, Akey CW (2005) A structure of the human apoptosome at 12.8 A resolution provides insights into this cell death platform. Structure 13(11):1725–1735. PubMed PMID: 16271896

Zamani MR, Aslani S, Salmaninejad A, Javan MR, Rezaei N (2016) PD-1/PD-L and autoimmunity: a growing relationship. Cell Immunol 10:27–41

Zhang X, Bennett V (1996) Identification of O-linked N-acetylglucosamine modification of ankyrinG isoforms targeted to nodes of Ranvier. J Biol Chem 271(49):31391–31398

Zheng X, Kuang Y, Lv W, Cao D, Sun Z, Sun X (2016) Genome-wide association study for muscle fat content and abdominal fat traits in common carp (*Cyprinus carpio*). PLoS ONE 11:e0169127

Zheng Y-L, Li B-S, Rudrabhatla P, Shukla V, Amin ND, Maric D, Kesavapany S, Kanungo J, Pareek TK, Takahashi S, Grant P, Kulkarni AB, Pant HC (2010) Phosphorylation of p27Kip1 at Thr187 by Cyclin-dependent Kinase 5 modulates neural stem cell differentiation. Mol Biol Cell 21:3601–3614

Zhou P, Chou J, Sanchez O, Yuan J, Wagner G (1999) Solution structure of Apaf-1 CARD and its interaction with caspase-9 CARD: A structural basis for specific adaptorycaspase interaction. Proc Natl Acad Sci 96:11265–11270

Zynda ER, Grimm MJ, Yuan M, Zhong L, Mace TA, Capitano M et al (2015) A role for the thermal environment in defining co-stimulation requirements for $CD4^+$ T cell activation. Cell Cycle 14:2340–2354

Chapter 7
Viruses

Abstract Viruses attack host cells using its machinery to infuse their genome and replicate themselves. Genome is the source of variation. Harmful, neutral, and beneficial sequences transfer between species and make the world more diverse. Viruses show amazing shapes, biophysical characteristics, and biochemical mechanisms. Studies of virus internalization and targeting strategies give prototypes for delivery vehicles in drug design. Our goals are to apply all knowledge for the benefit of people and find the means to avoid dangers.

Keywords Viral genome · Host · Entry complex · Capsid · Coat · Polymerase

Virus mechanism of action involves use of cell machinery of the attacked organism to infuse its genome and replicate itself. RNA or DNA genome of the virus in the form of single or double stranded highly packed molecule is enclosed by the protein capsid. Viruses exhibit amazing shapes. Their structure is extensively studied. Genome is the source of variation, and many sequences migrate into other organisms by a horizontal gene transfer. Host may use viruses for their functions such as transfer of nucleic acids or defense from other pathogens (Koonin and Dolja 2018). Harmful, neutral, and beneficial sequences transfer between species and make the world more diverse. Viruses also contain noncoding RNA fragments that regulate gene expression of cellular and viral proteins important for viral life cycle, pathogenesis, proliferation, escape of destruction, latency and reactivation (Chavez-Calvillo et al. 2018). Ability of viruses to mutate and appear in new forms presents a danger to living organisms as pathogens transform to become more virulent. Even one amino acid change can affect structure and ligand interaction (Kurochkina and Choekyi 2011). Emergence of mutant forms of pathogens is one of the sources of drug resistance since pathogens gain new properties and stop responding to existing drug. New viruses are being discovered (Chun et al. 2018; Ho et al. 2018; Nelson et al. 2018). Mechanisms of virus internalization, targeting and replication are used for the design of drug delivery vehicles (Singh et al. 2018; Sobhy et al. 2016), and we apply our knowledge to make them beneficial for people trying to avoid dangers.

N. Kurochkina, *Protein Structure and Modeling*,
https://doi.org/10.1007/978-981-13-6601-7_7

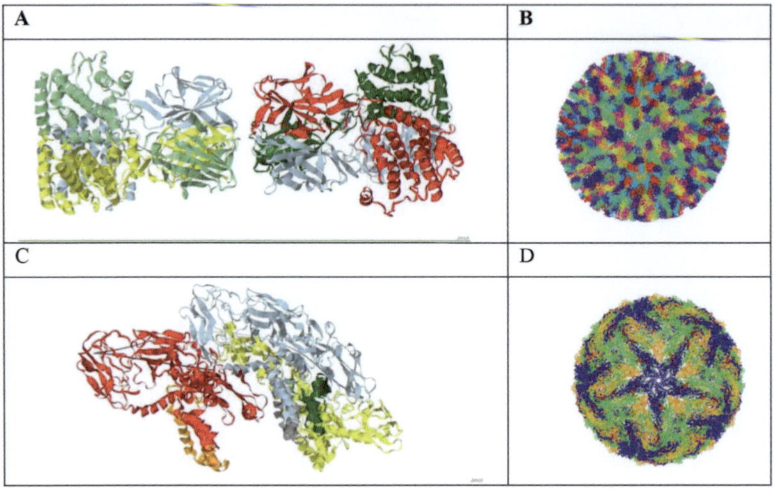

Fig. 7.1 *Bluetounge virus* (**A**) coat protein VP7 (1bvp); (**B**) core: VP3, VP7, inner layer (2btv) and *dengue virus* (**C**) components; (**D**) assembly (3j27)

Viral Proteins

The virus envelope glycoproteins mediate virus attachment and entry into host cells. In human immunodeficiency virus (HIV), entry complex in its prefusion form consists of three gpl20 surface glycoproteins and three membrane-anchored gp41 subunits. Binding to the receptor (CD44) induces a conformational change, formation of more stable trimer-of-hairpins structure, and fusion of the viral and cellular membranes (Liu et al. 2009). Similarly, the entry complexes of ISAV, IAV, IBV, ICV, and other viruses are organized. The capsid of icosahedral viruses is an assembly of jelly roll domains. This domain is a β barrel of two antiparallel sheets, each of four strands. Eight strands, with 12345678 order in amino acid sequence, have topology 1836 2745 in space. Double jelly roll major capsid protein, single jelly roll minor capsid protein, cysteine protease, and A type Atpase are proteins that involve virion morphogenesis. Many processes such as capsid assembly, virus maturation, and host cell entry depend on sequence of proteolytic steps and activation.

HIV-1 capsid with enclosed viral genome is cone shaped and includes hexameric lattice of 12 capsids protein pentameric modules (Fig. 7.1B). Structures of hexameric and pentameric capsid protein arrangements show that these two forms utilize different interfaces and flexibility (Mattei et al. 2016).

Refolding of the viral envelope glycoprotein GP41 into a fusion-active conformation stimulates HIV membrane fusion. In its fusion-active state, the gp41 ectodomain core exhibits a trimer of hairpins: three antiparallel C terminal helices associate with hydrophobic surface of the N terminal trimer. Therefore, a hexameric coiled coil of the gp120-gp41 complex is formed Fig. 7.2D). Alanine scanning

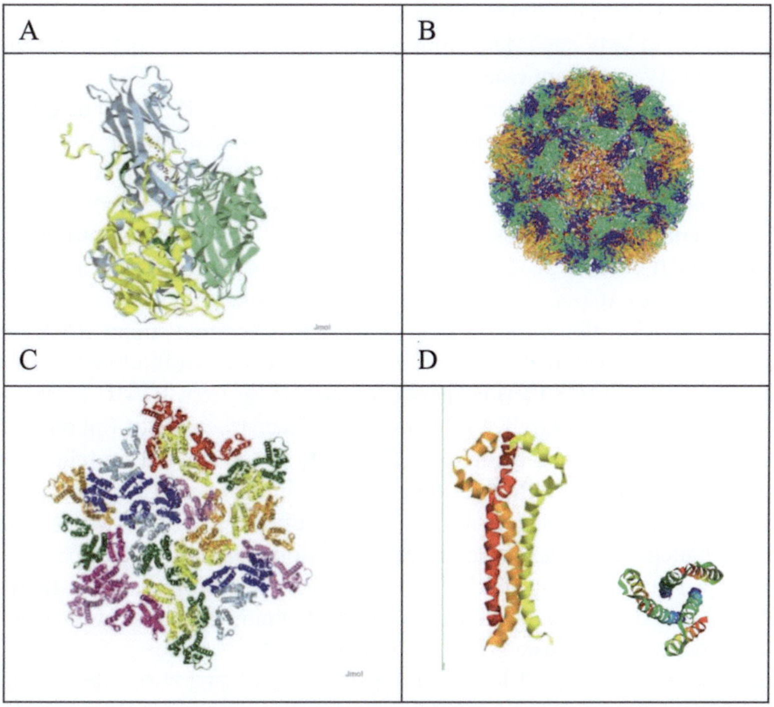

Fig. 7.2 HIV Virus. (**A**) Human rhinovirus type 14:human immunodeficiency virus type 1 V3 loop chimeric virus MN-III-2 (1k5m); coat protein VP1 (lightblue), chimera of HRV14 coat protein VP2 (P1B) and the V3 loop of HIV-1 GP120 (lightgreen), coat protein VP3 (yellow); coat protein VP4 (green); (**B**) HIV virus assembly(1k5m); (**C**) capsid protein P24 (5mcx); (**D**) envelope glycoprotein GP160 /6e8w/ and transmembrane glycoprotein (GP41) (1i5x)

mutagenesis shows that mostly hydrophobic residues are important for this interaction. Hydrophobic packing of specific helix/helix interfaces and lipid bilateral provides thermodynamically driven association (Liu et al. 2001).

A chimeric molecule that contains rhinovirus MN-III-2 with incorporated HIV-1 gp120 V3 loop (Fig. 7.2A) shows that HIV V3 loop can adopt multiple conformations in the context of the virus framework and possibly increase probability of an immune response of the neutralizing antibody (Ding et al. 2002).

HIV-1 envelope glycoprotein (Fig. 7.2D) contains epitopes of neutralizing antibodies in the membrane-proximal external region. This site in many viruses is targeted for the vaccine development. Together with its adjacent transmembrane domain, it is crystallized in micelles that mimic a lipid-bilayer membrane. The epitope site resides outside the lipid bilayer and folds as a trimer of gpl20 surface glycoproteins associated with three membrane-anchored gp41 subunits in prefusion conformation and interacting with phospholipid head groups (Fu et al. 2018).

Development of neutralizing antibodies against existing and newly emerging pathogens, particularly such dangerous as HIV and EBOLA which affect millions of people, is very important (Janus et al. 2015).

ISAV Infectious salmon anemia is caused by the virus (ISAV). The genome of ISAV consists of eight single-stranded RNA segments that encode for at least 10 proteins. Virus fusion mechanism involves engulfment into endocytic vesicles. The ISAVF fusion core protein is of class I viral fusion architecture (Cook et al. 2015). ISAV entry complex is composed of the ISAV F and ISAV hemagglutinin esterase (HE) proteins (Cook et al. 2015). As many other viruses, ISAV and HIV use glycoprotein fusion (Cook et al. 2015; 4xyp; Liu et al. 2009).

Rhinovirus (RV), the common cold virus, has a positive-strand RNA genome surrounded by nonenveloped, icosahedral capsid. The antigen-binding fragment of a neutralizing antibody (C5 Fab) triggers genome release from RV-B14. As a result, emptied viral particles neutralize virus infection. The structures of full and emptied particles show that binding of the 60 C5 Fabs to RV-B14 viral protein 3 induces significant conformational changes around holes in the capsid through which the viral RNA might exit (Dong et al. 2017). The difference in the two structures is mainly attributed to a "pocket factor" in the hydrophobic site of the VP1 jelly roll containing a fatty acid-like molecule that stabilizes the virus particle. In emptied capsid, two α-helices in VP2 related by twofold symmetry move away from each other and create a pore (Arnold and Rossmann 2014).

Bluetongue virus (Fig. 7.1A) is a large nonenveloped virus that contains a genome of ten double-stranded RNA sequences surrounded by protein layered surface (Zhang et al. 2010). It is transmitted by blood-feeding insects (*Culicoides sp.*) causing hemorrhage in livestock. Central hub domain contains sialic acid binding site. Triskelion shaped VP2 attachment trimer branches from the hub. VP5 membrane penetration trimer contains a central coiled-coil α-helical bundle, which is similar to the fusion proteins of many enveloped viruses (e.g., HIV, herpesviruses, vesicular stomatitis virus, and influenza virus) (Roy et al. 2017).

The *Dengue virus* (Fig. 7.1B) encodes three structural (C, prM, and E) and seven nonstructural proteins (NS1, NS2A, NS2B, NS3, NS4A, NS4B, and NS5) in its 5-capped positive-sense single-stranded RNAs. Capsid (C), precursor mature membrane protein (prM), and envelope (E) structural proteins form a capsid. Cleavage by a host-and virus-encoded proteases (NS2B-NS3) into converts polyprotein into active proteins. The nonstructural proteins direct viral RNA replication: synthesis of negative and positive strand RNAs and the addition of 5-cap to the positive sense RNAs (Teramoto et al. 2014). NS3 (helicase) and NS5 (MTase and polymerase) interdomain interactions are important for virus replication. NS5 dimerization interface and NS3/NS5 interface contain residues that are involved in specific interactions (Teramoto et al. 2017). Inhibition of the flaviviral protease is one of the major strategies in design of new therapies (Lai et al. 2013; Balasubramanian et al. 2016).

Ribonucleoprotein particle (RNP) is the site of genome packaging. Segmented genome fragments of the virus are each stored in a separate RNP. The RNP contains viral RNA-dependent RNA polymerase bound to the viral promoter, 5′ and 3′

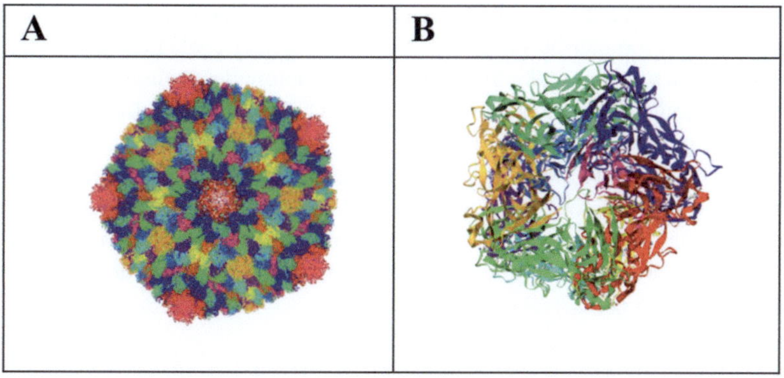

Fig. 7.3 DNA Viruses. (**A**) SPUTNIK virophage (3j26); (**B**) mavirus penton protein (6 g41)

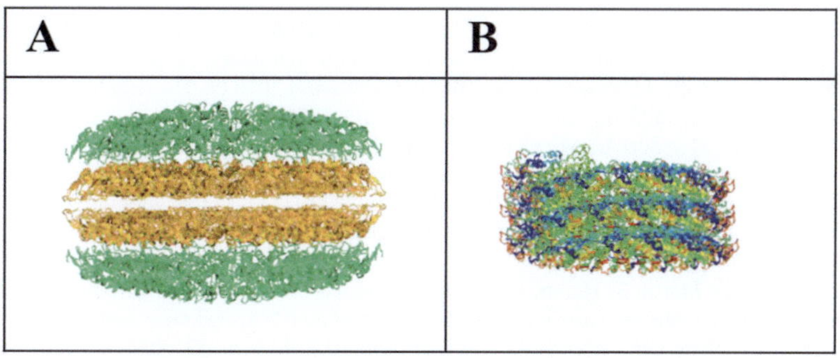

Fig. 7.4 TMV. (**A**) disk (1ie7); (**B**) spiral (3j06)

termini of the genome segment. The strands of viral RNA (vRNA) are coated by nucleoprotein (Pflug et al. 2017) (Nozawa et al. 2017).

DNA viruses, single stranded and double stranded, have many similar as well as distinct features with RNA viruses (Krupovic and Koonin 2014). Endonuclease and jelly roll fold versus protein primed type B DNA polymerase and capsid protein are just few examples.

Double stranded icosahedral DNA virus Sputnik (Fig. 7.3A) coassembles with mimivirus infecting amoeba. Its capsid consists of 260 trimeric and 12 pentameric subunits (pentons) in the lattice (triangulation number $T = 27$) and it is similar to mavirus (Fig. 7.3B) and Zamilon. These viruses infect freshwater, marine protozoa and algae. Double jelly roll major capsid protein, single jelly roll minor capsid protein, cysteine protease, and A type Atpase are conserved among virophages and define virion morphogenesis. Capsid assembly and virus maturation processes depend on sequential proteolytic steps (Born et al. 2018).

Tobacco mosaic virus (TMV) is a filamentous virus and is similar in its structure (Fig. 7.4) to bacterial flagella, cytoskeletal networks, muscle fibers, secretion systems. and F-actin assemblies. TMV coat protein contains a four alpha helical motif and can assemble as helical or disk arrangement of the repeated units. Nucleic acids of the virus genome can be observed inside the helical coat. Periodic repeats and flexibility are factors that make high resolution structure determination harder. In the four-layer aggregate disk structure, backbone residues form hydrogen bonds with other protein groups via water molecules.

In a disc structure, 17 subunits pack in one layer, whereas in a helix, 16 subunits. Conformation of each four helix subunit is very similar in the two arrangements. Water molecules are important for the obtaining stable conformation. Also, side chains of the hydrophobic core change rotameric states. RNA structure is unstable without the protein coat.

References

Balasubramanian A, Manzano M, Teramoto T, Pilankatta R, Padmanabhan R (2016) High-throughput screening for the identification of small-molecule inhibitors of the flaviviral protease. Antiviral Res 134:6–16. https://doi.org/10.1016/j.antiviral.2016.08.014. Epub 2016 Aug 15. PubMed PMID: 27539384; PubMed Central PMCID: PMC5065773

Born D, Reuter L, Mersdorf U, Mueller M, Fischer MG, Meinhart A, Reinstein J (2018) Capsid protein structure, self-assembly, and processing reveal morphogenesis of the marine virophage mavirus. Proc Natl Acad Sci USA 115(28):7332–7337. https://doi.org/10.1073/pnas.1805376115. Epub 2018 Jun 25. PubMed PMID: 29941605; PubMed Central PMCID: PMC6048507

Chavez-Calvillo G, Martin S, Hamm C, Sztuba-Solinska J (2018) The structure-to-function relationships of Gammaherpesvirus-encoded long non-coding RNAs and their contributions to viral pathogenesis. Noncoding RNA 4(4). pii: E24. doi: 10.3390/ncrna4040024. Review. PubMed PMID: 30261651

Chun J, Yang HE, Kim DH (2018) Identification and molecular characterization of a novel partitivirus from trichoderma atroviride NFCF394. Viruses 10(11). pii: E578.: 10.3390/v10110578. PubMed PMID: 30360465

Cook JD, Soto-Montoya H, Korpela MK, Lee JE (2015) Electrostatic architecture of the Infectious Salmon Anemia Virus (ISAV) Core fusion protein illustrates a Carboxyl-Carboxylate pH Sensor. J Biol Chem 290(30):18495–18504. https://doi.org/10.1074/jbc.M115.644781. Epub 2015 Jun 16. PubMed PMID: 26082488; PubMed Central PMCID: PMC4513110

Ding J, Smith AD, Geisler SC, Ma X, Arnold GF, Arnold E (2002) Crystal structure of a human rhinovirus that displays part of the HIV-1 V3 loop and induces neutralizing antibodies against HIV-1. Structure 10(7):999–1011. PubMed PMID: 12121655

Dong Y, Liu Y, Jiang W, Smith TJ, Xu Z, Rossmann MG (2017) Antibody-induced uncoating of human rhinovirus B14. Proc Natl Acad Sci USA 114(30):8017–8022. https://doi.org/10.1073/pnas.1707369114. Epub 2017 Jul 10. PubMed PMID: 28696310; PubMed Central PMCID: PMC5544331

Fu Q, Shaik MM, Cai Y, Ghantous F, Piai A, Peng H, Rits-Volloch S, Liu Z, Harrison SC, Seaman MS, Chen B, Chou JJ (2018) Structure of the membrane proximal external region of HIV-1 envelope glycoprotein. Proc Natl Acad Sci USA 115(38):E8892–E8899. https://doi.org/10.1073/pnas.1807259115. Epub 2018 Sep 5. PubMed PMID: 30185554; PubMed Central PMCID: PMC6156635

Ho KL, Gabrielsen M, Beh PL, Kueh CL, Thong QX, Streetley J, Tan WS, Bhella D (2018) Structure of the Macrobrachium rosenbergii nodavirus: A new genus within the Nodaviridae?, e3000038. PLoS Biol 16(10). https://doi.org/10.1371/journal.pbio.3000038. [Epub ahead of print] PubMed PMID: 30346944

Koonin EV, Dolja VV (2018) Metaviromics: a tectonic shift in understanding virus evolution. Virus Res. 246:A1–A3. https://doi.org/10.1016/j.virusres.2018.02.001

Krupovic M, Koonin EV(2014Jun 18) Evolution of eukaryotic single-stranded DNA viruses of the Bidnaviridae family from genes of four other groups of widely different viruses. Sci Rep.;4:5347. https://doi.org/10.1038/srep05347. PubMed PMID: 24939392; PubMed Central PMCID: PMC4061559

Kurochkina N, Choekyi T (2011) Helix-helix interfaces and ligand binding. J Theor Biol 283:92–102

Liu J, Deng Y, Dey AK, Moore JP, Lu M (2009) Structure of the HIV-1 gp41membrane-proximal ectodomain region in a putative prefusion conformation. Biochemistry 48(13):2915–2923. https://doi.org/10.1021/bi802303b. PubMed PMID: 19226163; PubMed Central PMCID: PMC2765501

Mattei S, Glass B, Hagen WJ, Kräusslich HG, Briggs JA (2016) The structure and flexibility of conical HIV-1 capsids determined within intact virions. Science 354(6318):1434–1437. PubMed PMID: 27980210

Nelson CA, Wilen CB, Dai YN, Orchard RC, Kim AS, Stegeman RA, Hsieh LL, Smith TJ, Virgin HW, Fremont DH (2018) Structural basis for murine norovirus engagement of bile acids and the CD300lf receptor. Proc Natl Acad Sci USA 115(39):E9201–E9210. https://doi.org/10.1073/pnas.1805797115. Epub 2018 Sep 7. PubMed PMID: 30194229; PubMed Central PMCID: PMC6166816

Nozawa K, Schneider TR, Cramer P (2017) Core mediator structure at 3.4 Å extends model of transcription initiation complex. Nature. https://doi.org/10.1038/nature22328. [Epub ahead of print]

Pflug A, Lukarska M, Resa-Infante P, Reich S, Cusack S (2017) Structural insights into RNA synthesis by the influenza virus transcription-replication machine. Virus Res 20. pii: S0168–1702(16)30782–1. https://doi.org/10.1016/j.virusres.2017.01.013. [Epub ahead of print]

Singh AK, Nguyen TH, Vidovszky MZ, Harrach B, Benko M, Kirwan A, Joshi L, Kilcoyne M, Berbis MÁ, Cañada FJ, Jiménez-Barbero J, Menéndez M, Wilson SS, Bromme BA, Smith JG, van Raaij MJ (2018) Structure and N-acetylglucosamine binding of the distal domain of mouse adenovirus 2 fibre. J Gen Virol 99(11):1494–1508. https://doi.org/10.1099/jgv.0.001145. Epub 2018 Oct 2. PubMed PMID: 30277856

Teramoto T, Balasubramanian A, Choi KH, Padmanabhan R (2017) Serotype-specific interactions among functional domains of dengue virus 2 nonstructural proteins (NS) 5 and NS3 are crucial for viral RNA replication. J Biol Chem 292(23):9465–9479. https://doi.org/10.1074/jbc.M117.775643. Epub 2017 Apr 10. PubMed PMID: 28396347; PubMed Central PMCID: PMC5465476

Teramoto T, Boonyasuppayakorn S, Handley M, Choi KH, Padmanabhan R (2014) Substitution of NS5 N-terminal domain of dengue virus type 2 RNA with type 4 domain caused impaired replication and emergence of adaptive mutants with enhanced fitness. J Biol Chem 289 (32):22385–22400. https://doi.org/10.1074/jbc.M114.584466. Epub 2014 Jun 5. PubMed PMID: 24904061; PubMed Central PMCID: PMC4139246

Chapter 8
Methods of Structure Determination

Abstract Structure of biological macromolecules and small molecules ligands is very important for our understanding of living organisms and processes. High resolution structures give us atomic positions of molecules and their complexes. Imaging of individual molecules and their dynamic behavior in cellular reactions and pathways is rich source of data that brings new insights about mechanisms of action. Experimental and theoretical models of biological molecules become more and more accurate and contribute to all areas of biomedical research.

Keywords Crystallography · NMR · Electron microscopy · Theoretical model · Structure

Studies of structural principles of protein molecule, its conformation and relationships between primary, secondary, tertiary, and quaternary structure raise and answer many important questions. High resolution structures, imaging of cellular processes in living cells, mapping of protein interactions and colocalization, advancements in proteomics and mass spectroscopy, development of noninvasive methods of medical diagnostics (Streckfus et al. 2012) bring into life many new applications of structural biology.

Crystallography

Discovery of X-rays by Roentgen and obtaining diffraction from molecular crystals allowed determining structure of many molecules including proteins and nucleic acids. Diffraction of proteins identified distinct patterns of polypeptide chain folding: fibers of α and β types. Diffraction patterns represent different types of molecular conformation. First protein crystal structures of myoglobin (1957) and hemoglobin (1960) were solved. Crystallography in 1950s and 1960s started by Max Perutz, John Kendrew and their colleagues brought new advancements in structural exploration of biological macromolecules (Schulz and Schirmer 1982; Lomelino et al.

N. Kurochkina, *Protein Structure and Modeling*,
https://doi.org/10.1007/978-981-13-6601-7_8

2018). Important applications of protein crystallography to medicine suggested new protein targets for structural studies: insulin, pepsin, lysozyme, chymotrypsin, trypsin, renin, HIV protease, and many other (Thomas et al. 2017).

Structure guided drug discovery exploded. Many inhibitors, activators, allosteric regulators of enzymes were discovered and newly designed for drug applications. Large protein complexes and organelles such as ribosome revealed detailed structure and mechanism studied by electron microscopy and crystallography (Dao et al. 2018).

Protein data bank (PDB) at the Brookhaven National Laboratory from the first few structures grew to hundreds thousands (Bernstein et al. 1977).

Many methods accompany crystallographic analysis: high throughput screening (HTS with fluorescence-based thermal shift measurements, ligand-based NMR, surface plasmon resonance, isothermal calorimetry, mass spectroscopy and other methods.

Electron Microscopy

Structures determined by crystallography as larger complexes are also viewed using cryo electron microscopy (Cryo EM). This method became quicker and less costly alternative to crystallography which with improvement in resolution is used for smaller molecules about 17 kDa as well. Use of direct electron detectors for low contrast flexible organic materials provides better treatment of sensitive to radiation samples in amorphous ice. Cryo EM is applied to view structure and dynamics of conformational transitions, particularly in large molecules of more than 100 kDa. High resolution structures of molecules in solvent exhibiting multiple conformational states are determined by Cryo EM and complement those available from crystallographic and NMR studies (Khoshouei et al. 2017).

Nuclear Magnetic Resonance, NMR

Amino acid residues undergo hydrogen exchange with solvent. NMR experiments allow measuring the time course at which the amino acid hydrogen atoms become protected from exchange with solvent. Tracing amide nitrogens involved in hydrogen bonds provides data on secondary and tertiary structure of the protein and kinetics of the folding (Alderson et al. 2017). Protein structures determined using NMR spectroscopy show ensemble of states the protein molecule exhibits in solution.

Folding of small proteins occurs on the timescale of milliseconds which is a rapid process in the life of cell. NMR provides data on transitions between folded and unfolded states. Use of NMR for folding experiments to accurately study model of protein folding and assess events and pathways of the folding process under native conditions applies hydrostatic pressure and includes temperature dependence (Charlier et al. 2018).

Imaging of organs and tissues is very important for medical diagnostics. Advancements in diagnostic methods bring new noninvasive methods with significant benefits to patients. Modern methods include main 4 non-invasive imaging techniques such as echocardiography, magnetic resonance imaging (MRI), multi-detector-computed tomography (MDCT), and nuclear imaging (positron emission tomography [PET] and single-photon emission-computed tomography [SPECT]). These new and established imaging techniques provide data on organs undergoing treatments, their anatomy and function. Imaging provides invaluable tool for accurate diagnostics and the assessment of the pathophysiology of organ malfunction (Bax et al. 2015).

Progress in solid state NMR allows to rapidly determine structure of large and small molecules applying isotope labeling strategies. This method overcomes requirements of sample crystallization or size (Martin et al. 2018).

Superresolution Microscopy and Mass Spectroscopy

New methods of super resolution imaging: photoactivated localization microscopy (PALM)/stochastic optical reconstruction microscopy (STORM), stimulated emission depletion microscopy (STED), single molecule guided Bayessian localization superresolution microscopy SIMBA), and expansion microscopy (ExM), significantly improved visualization of dynamic processes on molecular scale. STORM and STED improve resolution tenfold approximately from 250 to 25 nm (Rappoport 2018; Valm et al. 2017).

New imaging methods help understanding mechanisms of endocytosis. In nerve cells, important subjects studied by these methods are properties of excitatory and inhibitory synaptic transmission, molecular fusion, localization and movement of substances within cell, across the membrane and between cells, and cytoskeletal dynamics (Igarashi et al. 2018).

Structural illumination microscopy (SIM) improves resolution two fold but in contrast to STORM and STED is applicable to living cells. Making visualization in living cells possible advances our knowledge of dynamic cellular processes.

Theoretical Models

Growing amount of data becomes available from genetic and genomic studies as a result of sequencing of genomes of many organisms including those of extinct species. Human genome is source of our understanding of health and disease (Collins and Gottlieb 2018). Although methods for obtaining secondary and tertiary structure of proteins provide more and more data, there exists a large gap between number of sequences and number of three-dimensional structures determined for proteins, or between available primary and secondary/tertiary structure. New methods that provide tertiary structures of proteins help to reveal mechanisms of action of proteins and design drugs, nanostructures and nanomaterials. Accurate prediction of protein conformation has significant impact on all aspects and areas of biomedical research (Chou 2004). Cancer tumors exhibit high rate of mutations that gives abnormal isoforms of enzymes involved in disease progression (Halazonetis et al. 2008). New and existing therapies meet challengies as drug resistance presents new problems. Emerging pathogenic organisms produce mutant forms of proteins which were not found before and may not have readily available medicine to counteract increased virulence. Changes of the structure of proteins due to the influence of mutations become critical points of the analysis. Prediction of protein structure and its modifications contributes significantly to understanding of molecular function, mechanisms of protein action and interactions in multiple metabolic pathways. Mutagenesis as mechanism of new evolving molecules and organisms and as a tool of investigation of their function gives enormous amount of data as new proteins for which three-dimensional structure has to be determined and influence of changes has to be estimated become objects of studies and design of new therapies. It is very important to identify factors, which characterize structure, identify amino acids, which determine particular fold, and develop new methods for its accurate prediction.

Theoretical models close to experimentally determined structures are widely used in the analysis and design. Structure coordinates of the predicted molecules serve as preliminary models of crystallographic, NMR, and electron microscopy studies or as their quick alternatives. Development of a theoretical model is an iterative process in which the proposed model absorbs all available experimental data, goes through cycles of refinement, and serves as a source of new experimental design. We should not ignore principles of protein structure and importance of interfaces; otherwise, we will see the consequences in toxicity of conjugate molecules, such as immunotoxins, inaccurate drug delivery, and errors of colocalization.

Theoretical models save time and resources. Extensive use of theoretical models suggests new ways of preserving lives of animals used in biomedical research.

Protein Modeling is an essential part of experiment planning and interpretation and presentation of results.

Proteins as regular objects are well described by formal approaches and mathematical formulas. More and more new ways to look at protein structure are on the way. Will advancements in mathematics, physics, and other areas contribute? (Irving 2018; Sternberg 2018; Stephens 2018).

Modeling with Formula

(Copied with permission from Kurochkina "Approximation of the protein molecule with formula for spatial location of amino acids within secondary structure elements").

Chemical formulas describe covalent bonds of substances and atomic positions in space. Molecules also have class of interactions called van der Waals interactions. These interactions are particularly important for large macromolecules such as proteins and nucleic acids. Protein secondary structure and tertiary structure exhibit high degree of regularity demonstrated by existence of a number of hydrogen-bonded elements and tertiary structure motifs. These regularities originate from a very well organized network of contacts. In this work, a formula describing spatial location and van der Waals interactions of amino acid residues within secondary structure elements, which gives a description of the intramolecular and intermolecular contacts, is presented. A pair of α-helices forms an interface as a sequence of interacting pockets with side chains packed "knobs into holes" and stacked in regular layers. Interfaces of helical pairs are assembled into tertiary structure. This formula contains (1) a precise pattern of contacts of the helix-helix interface and (2) an arrangement of helix-helix interfaces combined to form a three-dimensional structure. In the description of protein motifs such as four-α-helix bundle, TIM-barrel, assembly of a virus into helical rod, circular arrangement of contacts is observed.

Examples of using this formula for protein modeling are shown in Appendix B.

Amino Acid Combinations and Fold/Structure Prediction

Amino acids that form hydrophobic core of helix-helix interfaces are important for the orientation of helices and overall fold of the molecule. Internal interface positions analogous to positions **a** and **d** in GCN4 leucine zipper nomenclature are recognized as major determinants of the core organization. Ankyrin repeat sequence adopts a fold of two antiparallel α-helices assembled in a superhelical structure as a contiguous spiral stack. Ankyrins and TIM-barrel proteins have identical and homologous combinations at positions **a** and **d** of parallel interfaces with negative interhelical angles which indicates common characteristics of their fold. Combinations at positions **a** and **d** of ankyrin antiparallel interfaces are homologous to those of the four-α-helix motif exhibiting positive interhelical angles. Location of characteristic amino acid combinations at the edges of helical surfaces results in a unique pattern of contacts formed by interacting helices of inner and outer row of ankyrin that distinguishes them from each other and from TIM-barrel proteins.

Combinations of Parallel Interfaces

Arrangement of ankyrin parallel α-helices is similar to the arrangement observed in TIM-barrel proteins. TIM-barrels are 8α/8β proteins with circular arrangement of secondary structure elements (Wierenga et al. 1991). This large group includes triosephosphate isomerase, pyruvate kinase, malate synthase, fructose-1,6-biphospate aldolase, xynalase and other proteins. Pairs of parallel α-helices in both ankyrin and TIM-barrel proteins exhibit negative interhelical angles. However, curvature direction of the helical row is opposite in the two proteins and correlates with distribution of contacts at the helical edges (Table 5.1). In addition, ankyrin has an antiparallel arrangement of α-helices carrying positive interhelical angles. Some of ankiryn repeat proteins such as gankyrin also contain a row of β-hairpins which resembles inner β-barrel of the TIM-barrel.

The determinants of helix–helix packing are described for the geometry of the interacting surfaces, interhelical angles, layered packing arrangements, amino acid patterns (Crick 1953; Efimov 1979; Gernert et al. 1995) and energetics of the regular structural elements such as four-α-helix, seven-helix, 8α/8β (Chou et al. 1991) and other types of associations. Helix-helix interactions play an important role in the protein three-dimensional structure (Carlacci et al. 1991), folding and stability of a protein, conductance of soluble and membrane protein channels (Schnell and Chou 2008), and helical assemblies (Kurochkina 2010). Amino acids located at core positions **a** and **d** of each α-helix contribute to specificity of helix-helix interactions (Kurochkina 2007, 2008). Distinct sequence combinations are present in leucine zipper coiled coil exhibiting positive interhelical angles and TIM barrel proteins with negative interhelical angles (Kurochkina 2008). Comparison of ankyrin and TIM-barrel proteins with respect to (a) amino acid combinations at helix-helix interfaces and (b) arrangement of helix-helix contacts on helix edges shows that these proteins have identical and homologous combinations at positions **a** and **d** of parallel interfaces with negative interhelical angles which indicates common characteristics of their fold. Combinations at positions **a** and **d** of antiparallel interfaces are homologous to those of the four-α-helix motif and exhibit positive interhelical angles (Table 8.1). Location of characteristic amino acid combinations at the edges of helical surfaces results in a unique pattern of contacts formed by interacting helices of inner and outer row of ankyrins that distinguishes them from each other and from TIM-barrel proteins (Table 5.1). These data contribute to understanding the relationship between amino acid sequence and structure of the protein and lead to the development of the reliable methods for tertiary structure prediction and design of new compounds.

Proteins, which contain the ankyrin repeat, TIM-barrel, and four-α-helix bundle motifs are listed in Table A.1.

Ankyrin repeat proteins contain repeated combinations of nonpolar amino acids at positions **a** and **d** of helix-helix interfaces (Table A.2). Polar amino acids (serine/threonine) at these positions are included if they exhibit the same atomic contacts as nonpolar amino acids. These combinations when compared with combinations

Table 8.1 Side chain rotamers of amino acids at positions **a, g, a', g'**. Amino acid combinations are of $\frac{AV}{LL}$ type in AnkyrinR and $\frac{LY}{LL}$ type in TRPVs

		a			a'		
Protein	PDB code	residue	χ_1	χ_2	residue	χ_1	χ_2
AnkyrinR	1n11	L408	101	-175	L423	146	−49
		L474	121	−164	L489	138	−60
		L507	147	−162	L522	113	−72
		L573	145	−164	L588	95	−60
		L639	149	−168	L654	95	−51
		L705	142	153	L720	73	−49
		L738	143	144	L753	66	−30
rTRPV2	2eta	L250	3	−162	L274	5	175
hTRPV2	2f37	L249	31	−116	L273	−26	150
TRPV4	3jxj	L311	48	−87	L335	−90	124
TRPV6	2rfa	L200	−85	140	L219	54	−82

previously found in a set of high resolution crystallographic and NMR structures of different motifs (*Kurochkina* 2008) show that similar combinations of aliphatic amino acids at positions **a** and **d** occur at helix/helix interfaces in parallel and antiparallel coiled coil, globular α-helical, TIM-barrel, αβ(including α/β, α + βand other) and other proteins.

Three-dimensional structure of ankyrin has common features with TIM-barrel proteins since each of the two rows of parallel α-helices exhibits negative interhelical angles (Fig. 1.23). It also carries characteristics of the four-α-helix motif as a pair of two-helical structures each characterized by positive interhelical angles of antiparallel interfaces and arranged parallel to each other. Significant differences between TIM barrel protein and ankyrin include curvature direction of the helical row, which is opposite in the two proteins, and circular versus spiral arrangement of α-helices. Combinations of the inner row and outer row of ankyrin repeat proteins are most similar to combinations of the TIM-barrel proteins. This finding is consistent with the fact that both proteins exhibit negative interhelical angles of parallel interfaces and short surface of the interface extending to one or two helical turns. The most conserved combination of amino acids contains proline or alanine (position 5 of the consensus sequence) of one helix and histidine or alanine (position 7) of another helix at positions **a** and **a'** combined with alanine (position 10) and valine/cysteine (position 8) at positions **d** and **d'** (Table A.2, Fig. 8.1). As a result, this combination $\frac{x\,\beta'}{\beta'\beta}$ belongs to group $\frac{\gamma\,\beta'}{\beta'\beta}$ if x is V/C/T or group $\frac{\delta\,\beta'}{\beta'\beta}$ if x is L/I. When this combination is viewed in the coordinate system of the layer (Kurochkina 2008) regarding interlayer contacts, it can be seen that proline and histidine direct Cβ atom toward the upper layer mimicking alanine. That explains substitutions of proline and histidine to alanine at these positions of the consensus sequence. Substitutions of proline for alanine are found for example in B-cell lymphoma 3- encoded protein,

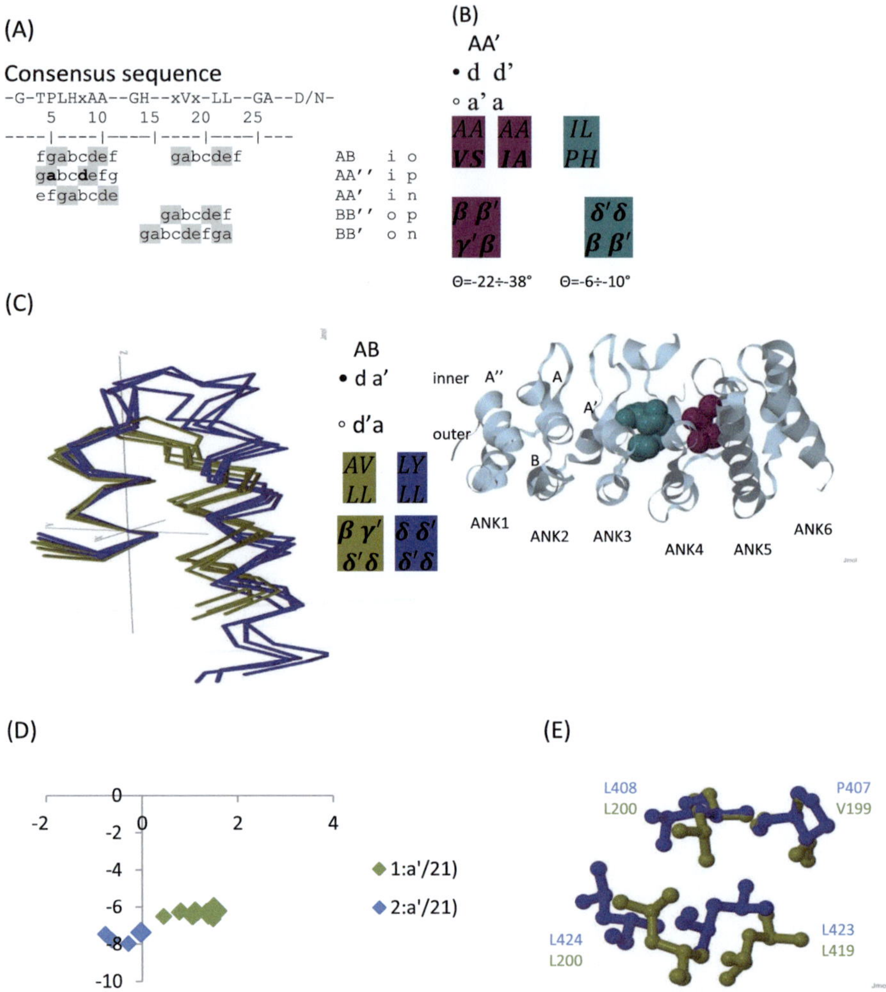

Fig. 8.1 Structure and specific amino acid combinations of ankyrin repeats. (**A**) Consensus sequence of a pair of antiparallel helices AB, and assigned positions **a** and **d** for interactions of inner row helix A (i) with preceding A' (p), antiparallel outer row B (o), and following A" (n) helices. Within one repeat, helices A and B form antiparallel interface involving amino acids at positions g, a, d, e (5, 6, 9, and 10) and positions g', a', d', e' (17, 18, 21, 22); x – hydrophobic residue. (**B**) TRPV1 Ankyrin repeats: two distinct combinations of amino acids at positions **a** and **d** of helix A and positions **a'** and **d'** of helix B, $\frac{AA}{VS}$ producing larger twist, and $\frac{IL}{PH}$ with smaller twist; Θ – interhelical angle. (**C**) α-carbon backbone of the ankyrin repeats superimposed at positions **a** of helices A in the coordinate system (*14*); clustering of the two distinct amino acid combinations of antiparallel interface given by positions **a** and **d** of helix A and positions **a'** and **d'** of helix B: $\frac{AV}{LL}$ (olive) and $\frac{LY}{LL}$ (blue) types. (**D**) Clustering of positions **a** of the two types of interfaces in the XY plane. (**E**) Side chain rotamers of amino acids at positions **a**, **g**, **a'**, **g'**; ankyrinR /1n11/ (blue) and TRPV6 /2rfa/ (olive) ankyrin repeats

PYK2-associated protein, and cyclin-dependent kinase 6 inhibitor.Combinations of this group are present in TIM barrel proteins. The combination is identical in several TIM barrel proteins: *Thermoproteus tenax* fructose-1,6-biphosphate aldolase, pyruvate kinases, and ankyrin repeat protein p18 cyclin-dependent kinase 6 inhibitor. In this combination, substitutions of histidine and proline to alanine are observed. Although both histidine and alanine are in good agreement at this position, histidines of ankyrin repeats of p18 were shown to be important for the stability of the molecule (Venkataramani et al. 2002). Combination of the group $\begin{smallmatrix}\delta\,\delta'\\ \beta'\beta\end{smallmatrix}$ is rare and occurs in transient receptor potential cation channels (TRPV). Isoleucine has (−, t) conformation, while leucine being at (t, +) conformation imitates alanine. Combination of the group $\begin{smallmatrix}\delta\,\gamma'\\ \beta'\beta\end{smallmatrix}$ is also rare and occurs in TRPV and NOTCH proteins. Combinations of the outer row have similarities with TIM-barrel, glutathione transferase and other proteins. Human AnkyrinR combinations are identical to combination found in *Arabidopsis thaliana* glutathione s-transferase zeta. Similar combinations are present in *Mus musculus* Notch1 protein, *Homo sapiens* pyruvate kinase, *Bacillus subtilis* L-Ala-D/L-Glu epimerase and glutathione transferases. Similarity of ankyrin combinations with *Mycobacterium tuberculosis* 2-isopropylmalate synthase, *Pseudomonas fluorescens* glutathione transferase A, and *Aeropyrum pernix* deoxyribose phosphate aldolase can be seen.

We can see that Phe can be equivalent to Ala or Leu. Aromatic residues can be accommodated at helix interface core positions, for example, in vimentin.

Combinations of antiparallel interfaces between inner row and outer row of ankyrin repeats have positive interhelical angles and are identical or similar to combinations of the four-helix bundle proteins such as TF-1 cell apoptosis-related protein 15 and annexin. All identities and similarities of the inner and outer rows of ankyrins are consistent with interhelical angles. Parallel interfaces that exhibit negative interhelical angles carry characteristic amino acid combinations, which distinguish them from interfaces with positive interhelical angles. Antiparallel interfaces exhibit positive interhelical angles and form several groups of combinations characteristic for this fold. The parallel combinations and groups of the inner row and outer row show symmetric arrangement of amino acids at positions **a**, **a**' but asymmetric arrangement of amino acids at positions **d**, **d**'. Therefore, they have the same type of helix-helix interface as TIM-barrel proteins and other proteins with negative interhelical angles such as alcohol dehydrogenase, phosphotriesterase, and isocitrate dehydrogenase. All groups of the inner row differ from groups of the outer row. As a result, they differ from each other by interhelical distance, smaller for the former and larger for the latter (Fig. 5.4). Most of the residues at positions **a** and **d** belongs to a consensus sequence and are important structural determinants.

Ankyrins have identical or similar amino acid combinations at positions **a** and **d** with TIM-barrel, four-α-helix, and other proteins. These combinations are associated with common features of their threedimensional structure. Furthemore, location of characteristic amino acid combinations at the edges of helical surfaces and consequently a pattern of contacts uniquely identifies particular protein fold.

Symmetric **a**, **a**' arrangements of amino acids combined with asymmetric **d**, **d**' arrangements that is a distinguished characteristic of TIM-barrel sequence is also present at parallel interfaces of ankyrin as well as glutathione s-transferase and other proteins with identical type of helix-helix interface. Similarities of sequence combinations that can be seen between ankyrin and four-α-helix motif reflect their common features such as interhelical angles and modular organization of homologous repeat units assembled as dimers of dimers. Most consensus residues of ankyrin repeats are located at helix-helix interfaces and fill positions **a** and **d** as well as **e** and **g**. The strong conservation of these residues suggests their role as providers of conserved structural features and uniform propagation of the repeat units. Amino acids at positions **a** and **d** and to some extent **e** and **g** of ankyrin are important for the conserved fold and stability of the molecule. The distinctive feature of the ankyrin repeats, the curvature of the repeat stack, is attributed to the relative side chain volumes of the core residues at positions 10, 17, and 18 (Michaely et al. 2002) that is consistent with the presented work showing that amino acid combinations at helix-helix interfaces and interhelical distances differ for the inner type and outer type of the interface. In addition, location of amino acid combinations at the particular edges of the helical surfaces is important for the direction and extent of curvature. Morphological features of each type of structural motif are consistent with similarities and differences in their amino acid composition at critically important core positions. This work also raises a question about similarities and differences in other repeat containing molecules. Multiple two-helical repeats contiguously stacked can be observed for instance in tetratricopeptide and armadillo proteins. Tetratricopeptide motif exhibits positive interhelical angles of parallel and antiparallel interfaces, whereas armadillo repeat carries both positive and negative interhelical angles of parallel interfaces combined with positive angles of antiparallel interfaces. Taken together, the results of this study provide the basis for future investigation of the key determinants of the structural characteristics for these proteins.

This approach to the investigation of the relationship between amino acid sequence and structure of the protein leads to the development of the reliable methods for tertiary structure prediction and design of new compounds. This problem has significant impact on many research areas such as protein structure, stability, mutagenesis, and protein-ligand interactions.

Fold Recognition

Helix-helix interfaces of helical assemblies carry specific amino acid combinations that determine fold of the molecule. Distribution of characteristic combinations at particular helical edges is also important. Based on these key factors, a new method is developed for the prediction of three-dimensional structure of helical assemblies. Parameters of each helix-helix interface that correspond to certain amino acid combination are used to generate coordinates of the two interacting α-helices, translate and rotate one helix relative to another and achieve the structure consistent with the interface type. This method was applied for structure prediction of ankyrin repeats (Table A.3) and demonstrated reproduced specific features of transient receptor proteins for not only interfaces that carry consensus sequence but also those that deviate from the pattern (Hall et al. 2018). Predicted models are in good agreement with crystallographic structures and exhibit root mean square deviation of α-carbon positions of the model from x-ray structure in the range of 0.4–2.8 Angstrom.

Helical assemblies, structural units of many proteins, play important roles in mediating protein-protein interactions and cell signaling (Andrade et al. 2001; Li et al. 2006; Teweri et al. 2010; Sawyer et al. 2013). Involvement in vital physiological processes including cell cycle control, transcription regulation, development, and sensory transduction, links these molecules to cancer, schizophrenia, Alzeimer's, Huntington and other diseases (Utreras et al. 2013; Lishko et al. 2007; Sanders et al. 2014; Sun et al. 2014; Zhang et al. 2014). Designed Ankyrin Repeat Proteins (DARPins) are promising drug candidates able to bind their targets with high affinity (Stumpp et al. 2008). Unique elastic properties of superhelical structures make them attractive scaffolds for bio-inspired design of nanomaterials (Lee et al. 2006). Structure of the helical assembly contains a pair of antiparallel helices stacked in a parallel fashion to form a spiral, helix of helices. Each type of the repeat - Ankyrin, Armadillo/HEAT, and tetratricopeptide – possesses characteristic properties and unique amino acid composition (Andrade et al. 2001; Li et al. 2006; Teweri, et al. 2010; Sawyer et al. 2013). Consensus sequences were derived for the repeats (Mosavi et al. 2002) although some repeats show deviations from the pattern (Gaudet 2008).

Amino acid combinations characteristic for each type of helix-helix interface correlate with the arrangement of α-helices in proteins (Kurochkina 2008, Kurochkina and T. Choekyi 2011). Location of specific combinations on the helical edges contributes to the shape of the assembly (Kurochkina 2010). Specific amino acid combinations of helix helix interfaces and their distribution at the helical edges are used to predict three-dimensional structure of the assemblies. The approach of treating amino acid sequence in three dimensions allows more accurate prediction of the repeats compared to the prediction using linear procedures. Parameters of each helix-helix interface that correspond to certain amino acid combination are used to generate coordinates of the two interacting α-helices, translate and rotate one helix relative to another and achieve the structure consistent with the interface type. This method was applied for the prediction of ankyrin repeats structure. Produced models are in good agreement with experimentally determined structures. There exists a large gap between number of three-dimensional structures and number of sequences determined for these proteins. New method provides tertiary structures of many proteins that are necessary to reveal mechanisms of action and design drugs, nanostructures and nanomaterials.

Helix-helix interface is formed by amino acids located mainly at the conserved core positions **a** and **d** and adjacent positions **e** and **g** in leucine zipper nomenclature (Hodges et al. 1972). In ankyrin repeat molecule, each inner row helix

(A) participates in antiparallel interface with outer row helix (B) and two adjacent parallel interfaces, one with preceding (A') and one with following (A") helices. We observe AB, AA', and AA" interfaces in the inner row and AB, BB' and BB" interfaces in the outer row (Fig. 8.1). Contacts between the neighboring outer row helices BB' and BB" are less tight, even absent and may reappear upon ligand binding or mechanical stress.

Assignment of positions **a**, **d**, **e**, and **g** to each helix-helix interface (Fig 8.1A) demonstrates contacts of the central helix with three surrounding helices that follow pattern repeated at each ankyrin repeat unit. TIM-barrel proteins utilize edges (i, i + 1) and (i + 5, i + 6) to contact preceding and following helices whereas 4-α-helix bundle proteins use (i, i + 1) and (i + 2, i + 3) edges (Kurochkina 2008).

Amino acid combinations at antiparallel AB interface cluster into several groups. Most frequent combinations are $\frac{AV}{LL}$ ($\frac{AI}{LL}$) and $\frac{AA}{LL}$. Combinations of the N- and C-terminal repeats show more variation. Distribution of amino acid combinations at 159 antiparallel interfaces of 27 proteins shows amazing conservation of $\frac{AV}{LL}$ type: 50% of all interfaces belong to this group (Fig. 8.2). Parallel interfaces AA' and AA" frequently contain $\frac{LA}{HP}$, $\frac{LV}{HP}$, and $\frac{VA}{AA}$. Method generates all possible amino acid combinations at all helix-helix pairs starting with predicted α-helices arranged as A, B, A', A" unit with contacts following pattern of contacts observed. Combinations are compared with frequently observed combinations and score based on similarity is calculated. The score is used to generate tertiary structure with predicted helix-helix interfaces. Structure of each helix-helix interface is calculated using average parameters of the interface type. If the predicted interface has one of the frequently observed combinations, structure coordinates are assigned according to the average position of amino acids at this interface. If the predicted interface has rare amino acid combination, coordinates are assigned according to the structure of the most similar combination. Parameters of each interface type include displacement of one helix relative to another that allow to place two interacting helices in relative orientation consistent with their interface type. Differences in parameters between the two types of antiparallel interfaces, for example, are seen for $\frac{AV}{LL}$ and $\frac{LY}{LL}$ combinations (Fig 8.1C, D). Positions **a'** of the two types are shown in the coordinate plane so that α-carbon of amino acid at position **a** is at the origin, peptide group between residue at position **a** and residue at position **g** preceding **a** is in XZ plane, vector from C_α at position **g** to C_β at position **a** is parallel to X axis, and negative end of the Y-axis points toward interacting helix (Kurochkina 2008). Matrix transformations are used to rotate/translate one of the two interacting helices to the origin of the coordinate system using α-carbon positions **a**, **d**, and **g**, and then to assign the second helix average coordinates of its interface type. This procedure treats two interacting α-helices as rigid bodies. Side chain rotamers reflect geometry of the particular amino acid combination (Table 8.1, Fig 8.1E) and are assigned using template coordinates of the two interacting helices. After coordinates of the interface are

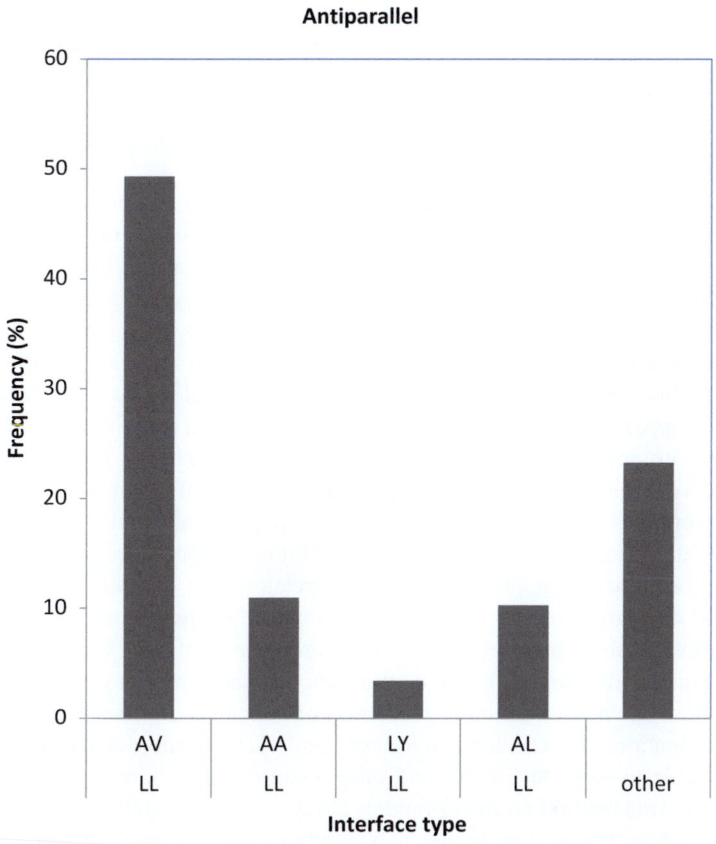

Fig. 8.2 Frequency of amino acid combinations at positions **a** and **d** of ankyrin repeats antiparallel interfaces

assigned to each pair of α-helices, energy minimization is used to adjust positions of all atoms.

The described method is applied for structure prediction of helical assemblies, such as ankyrin repeats proteins, and can be illustrated for a group of transient receptor potential vanilloid ankyrin domain (TRPV). TRPV channels, their pore structure, oligomerization, regulation, and other properties show similarities to voltage gated sodium/potassium channels (Long et al. 2007; Kedei et al. 2001). Structure of TRPV ankyrin domain comprises six repeats, ANK1–6. Repeats ANK1–4 follow canonical consensus sequence, whereas repeats ANK5–6 deviate. This deviation in amino acid sequence is reflected in change of canonical structure of the repeats. The twist angle of canonical repeats ANK1–4 is small, approximately 6°, and increases in ANK5–6 to 22–38° (Phelps et al. 2008). The twist angle between repeats (repeat AB and repeat A'B′, for instance) is a combination of two interhelical angles, one between inner row helices, such as A and A', and another between outer row helices, B and B′. Interhelical angles show good correlation with amino acids at positions **a** and **d** (Kurochkina 2008, Kurochkina and Choekyi 2011). The question is whether interhelical angles in TRPV channels also show correlation with amino acid sequence of the helix-helix interfaces. Indeed, amino acid combinations at the parallel interfaces of ANK1–4 are typical for ankyrin repeats, whereas combinations at ANK4–5 differ and cluster into two groups, $\frac{\beta\,\beta'}{\gamma'\beta}\left(\frac{AA}{VS}, \frac{AA}{IA}\right)$ and $\frac{\beta\,\delta'}{\gamma'\beta}\left(\frac{AI}{VS}\right)$ (Fig. 8.1C). Although ANK4–5 combinations differ from other ankyrin repeats combinations, they are identical to combinations found in several TIM-barrel proteins such as fructose 1, 6-biphosphate aldolase, pyruvate kinase, dihydrodipicolinate synthase, and isopropylmalate dehydrogenase, and exhibit close values of interhelical angles.

Similarities in amino acid composition of helix-helix interfaces exist in many proteins. What can we conclude when we observe identical amino acid combinations in leucine zipper and β-lactamase? The answer is that both proteins contain a pair of α-helices with positive interhelical angles. Another example is presence of identical amino acid combinations in ankyrin repeats and TIM barrel proteins. Although types of the assemblies differ in these structural motifs, row of parallel α-helices carry negative interhelical angles in both proteins. Not only sign of the angle, positive or negative, but also range of angle values correlates with composition. The larger angles for one group of combinations and smaller for another demonstrate differences in twist values between consecutive repeat units of TRPV molecules. Conserved amino acids adopt different conformational states in each type of amino acid combination. Method of structure prediction of helical assemblies, based on amino acid combinations at helix-helix interfaces and their location at particular helical edges, is able to reproduce these and other features in proteins containing helical assemblies. This method produces models close to experimentally determined structures that can be widely used in the analysis and design. Structure coordinates of the predicted molecules can serve as preliminary models of crystallographic, NMR, and electron microscopy studies or as their quick alternatives. Helical assemblies, from one pair of α-helices to multiple repeats, are essential components of large number of protein structural motifs including leucine zipper, ankyrin, ARM/HEAT, LRR, TIM-barrel, 4-helix bundles, transmembrane segments and Rossmann fold.

Specific amino acid combinations and their location at helical edges are key factors of the fold of helical assemblies.

Amino acid combinations Repeated combinations of nonpolar amino acids at positions **a** and **d** of helix-helix interfaces of proteins are identified. These combinations are compared with combinations previously found in a set of high resolution crystallographic and NMR structures of different fold containing helix-helix interfaces of parallel/antiparallel orientation and positive/negative interhelical angles (Kurochkina 2007, 2008). Repeated combinations of aliphatic amino acids at positions **a** and **d** of helix-helix interfaces occur in parallel and antiparallel coiled coil, globular α-helical, TIM-barrel, αβ (including α/β and α + β) and other proteins. Characteristic combinations of interfaces with positive and negative interhelical angles predict the type of interface (Kurochkina and Choekyi 2011).

References

Alderson TR, Charlier C, Torchia DA, Anfinrud P, Bax A (2017) Monitoring hydrogen exchange during protein folding by fast pressure jump NMR spectroscopy. J Am Chem Soc 139 (32):11036–11039. doi: https://doi.org/10.1021/jacs.7b06676. Epub 2017 Aug 7. PubMed PMID: 28766333; PubMed Central PMCID: PMC5586491

Bernstein FC, Koetzle TF, Williams GJB, Meyer EF Jr, Brice MD, Rodgers JR, Kennard O, Shimanouchi T, Tasumi M (1977) The protein data bank: a computer-based archival file for macromolecular structures. J Mol Biol 112:535

Charlier C, Alderson TR, Courtney JM, Ying J, Anfinrud P, Bax A (2018) Study of protein folding under native conditions by rapidly switching the hydrostatic pressure inside an NMR sample cell. Proc Natl Acad Sci USA 115(18):E4169–E4178. https://doi.org/10.1073/pnas.1803642115. Epub 2018 Apr 16. PubMed PMID: 29666248; PubMed Central PMCID: PMC5939115

Chou KC (2004) Review: Structural bioinformatics and its impact to biomedical science. Curr Med Chem 11:2105–2134

Collins FS, Gottlieb S (2018) The next phase of human gene-therapy oversight. N Engl J Med doi: 10.1056/NEJMp1810628. [Epub ahead of print] PubMed PMID: 30110242

Crick F (1953) Acta Crystallogr 6:689

Dao EH, Poitevin F, Sierra RG, Gati C, Rao Y, Ciftci HI, Aksit F, McGurk A, Obrinski T, Mgbam P, Hayes B, DE Lichtenberg C, Pardo-Avila F, Corsepius N, Zhang L, Seaberg MH, Hunter MS, Liang M, Koglin JE, Wakatsuki S, Demirci H (2018) Structure of the 30S ribosomal decoding complex at ambient temperature. RNA. pii: rna.067660.118. doi: 10.1261/rna.067660.118. [Epub ahead of print] PubMed PMID: 30139800

Efimov AV (1979) Packing of alpha-helices in globular proteins. Layer-structure of globin hydrophobic cores. J Mol Biol 134(1):23–40

Gaudet R (2008) A primer on ankyrin repeat function in TRP channels and beyond. Mol Biosyst 4:372–379

Gernert KM, Surles MC, Labean TH, Richardson JS, Richardson DC (1995) Alacoil: a very tight, antiparallel coiled-coil of helices. Protein Science 4:2252–2260

Halazonetis TD, Gorgoulis VG, Bartek J (2008) An oncogene-induced DNA damage model for cancer development. Science 319:1352–1355

Hall BE, Prochazkova M, Sapio MR, Minetos P, Kurochkina N, Binukumar BK, Amin ND, Terse A, Joseph J, Raithel SJ, Mannes AJ, Pant HC, Chung MK, Iadarola MJ, Kulkarni AB (2018) Phosphorylation of the Transient Receptor Potential Ankyrin 1 by Cyclin-dependent Kinase 5 affects Chemo-nociception. Sci Rep 8(1):1177. https://doi.org/10.1038/s41598-018-19532-6. PubMed PMID: 29352128; PubMed Central PMCID: PMC5775258

Hodges RS, Sodek J, Smillie LB et al (1972) Amino-acid sequence of rabbit skeletal tropomyosin and its coiled-coil structure. Proc Natl Acad Sci 69:3800–3804

Igarashi M, Nozumi M, Wu LG, Cella Zanacchi F, Katona I, Barna L, Xu P, Zhang M, Xue F, Boyden E (2018) New observations in neuroscience using superresolution microscopy. J Neurosci 38(44):9459–9467. https://doi.org/10.1523/JNEUROSCI.1678-18.2018. PubMed PMID: 30381437

Kedei N, Szabo T, Lile JD, Treanor JJ, Olah Z, Iadarola MJ, Blumberg PM (2001) Analysis of the native quaternary structure of vanilloid receptor 1. J Biol Chem 276:28613–28619

Khoshouei M, Radjainia M, Baumeister W, Danev R (2017) Cryo-EM structure of haemoglobin at 3.2 Å determined with the volta phase plate. Nat Commun 8:16099. https://doi.org/10.1038/ncomms16099

Kurochkina N (2007) J Theor Biol 247:110–121

Kurochkina N (2008) J Theor Biol 255:188–198

Kurochkina N, Choekyi T (2011) Helix-helix interfaces and ligand binding. J Theor Biol 283:92–102
Lee G, Abdi K, Jiang Y, Michaely P, Bennett V, Marszalek PE (2006) Nanospring behaviour of ankyrin repeats. Nature 440:246–249
Lishko PV, Procko E, Jin X, Phelps CB, Gaudet R (2007) The ankyrin repeats of TRPV1 bind multiple ligands and modulate channel sensitivity. Neuron 54:905–918
Lomelino CL, Andring JT, McKenna R (2018) Crystallography and Its Impact on Carbonic Anhydrase Research. Int J Med Chem 2018:9419521. https://doi.org/10.1155/2018/9419521. eCollection 2018. Review. PubMed PMID: 30302289; PubMed Central PMCID: PMC6158936
Long SB, Tao X, Campbell EB, MacKinnon R (2007) Atomic structure of a voltage-dependent K+ channel in a lipid membrane-like environment. Nature 450:376–382
Martin RW, Kelly JE, Kelz JI. Advances in instrumentation and methodology for solid-state NMR of biological assemblies. J Struct Biol. 2018. pii: S1047-8477(18)30253–30253. https://doi.org/10.1016/j.jsb.2018.09.003. [Epub ahead of print] PubMed PMID: 30205196
Michaely P, Tomchick DR, Machius M, Anderson RGW (2002) Crystal structure of a 12 ANK repeat stack from human ankyrinR. EMBO J 21:6387–6396
Mosavi LK, Minor DL, Peng Z (2002) Consensus-derived structural determinants of the ankyrin repeat motif. Proc Natl Acad Sci 99:16031–16034
Phelps CB, Huang RJ, Lishko PV, Wang RR, Gaudet R (2008) Structural analyses of the Ankyrin Repeat Domain of TRPV6 and related TRPV ion channels. Biochemistry 47:2476–2484
Sanders SS, Mui KKN, Sutton LM, Hayden MR (2014) Identification of binding sites in Huntingtin for the Huntingtin Interacting Proteins HIP14 and HIP14L. PlosOne 28:e90669
Sawyer N, Chen J, Regan L (2013) All repeats are not equal: a module-based approach to guide repeat protein design. J Mol Biol 425:1826–1838
Schnell JR, Chou JJ (2008) Structure and mechanism of the M2 proton channel of influenza A virus. Nature 451:591–595
Schulz GE, Schirmer RH (1982) Principles of protein structure. Ed: Cantor, C. R.,
Streckfus CF, Arreola D, Edwards C, Bigler L (2012) Salivary protein profiles among HER2/neu-receptor-positive and -negative breast cancer patients: support for using salivary protein profiles for modeling breast cancer progression. J Oncol 2012:413256. https://doi.org/10.1155/2012/413256. Epub 2012 Apr 10. PubMed PMID: 22570650; PubMed Central PMCID: PMC3335259
Stumpp MT, Binz HK, Amstutz P (2008) DARPins: a new generation of protein therapeutics. Drug Discov Today 13:695–701
Teweri R, E Bailes, KA Bunting, JC Coates (2010) Trends Cell. Biol 20:470
Thomas SE, Mendes V, Kim SY, Malhotra S, Ochoa-Montaño B, Blaszczyk M, Blundell TL (2017) Structural biology and the design of new therapeutics: from HIV and cancer to mycobacterial infections: a paper dedicated to John Kendrew. J Mol Biol 2017 Jun 23. pii: S0022-2836(17)30315–7. https://doi.org/10.1016/j.jmb.2017.06.014. [Epub ahead of print]
Utreras E, Prochaskova M, Terse A, Gross J, Keller J, Iadarola MJ, Kulkarni AB (2013) TGF-β1 sensitizes TRPV1 through Cdk5 signaling in odontoblast-like cells. Mol Pain 9:24–38
Valm AM, Cohen S, Legant WR, Melunis J, Hershberg U, Wait E, Cohen AR, Davidson MW, Betzig E, Lippincott-Schwartz J (2017) Applying systems-level spectral imaging and analysis to reveal the organelle interactome. Nature 546(7656):162–167. https://doi.org/10.1038/nature22369. Epub 2017 May 24. PubMed PMID: 28538724; PubMed Central PMCID: PMC5536967
Venkataramani RN, MacLachlan TK, Chai X, El-Deiry WS, Marmorstein R (2002) Structure-based Design of p18INK4cProteins with Increased Thermodynamic Stability and Cell Cycle Inhibitory Activity. J Biol Chem 277:48827–48833

Wierenga RK, Noble MEM, Vriend G, Naughe S, Hol WGJ (1991) Refined 1.83 Angstrom structure of trypanosomal triosephospate isomerase, crystallized in presence of 2.4M- Ammonium sulphate. A comparison with the structure of the trypanosomal triosephosphate isomerase-glycerol-3-phosphate complex. J Mol Biol 220:995–1015

Zhang C, Caj J, Zhang J, Li Z, Guo Z, Zhang X, Lu W, Zhang Y, Yuan A, Yu S, Fang Y (2014) Pharacogenetic effects of dopamine transporter gene polymorphisms on response to chlorpromazine and clozapine and on extrapyramidal syndrome in schizophrenia. Prog Neuropsychopharmacol Biol Physciatry 50:110

Appendices

Appendix A

Mathematical Formulas

Interhelical angles Angle between two α-helices (Fig. 1.14) is defined as an angle between their axes (Kurochkina 2008) if distant helix rotates relative to proximal helix clockwise (negative direction) or counterclockwise (positive direction). Viewer sees proximal helix in front of the distant helix. The angle values agree with those calculated by Chothia and coworkers (Chothia et al. 1981). Interhelical angle α is calculated as an angle between vector X (x_1, y_1, z_1) and vector Y (y_1, y_2, y_3):

$$\alpha = \cos^{-1}(x_1x_2 + y_1y_2 + z_1z_2)/\left(\sqrt{x_1^2 + y_1^2 + z_1^2} * \sqrt{x_2^2 + y_2^2 + z_2^2}\right)$$

X is the vector via ends of the first helix connecting centers of mass of the first four residues the last four residues of the helix,

Y is the vector via ends of the second helix.

Angles between the edges of the helix surface Two consecutive amino acids, amino acid i and amino acid i+1, form (i, i+1) edge on the surface of the helix (Fig. 5.6). Angle between any two edges, (i, i+1) and (i+1, i+2), (i+1, i+2) and (i+2, i +3), and so on, is calculated. The angle θ between the two edges, (i, i+1) and (i+1, i +2), for example, designated as vectors $v1$ (x_1, y_1, z_1) and $v2$ (x_2, y_2, z_2) correspondingly, is calculated as

$$\theta = \cos^{-1}(x_1x_2 + y_1y_2)/(|v1||v2|)$$

where

N. Kurochkina, *Protein Structure and Modeling*,
https://doi.org/10.1007/978-981-13-6601-7

$$| v_1 |= \sqrt{x_1^2 + y_1^2 + z_1^2}$$

and

$$| v2 |= \sqrt{x_2^2 + y_2^2 + z_2^2}$$

Angles between the planes containing edges of the helix surface Two consecutive amino acids, amino acid i and amino acid i+1, form (i, i+1) edge on the surface of the helix (Fig. 5.6). Peptide group atoms $C_\alpha^i, C^i, \ N^{i+1}, \ C_\alpha^{i+1}, \ C_\beta^{i+1}$ are in one plane and include the edge (i, i+1). Therefore, we can write an equation for this plane:

$$\begin{vmatrix} x & y & z \\ A_1 & A_2 & A_3 \\ B_1 & B_2 & B_3 \end{vmatrix} = 0,$$

where

A_1, A_2, and A_3 are components of the vector $C_\alpha^i - C_\alpha^{i+1}$,

B_1, B_2, and B_3 are components of the vector $C_\beta^{i+1} - C_\alpha^{i+1}$, and $\begin{vmatrix} A_2 & A_3 \\ B_2 & B_3 \end{vmatrix}, \begin{vmatrix} A_1 & A_3 \\ B_1 & B_3 \end{vmatrix}, and \begin{vmatrix} A_2 & A_3 \\ B_2 & B_3 \end{vmatrix}$ are components of the normal vector N to this plane.

An angle between normal vectors N_1 (x_1, y_1, z_1) and N_2 (x_2, y_2, z_2) is the angle θ between the two planes, one containing edge (i, i+1), another containing edge (i+1, i +2), is calculated as:

$$\theta = \cos^{-1}(x_1x_2 + y_1y_2 + z_1z_2) \ / \ \left(\sqrt{x_1^2 + y_1^2 + z_1^2} * \sqrt{x_2^2 + y_2^2 + z_2^2}\right)$$

Similarly, angles for other pairs of planes of the helix surface are calculated.

Interior angles of the n-sided polygon $\theta_{ideal} = 180°$~*(n-2)/n.

The amount of area of the amino acid covered by contact with another helix Interface area change is the amount of surface area that amino acid buries in contact with another α-helix. It is calculated as

$$\Delta A = (A_i - A_c) * 100\% / A_i$$

where

A_i is the total solvent accessible area of the amino acid in standalone α-helix, i. e. single α-helix transferred from protein interior into solvent;

A_c is the solvent accessible area of amino acid in contact with another α-helix.

Difference A_i-A_c is the amount of surface area covered by contact due to association with another α-helix for each amino acid.

Coordinate system of the peptide group between two residues i and i+1 (Kurochkina N. 2008) is used for comparison of the helix interfaces. The cross-section of the helix (Fig. 1.23) is in XY plane whereas helix axis goes along the Z axis.

Resources

Molecular graphics and matrix transformation operations were performed with JMOL, an open-source Java viewer for chemical structures in 3D (http://www.jmol.org).

Protein data set. Protein crystallographic structures are used from the Protein Data Bank (PDB) (Table A.1) (Bernstein et al. 1977).

Helix-helix interface positions Helix positions in leucine zipper nomenclature are named by letters **a-g**. At helix-helix interface, core is occupied mainly by positions **a** and **d**, and amino acids at these two positions bury the largest amount of surface area. Flanking positions **e** and **g** contain polar residues and bury smaller amount of surface area (O'Shea et al. 1991; Krylov et al. 1994).

Similarly, the assignment of interface positions is used for each pair of interacting helices including parallel and antiparallel arrangements. Heptad repeat is subdivided into "3" and "4" types of pockets. If helix-helix interface is less than one turn of the helix, it is formed either by "3" or "4" type of pocket making ½ turn of a helix. Assignment of positions to helix interfaces is done by (1) calculation of helix-helix contacts, (2) estimation of accessible surface area change of each amino acid at helix-helix interface, and (3) visual inspection (Table A.2).

Table A.1 List of proteins and protein data bank entries for crystallographic/NMR structures

Protein	Source	PDB code
Ankyrin repeat prediction		
1Ankyrin-R (D34 region)	*H. sapiens*	1n11
2Gankyrin	*H. sapiens*	1uoh
3Gankyrin	*S. cerevisiae*	1ixv, 1wg0
[93]Gankyrin	*M. musculus*	2dvw 2dwz
4PYK2-associated protein β	*H. sapiens*	1nfi
5Cyclin-dependent kinase 6 inhibitor	*H. sapiens*	1mx2, 1ihb
6Cyclin-dependent kinase 4 inhibitor B p15	*M. musculus*	1d9s
7Ankyrin domain of GABPβ1	*M. musculus*	1awc
8Notch 1 ankyrin domain	*H. sapiens*	1yyh, 2f8y

(continued)

Table A.1 (continued)

Protein	Source	PDB code
9B-cell lymphoma 3-encoded protein	*H. sapiens*	1k1a
10Ser/Thr protein phosphatase pp1-β	*G. gallus*	1s70
11PYK2-associated protein β	*M. musculus*	1dcq
12Notch 1 protein	*M. musculus*	2qc9
13Notch receptor	*D. melanogaster*	1ot8
14Consensus ankyrin repeat protein SANK	*Designed*	1mj0
	M. musculus	2v4h
151D5 DARPIN	*H. sapiens*	2xai
16Ankyrin repeat and SOCS box protein 9	*L. pneumophila*	2aja
17Q5ZSV0	*R. norvegicus*	
18	*C. parvum*	2w4s
19RNA-binding domain	*R. norvegicus*	2nyj, 2pnn
20TRPV1	*R. norvegicus*	2eta, 2etb
21TRPV2	*H. sapiens*	2f37
22TRPV2	*G. gallus*	3jxj
23TRPV4	*M. musculus*	2rfa
24TRPV6	*H. sapiens*	6bo8
25 TRPV6	*H. sapiens*	3lvq
26ARF-ARFGAP complex	*V. virus*	3kea
27Host range protein K1	*T. volcanium*	2rfm
28TV1425	*H. sapiens*	3b95, 3b7b
29G9A/GLP histone methyltransferase	*H. sapiens*	3f6q
30Integrin-linked protein kinase	*H. sapiens*	3c5r
31BARD1	*H. sapiens*	3eu9
32Huntingtin-interacting protein 14	*L. major*	3ljn
33Ankyrin repeat protein	*H. sapiens*	3d9h
34ASB9	*A. thaliana*	3deo
35CPSRP43	*S. cereviciae*	1sw6
[90] Regulatory protein SWI6	*R. norvegicus*	1myo
[91] Myotrophin	*H. sapiens*	2b0o
[92] UPLC1 gap domain	*E. coli*	2l6b
[94] NR1C	*H. sapiens*	2vge
[95] IASPP	*H. sapiens*	2xzd
[96] DARPIN	*H. sapiens*	3ehq 3ehr
[97] Osteoclast stimulating factor 1	*R. norvegicus*	
[109]Osteoclast stimulating factor 1	*H. sapiens*	1dc2
[168]Cyclin-dependent kinase 4 inhibitor A p16		
Tetratricorepeat		
36Protein Phosphatase 5	*17Q5ZSV0*	17Q5ZSV0
37Mitotic checkpoint Ser/Thr-protein	*H. sapiens*	2wvi
38RAS-related c3 botulinum toxin substrate	*H. sapiens*	1e96
39Lipoprotein NLPI	*E. coli*	1xnf

(continued)

Table A.1 (continued)

Protein	Source	PDB code
40PEX5	*H. sapiens*	1fch
41Cyclophilin 40	*B. taurus*	1ihg, 1iip
42Prolyl 4-hydroxylase alpha-1 subunit	*H. sapiens*	1tjc
43Neutrophil cytosol factor 2	*H. sapiens*	1wm5
44PDCD6-interacting protein	*H. sapiens*	2oev
45Stress-induced-phosphoprotein 1	*H. sapiens*	3esk, 3fwv
46Epsilon-coat protein	*B. taurus*	3mkr
[98]Type 4 fimbrial biogenesis protein PILF	*P. aeruginosa*	2fi7
[100]STRAP	*H. sapiens*	2xvs
[101]STRAP	*M. musculus*	4abn
[102]Mitochondrial fission protein 1	*S. cereviciae*	3o48
[108]Formate-dependent nitrite reductase	*E. coli*	2e2e
[110]70 KDa peptidylprolyl isomerase	*P.falciparum*	2fbn
[111]EMBR	*M. tuberculosis*	2fez 2ff4
[123]Suppressor of g2 allele of SKP1 homolog	*H. sapiens*	1rl1
[124] P67PHOX	*H. sapiens*	1wm5
[125]Mitochondrial import receptor subunit Tom20-3	*A. thaliana*	1zu2
[126]Smooth muscle cell associated protein-1	*H. sapiens*	2dba
[127]BF1650	*B. fragilis*	2kc7
[128]MRR121A	*M. maripaludis*	2kck 2kcl
[129]HOP	*H. sapiens*	1elr
[130]SRR115C	*S. ruber*	2kcv?3ma5
[135]Transcriptional regulatory protein COMA?	*B. subtilis*	3ulq 3q15
[137]TAH1	*S. cereviciae*	2l6j
[138]Small glutamine-rich protein 2	*S. cereviciae*	2lxb 2lxc
[139]Secretion translocator chaperone SYCD	*Y. enterocolitica*	2vgx
[140] SGTA protein	*H. sapiens*	2vyi
[141]MAMA AMB-1 C2221	*M. magnetotacticum*	3as4
[142]ALGK	*P. fluorescens*	3e4b
[143]Anaphase-promoting complex subunit 7	*H. sapiens*	3ffl
[144]Chaperone protein IpgC	*S. flexneri*	3gz1 3gyz
[145]WD and tetratricopeptide repeats protein 1	*H. sapiens*	3i7n
[146]Possible protein of nuclear scaffold	*E. cuniculi*	3kae
[147] Programmed cell death protein 10	*H. sapiens*	3l8j
[148] alpha/epsilon-COP?	*B. Taurus*	3mkr
[149]NRO1/ETT1	*S. pombe*	3qtm
[150]G-protein-signaling modulator 2	*H. sapiens*	3sf4
[151]O-GlcNAc transferase OGT	*H. sapiens*	1w3b, 4ay5
[156]TOM70	*S. cereviciae*	2gw1
[165]Cysteine-rich protein	*H. pylori*	1ouv
Pentatricopeptide		
AT2G32230	*A. thaliana*	4g23

(continued)

Table A.1 (continued)

Protein	Source	PDB code
Armadillo/HEAT repeat		
[47]Importin alpha 2	*M. musculus*	1iq1
[48]FANCF	*H. sapiens*	2iqc
[105]β-catenin	*H. sapiens*	1g3j 1qz7
[106]β-catenin 2bct	*M. musculus*	1i7w 1i7x
[99]Adenomatous polyposis coli protein	*H. sapiens*	3au3
[103] Importin alpha	*S. cereviciae*	1un0 1bk5
[104]ENTH domain of rat epsin 1	*R. norvegicus*	1edu
[105]Inositol 1,4,5-trisphosphate receptor type 1	*M. musculus*	1n4k
[106] Plakophilin 1	*H. sapiens*	1xm9
[107]MDIA1	*M. musculus*	2bap, 1z2c
[112]RCD-1	*H. sapiens*	2fv2
[113]Diaphanous protein homolog 1	*M. musculus*	2f31
[114]General vesicular transport factor P115	*H. sapiens*	2w3c
[115]SYS-1 protein	*C. elegans*	3c2g
[116] Phospholipase A2-activating protein	*H. sapiens*	3ebb
[117] Protein diaphanous homolog 1?	*M. musculus*	3eg5
[118]PUL	*S. cereviciae*	3gae, 3l3f
[119]General vesicular transport factor P115	*B. taurus*	3grl
[120]Flagellar motor switch protein FLIG	*A. aeolicus*	3hjl
[121]δ-catenin	*H. sapiens*	3l6x, 3l6y
[122]Ubiquitin conjugation factor E4	*S. cereviciae*	3m62
[131]UNC-45	*D. melanogaster*	3now
[132]Nucleotide exchange factor SIL1	*S. cereviciae*	3qml
[133]Inositol 1,4,5-trisphosphate receptor type 1	*R. norvegicus*	3t8s
[134]Apoptosis inhibitor 5	*H. sapiens*	3u0r
[136]Diaphanous protein	*E. Histolytica*	4dvg
[153]Importin beta	*M. musculus*	1gcj 1qgk
[154]Importin 13	*H. sapiens*	2xwu 2x19
[155]PP2A	*H. sapiens*	1b3u
[169]TOM1	*H. sapiens*	1elk
LRR		
[157]Internalin A	*L. monocytogenes*	2omt 1h6t
[158]Ribonuclease inhibitor	*B. taurus*	1dfj 2bnh
[159]PGIP	*P. vulgaris*	1ogq
[160] Internalin B	*L. monocytogenes*	1d0b
[161] Outer Arm dynein	*C. reinhardtii*	1ds9
[162] Nuclear RNA export factor 1	*H. sapiens*	1fo1 1ft8
[163] Leucine rich repeat variant	*A. vinelandii*	1lrv
[164] Outer arm dynein light chain 1	*C. reinhardtii*	1m9l
[166]Ribonuclease inhibitor	*H. sapiens*	1z7x
[167]RAN GTPASE-activating protein 1	*S. pombe*	2ca6

(continued)

Table A.1 (continued)

Protein	Source	PDB code
Four-α-helix bundle		
50Myohemerythrin	*T. zostericola*	2mhr
51Hemerythrin	*Thermiste discrita*	2hmq
52Coat protein	Tobacco mosaic virus	2tmv, 1ei7
53TF-1 cell apoptosis-related protein 15	*H. sapiens*	3ajm, 3l8j
54Annexin	*H. sapiens*	1axn
55Apoferritin	*E. caballus*	1hrs
56Ferritin	*H. sapiens*	1fha
TIM-barrel		
57Triose phosphate isomerase	*T. brucei*	5tim
58 :	*L. mexicana*	1n55
59 :	*P.falciparum*	1o5x
60 :	*S. cereviciae*	7tim
61Fructose-1,6-biphosphate aldolase	*T. aquaticus*	1rvg
62 :	*D. melanogaster*	1fba
63 :	*H. sapiens*	1ald
64 :	*T. tenax*	1ojx
65 :	*E. coli*	2coa
66 :	*O. cuniculus*	3vb4, 3dft
67Enolase	*T. brucei*	1oep
68 :	*S. cereviciae*	3enl
69 :	*H. sapiens*	1te6
70Malate synthase	*E. coli*	1p7t
71 :	*M. tuberculosis*	1n8w
72Pyruvate kinase	*O. cuniculus*	1pkn
73 :	*H. sapiens*	1liu
74Xylanase inhibitor protein I	*T. aestivum*	1te1, 1om0
75Xylanase	*A. nidulans*	1hb2, 2ivj
76Xylanase	*P. simplicissimum*	1bg4
77Endoglucanase	*T. maritime*	3mmu
78Haementhin	*S. multiflorus*	3d5h
79Chromate reductase	*T. scotoductus*	3hgj
80L-Ala-D/L-Glu epimerase	*B. subtilis*	1tkk
812-Isopropylmalate synthase	*M. tuberculosis*	1sr9
82Deoxyribose phosphate aldolase	*A. pernix*	1n7k
83Deoxyribose phosphate aldolase	*E. coli*	1p1x
84D-Hydantoinase	*Thermus sp*	1gkp
Glutathione s-transferase		
85Glutathione s-transferase zeta	*Arabidopsis thaliana*	1e6b
86Glutathione s-transferase phi	*Arabidopsis thaliana*	1gnw
87Glutathione s-transferase GTT2	*S. cereviciae*	3ibh
88Glutathione s-transferase A	*P. fluorescens*	3lxt
89Glutathione s-transferase zeta 1-1	*M. musculus*	2cz2

Table A.2 Combinations at positions a and d of ankyrin repeats that are identical or similar to those of other proteins. Numbers to the right of the amino acid combination show protein as listed in Table A.1. Negative angle is given as (–)

Layer	Atomic groups	Amino acids at positions **a** and **d**	Angle	Row
"3" part Parallel • d d' ◦ a' a Ank	$\gamma\ \beta'$ $\beta'\ \beta$	VA[5,64,72,73] VA[1] TA[84] CA[1] AC[10] AA HP AA HP HP	-	Inner
	$\delta\ \beta'$ $\beta'\ \beta$	IA[1,10,29] LA[1,2,5,6,8,12−16,34,93] LA[5,35] IA[14,24] HP HP HA HA	-	inner
	$\delta\ \delta'$ $\beta'\ \beta$	LI[20−22,24,33] HP		
	$\delta\ \gamma'$ $\beta'\ \beta$	LV[8,9,12,13,20,21,34] IV[1,3] LC[33] IC[2,11,93] HP HP HP HP LV[9] HA IL[161] IL MS[1] HP VC[2,93] VC[9] HP HA VL[3] FA WA[2,93] YA[2,30,31,93] RA[2,93] WC[3] LC[3] RV[3] CA HP HA HP HC HP HV[9] HP SV[9] IA		
"3" part Parallel • d d' ◦ a' a ARM		LP[103] FT[103] AI AL		outer
"3" part Parallel • d d' ◦ a' a ARM		AN[103] AN[103] CY[103] AT TG AS		inner
"4" part Parallel ◦ a' a • d d' LRR		LA[159] LA[157] LS[157] PA LV IS LA[157] LS[157] IP IP LG[157] IP		

(continued)

Table A.2 (continued)

"4" part Parallel ∘ a' a • d d' LRR		LA^{159} $LA^{157,161}$ LS^{157} PA LV IS LA^{157} LS^{157} IP IP LG^{157} IT		
"3" part Anti- Parallel • d a' ∘ d' a	$\beta\ \delta'$ $\delta'\gamma$	AL^{55} AL^{5} AL^{55} LV LI II		
	$\beta\ \gamma'$ $\delta'\delta$	$AV^{1-4,15,20,21,24,27-31,34,35,90,91,93,95-97}$ LL $AI^{,3,10,21,22,28,31,35}$ AT^{1} AS^{3} LL LL LL $AV^{?27,33,94}$ AV^{93} $AI^{2,?4}$ LY LI LM LM LL		
	$\delta\ \gamma'$ $\gamma'\delta$	LV^{106} LI^{153} IM^{153} LV^{153} IL II VL IM		
	$\beta\ \delta'$ $\delta'\delta$	$AL^{92,24}$ LL		
	$\delta\beta'$ $\delta'\delta$	LA^{155} LL		
	$\beta\ \beta'$ $\delta'\delta$	$AA^{1,2,3,8,10,24,28,29,30,33,93}$ $AS^{3,103?}$ LL LL		
	$\gamma\ \beta'$ $\delta'\delta$	VA^{1} LL		
	$\delta\ \gamma'$ $\delta'\delta$	$IV^{103,47}$ $LI^{103,106}$ LV^{153} II^{155} II^{47} II^{47} LL LL LL LL FL YL		
	$\delta\ \gamma'$ $\delta'\gamma$	$LV^{47?}$ LI^{47} LI IV		
	$\gamma\ \gamma'$ $\delta'\delta$	VV^{103} FL		
	$\delta\ \gamma'$ $\delta'\gamma$	LI^{47} LI^{47} YV II		
	$\delta\ \delta'$ $\delta'\delta$	IL^{103} LL^{106} LM ML		
	$\gamma\ \delta'$ $\delta'\delta$	SL^{153} IL		

(continued)

Table A.2 (continued)

		AV^{91} TW^{153} YF VF AD^{92} IL AK^{3} FE		
"4" part Anti- Parallel ∘ d a' • d' a		FL^{47} FL^{47} SL AL		

Ankyrin consensus sequences
 	
`123456789012345678901234567890`
`-G-TPLHψAA--GH--ψV-ψLL--GA--DIN--- /1n11/`
`H1 H2`
`NGRTPLHψAARNGHLEψVKLLLEADVNAKDK--- /1noq/`
`-G-TPLHψA---G---ψV-ψLL--GA--------/2eta 2f37/`				

Table A.3 Predicted TRPA1 combinations

Layer	Atomic groups	Amino acids at positions **a** and **d**	Angle	Row
"3" part Parallel • d d' ∘ a' a	$\gamma\,\beta'$ $\beta'\beta$	$VA^{5,64,72,73}$ VA^{1} TA^{84} CA^{1} AC^{10} AA HP AA HP HP	-	Inner
	$\delta\,\beta'$ $\beta'\beta$	$IA^{1,10,29}$ $LA^{1,2,5,6,8,12-16,34}$ $LA^{5,35}$ $IA^{14,24}$ HP HP HA HA	-	inner
	$\delta\,\delta'$ $\beta'\beta$	$LI^{20-22,24,33}$ HP		
	$\delta\,\gamma'$ $\beta'\beta$	$LV^{8,9,12,13,20,21,34}$ $IV^{1,3}$ LC^{33} $IC^{2,11}$ HP HP HP HP		
		WA^{Ink2-1} $IA^{Ank2-3,}$ SV^{Ank3-4} $LA^{Ank5-6,}$ HP HP HA HP		

(continued)

Table A.3 (continued)

	$\delta\ \beta'$ $\gamma'\ \gamma$	LA [73] IA [80] LA [1,85] IA [86,87] IA [8,24] IA [12] VV VV VI VT SV SV	-	outer
	$\gamma\ \beta'$ $\gamma'\ \gamma$	VA [1] VA [12] VA [81] SA [82] TT VT TV VV	-	
	$\gamma\ \delta'$ $\gamma'\ \gamma$	VL [1,6] VV	-	
	$\beta\ \delta'$ $\varepsilon'\ \gamma$	AL [1,89] AL [1] MV MT	-	
	$\gamma\ \gamma'$ $\gamma'\ \beta$	VV [1] VV [73] TC [83] VA IA IA	-	
"3" part Anti-Parallel • d a' ◦ d' a	$\beta\ \delta'$ $\delta'\gamma$	AL [5] LI		
	$\beta\ \gamma'$ $\delta'\delta$	AV [1,2,4,15,20,21,24,27–31,34,35,90,91,95–97] AV [33,94] LL LI AI [1,10,21,22,28,31,35] LL	+	
		AV [Ank2,Ank9,Ank10,Ank14] AI [Ank3,Ank7] TL [Ank4] LL LL LL AC [Ank5] AD [Ank11] AM [Ank13] LV LL IL		

Appendix B

For a pair of interacting α-helices I and J, there are several possible ways to arrange amino acids of one helix to interact with amino acids of another helix. Interaction of two α-helices results in contacts between contiguous pockets situated in two neighboring rows where two pockets in one row are separated by seven amino acid residues. If we consider parallel packing of helices, pocket (i, i+3, i+4, i+7) contains side chain j+3, whereas pocket (j, j+3, j+4, j+7) contains side chain i+3. In the next interacting pair (i+4, i+7, i+8, i+11) and (j+4, j+7, j+8, j+11), pockets exchange side chains of residues i+7 and j+7. After that, interaction pattern is repeated starting from amino acid i+7 for helix I or amino acid j+7 for helix J. Interactions in one repeat can be described by the following formula for parallel (B.1) and anti-parallel (B.2) types of helix-helix interaction:

$$\begin{array}{llll} j+4+7n & i+3+7n & j+3+7n & i+4+7n \\ i-1+7n & j+7n & i+7n & j-1+7n \end{array} \quad (B.1)$$

$$\begin{array}{llll} j-4-7n & i+3+7n & j-3-7n & i+4+7n \\ i-1+7n & j-7n & i+7n & j+1-7n \end{array} \quad (B.2)$$

where n=. . .,-2,-1,0,1,2,. . . is a repeat number, and i and j are amino acids of the helices I and J correspondingly.

In this representation, side chain layers are stacked organized by ±3 and ±4 contacts. Each layer is composed of two amino acid residues from each helix, mainly with (i, i±1), (j, j±1) contacts. More close contacts occur between the residues from opposite helices compared to interaction of the neighboring side chains from the same helix in one layer. For instance, side chains i, j interact with each other while side chains i, i+1 (or j, j+1) are separated by larger distance. As a result, i, i+1 (or j, j +1) side chains usually do not form one hydrophobic cluster but rather are separated into two. Within one layer, side chain i interacts with side chains j and j+1 while side chain j interacts with side chains i and i+1. Two types of interactions [18, 19] are formed along the helix-helix interface: "4" part of the repeat (layers 1, 0) and "3" part of the repeat (layers 0, -1):

j+4	i+3	j+3	i+4	1	↑ +7n	j-4	i+3	j-3	i+4	↑ i+7n, j-7n
i-1	j	i	j-1	0		i-1	j	i	j+1	
j-3	i-4	j-4	i-3	-1	↓ -7n	j+3	i-4	j+4	i-3	↓ i-7n, j+7n

(B.3)

Sequence of the 3-4 repeats is assembled as an interface (B.3). Therefore, the packing of α-helices results in a scheme of contacts, which can be described by a set of rules. This scheme of contacts is shown for parallel type of helix-helix interaction of leucine zipper and for antiparallel type of interaction of cytochrome b (Table B.1).

Helix-helix pairs are assembled to form a structure (Table B.2). Binding of a helix or a pair of helices results in a new motif.

Four-α-helix motif consists of four α-helices. One of its frequently observed types, all antiparallel, can be viewed as two repeats of two antiparallel helices stacked antiparallel to each other. Let us designate the helices as I, J, K, and L. Four-α-helix arrangement, in which helices follow the order H1, H2, H3, H4 in protein sequence and form IJ, IK, JL, and KL antiparallel interfaces (Fig. B.1), occurs as left-handed (I=H1, J=H4, K=H3, L=H2) or right-handed (I=H1, J=H2, K=H3, L=H4) structure. Four more arrangements are possible (Fig. B.1A). Comparison of the amino acid sequence and three-dimensional structure of the two topologies (Fig. B.1B, C) shows how distribution of amino acids at positions **a** and **d** distinguishes one type from another. For each of these topological arrangements formula description shows repeated pattern of contacts characteristic for the four-α-helix motif (Table B.2A).

Table B.1 Scheme of contacts for interacting α-helices I (amino acids i ± 1, i ± 3, i ± 4) and J (amino acids j ± 1, j ± 3, j ± 4), positioned (A) parallel and (B) anti-parallel to each other. Corresponding positions a through g in (C) parallel and (D) antiparallel arrangements of helices. (E) Example of the scheme of contacts for the parallel interface of leucine zipper (pdb code 2zta); helix I (A subunit) and helix J (B subunit, primed numbers for amino acid residues) (F) Example of the scheme of contacts for the antiparallel interface of cytochrome c' molecule (pdb code 1bbh); helix I and helix J as helices 1 and 2 correspondingly. One heptad "3-4" repeat is boxed

(A)

	↑	↑	
	I	J	
j+4	**i+3**	**j+3**	**i+4**
i-1	**j**	**i**	**j-1**
j-3	**i-4**	**j-4**	**i-3**

(B)

	↑	↓	
	I	J	
j-4	**i+3**	**j-3**	**i+4**
i-1	**j**	**i**	**j+1**
j+3	**i-4**	**j+4**	**i-3**

(C)

e′	d	d′	e
g	a′	a	g′
e′	d	d′	e

(D)

g′	d	a′	e
g	d′	a	e′
g′	d	a′	e

(E)

29L	30′V	30V	29′L
27′K	26L	26′L	27K
22E	23′V	23V	22′E
20′E	19L	19′L	20E
15K	**16′N**	**16N**	**15′K**
13′L	**12L**	**12′L**	**13L**
8K	**9′V**	**9V**	**8′K**
6′E	5L	5′L	6E
1R	2′M	2M	1′R

(F)

38Q	**29N**	**39V**	**30L**
25K	**42A**	**26I**	**43A**
45V	**22N**	**46I**	**23M**
18F	49I	19M	50A

Helical assemblies (ankyrin, ARM/HEAT, PUM, TPR, PPR, LRR, TAL) comprise one or two rows of α-helices organized as a toroid. Ankyrin repeats consist of the two antiparallel rows of parallel helices. The repeated motif is a pair of antiparallel α-helices stacked parallel to each other and, therefore, contains four-α-helix motif. Let us consider a fragment containing three repeats with a total of six helices designated as I, J, and K in the inner row and L, M, and N in the outer row. Pairs IL, JM, and KN are antiparallel to each other. Each pair forms a helix-turn-helix motif. Inner row (IJ, JK) and outer row (LM, MN) interfaces are parallel. Two groups of helices, I, J, L, M and J, K, M, N, each comprises four-α-helix motif with two parallel and two antiparallel interfaces. If the inner row is compared with the outer row, we can see that direction of the assembly differs (Table. B.2). This difference can be explained by the two distinct patterns of contacts [21] and is clearly demonstrated by the formula: in contacts with preceding and following helices, edges (j-1, j) and (j+1, j+2) are utilized on each inner row helix, whereas edges (m-1, m) and

Table B.2 Description of an α-helical protein by the formula for spatial location of amino acid residues within secondary structure elements. (A) Four-α-helix motif; helices I, J, K, and L. (B) Assembly of the four-α-helix subunits into a helical rod: ankyrin repeats (C) Assembly of the barrel: TIM-barrel motif; helices I through P /1o5x/

(A)

I,J	I,L	K,L	J,K
j-4 i+3 j-3 i+4	l-4 i+4 l-3 i+5	l-5 k+3 l-4 k+4	j-3 k+4 j-4 k+5
i-1 j i j+1	**i l i+1 l+1**	**k-1 l-1 k l**	**k j+1 k+1 j+2**
j+3 i-4 j+4 i-3	l+3 i-3 l+4 i-2	l+2 k-4 l+3 k-3	j+4 k-3 j+5 k-2

Themiste zostericola myohemerythrin

34D 43N 35C 44L	35C 96V 36I 97V	86G 95N 87L 96V	87L 44L 88S 45A
46T 31G 47L 32I	99C 32I 100K 33F	98Y 83K 99C 84I	47L 84I 48V 85G
27K 50V 28I 51T	**28I103L 29F 104V**	**79D 102W 80F 103L**	**80F 51T 81L 52T**
53N 24E 54H 25H	106H 25H 107I 26K	105N 76M 106H 77H	54H 77H 55F 78K
20Q 57H 21L 58E			

(B)

Ankyrin inner row	(i-1, i) and (i+1, i+2)	
IJ	JM	JL
j+6 i+2 j+7 i+1	j+7 m-3 j+8 m-2	j+8 l+4 j+9 l+5
i-1 j+3 i-2 j+4	m j+4 m+1 j+5	l+1 j+5 l+2 j+6
j-1 i-5 j i-6	**j m+4 j+1 m+5**	**j+1 l-3 j+2 l-2**
	447A 449H 448G 450T	
413F 443M 412S 444A	452V 444A 453A 445A	477A 445A 476C 446R
439T 409H 440P 408L	**440P 456L 441L 457L**	**441L 473P 442H 472T**

Ankyrin outer row	(i-1, i) and (i-3, i-2)	
LM	JM	MN
l+7 m+3 l+8 m+4	j+7 m-3 j+8 m-2	n+2 m+1 n+1 m+2
m-1 l+3 m l+4	m j+4 m+1 j+5	**m-3 n-2 m-2 n-3**
l m-4 l+1 m-3	**j m+4 j+1 m+5**	n-5 m-6 n-6 m-5
458Q 424L 459N 423L	447A 449H 448G 450T	456L 492N 457L 491E
421K 455Y 420V 456L	**452V 444A 453A 445A**	489L 453A 488L 454K
451E 417L 452V 416H	440P 456L 441L 457L	**449H 485M 450T 484N**

(C)

TIM barrel: *Plasmodium falciparum* triosephosphate isomerase

I,J	JK	KL	LM
j+4 i+1 j+3 i+2	k+4 j+1 k+3 j+2	l+4 k+1 l+3 k+2	m+4 l+1 m+3 l+2
i-3 j i-2 j-1	**j-3 k j-2 k-1**	**k-3 l k-2 l-1**	**l-3 m l-2 m-1**
j-3 i-6 j-4 i-5	k-3 j-6 k-4 j-5	l-3 k-6 l-4 k-5	m-3 l-6 m-4 l-5
55L 24T 54L 25N			
20I 51T 21K 50H	48Y 82I 49D 81E	**79S 116S 80A 115A**	**113L 142V 114Q 141E**
48Y 17L 47H 18E			

MN	NO	OP	PI
n+4 m+1 n+3 m+2	o+4 n+1 o+3 n+2	p+4 o+1 p+3 o+2	i+4 p+1 i+3 p+2
m-3 n m-2 n-1	**n-3 o n-2 o-1**	**o-3 p o-2 p-1**	**p-3 i p-2 i-1**
n-3 m-6 n-4 m-5	o-3 n-6 o-4 n-5	p-3 o-6 p-4 o-5	i-3 p-6 i-4 p-5
146Q 191I 147V 190K			
188I 143I 187E 144T		247A 221I 246S 222Q	
139T 184V 140I 183L	**181A 220L 182Q219S**	**217C 243I 218S 242D**	**240T 23L 241V 22S**

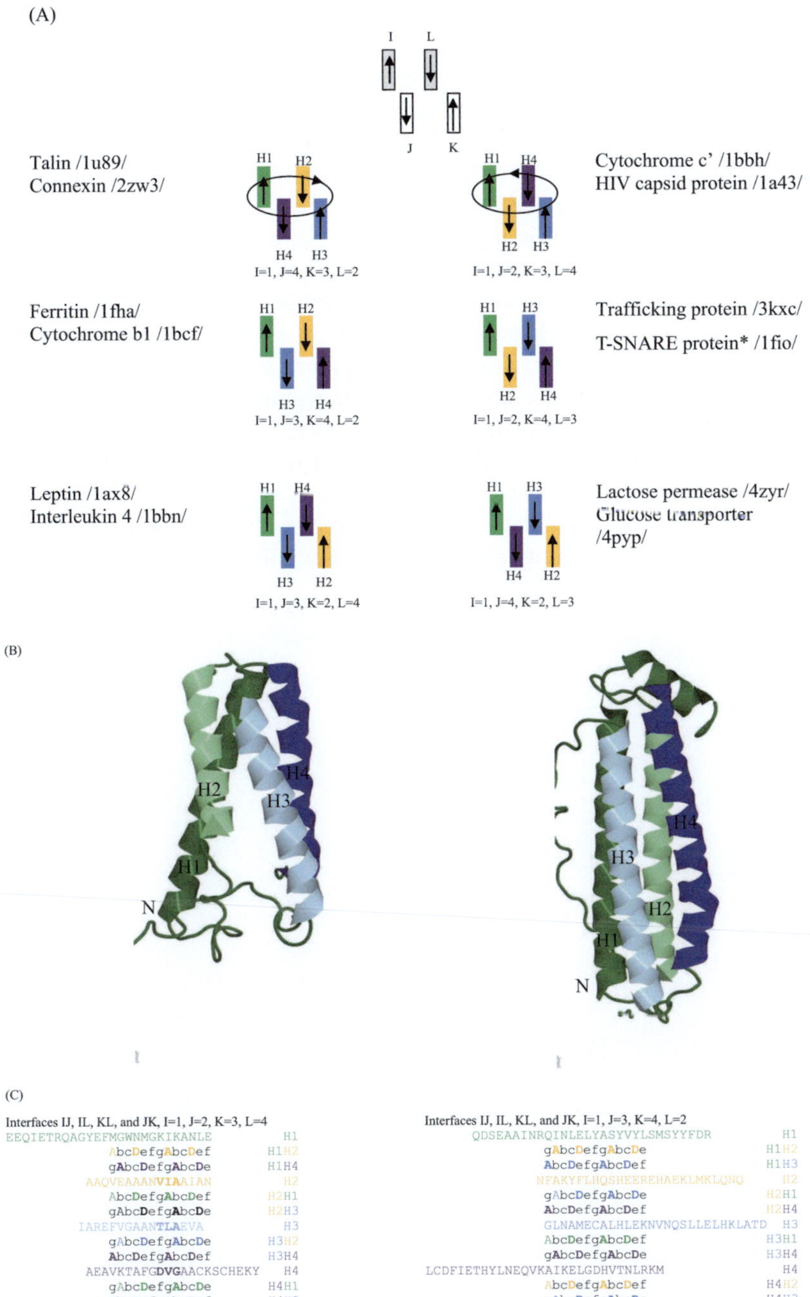

Fig. B.1 Three-dimensional structure of the antiparallel four-α-helix motif. (A) Six topological arrangements. Helices I, J, K, and L form IJ, IL, JK, and KL interfaces; helices J and K are in front of the helices I and L. For left-handed (left) and right-handed (right) topologies clockwise and counterclockwise direction of helices are indicated by arrows. One example (*) contains parallel arrangement of helices. Structure (B) and amino acid sequence (C) of the *C. vinosum* cytochrome c' /1bbh/ and *H. sapiens* ferritin /1fha/. Positions designations **a** through **g** of each interface are in leucine zipper nomenclature. Positions **a** and **d** involved in interfaces are capitalized and colored as interacting helix (m-3, m-2) are involved on each outer row helix (Table B.2B). As a result, for inner and outer rows of helices, there exists the characteristic repeated pattern of contacts. Each type of helical repeats, such as ARM/HEAT, PUM, TPR, PPR, LRR, TAL, exhibits a unique pattern and direction of the assembly correlates with the arrangement of contacts on the surface of the helix.

A cross section of the central pocket of the four-α-helix motif shows that each α-helix participates in two interfaces so that each of these interfaces involves one central and one side pocket. In total, α-helix can utilize all of its four central pockets together with four side pockets as observed in an assembly of the four-α-helical subunits of a tobacco mosaic virus coat protein into a helical rod. Each subunit is surrounded by four other subunits, two subunits within the same helical turn of the $16.\bar{3}$ subunits, one subunit of the previous helical turn, and one subunit of the following helical turn. In a circular arrangement of a TIM-barrel structure, each α-helix makes contacts with two other helices and a β-sheet. Therefore, three central pockets are involved in contacts within the motif and the forth central pocket contributes to the intersubunit or interdomain assemblies (Table B.1).

References

Abaan OD, Toretsky JA (2008) PTPL1: a large phosphatase with a split personality. Cancer Metastasis Rev 27:205–214

Abad MA, Medina B, Santamaria A, Zou J, Plasberg-Hill C, Madhumalar A et al (2014) Structural basis for microtubule recognition by the human kinetochore Ska complex. Nat Commun 5:2964. https://doi.org/10.1038/ncomms3964

Abil Z, Gumy LF, Zhao H, Hoogenraad CC (2017) Inducible control of mRNA transport using reprogrammable RNA-binding proteins. ACS Synth Biol 6(6):950–956. https://doi.org/10.1021/acssynbio.7b00025. Epub 2017 Mar 8. PubMed PMID: 28260376; PubMed Central PMCID: PMC5477001

Achilonu I, Gildenhuys S, Fisher L, Burke J, Fanucchi S, Sewell BT, Fernandes M, Dirr HW (2010) The role of a topologically conserved isoleucine in glutathione transferase structure, stability and function. Acta Crystallogr Sect F Struct Biol Cryst Commun 66(Pt 7):776–780. https://doi.org/10.1107/S1744309110019135. Epub 2010 Jun 23. PubMed PMID: 20606271; PubMed Central PMCID: PMC2898459

Adam A, Haberhauer G, Wölper C (2017) Bio-inspired herringbone foldamers: strategy for changing the structure of helices. J Org Chem 82(8):4203–4215. https://doi.org/10.1021/acs.joc.7b00185. Epub 2017 Apr 12

Adam C, Peters AD, Lizio MG, Whitehead GFS, Diemer V, Cooper JA, Cockroft SL, Clayden J, Webb SJ (2018) The Role of terminal functionality in the membrane and antibacterial activity of peptaibol-mimetic aib foldamers. Chemistry 24(9):2249–2256. https://doi.org/10.1002/chem.201705299. Epub 2018 Jan 17. PubMed PMID: 29210477

Adler, I A new look at geometry

Aggeli A, Nyrkova IA, Bell M, Harding R, Carrick L, McLeish TCB, Semenov AN, Boden N (2001) Hierarchical self-assembly of chiral rod-like molecules as a model for peptide beta -sheet tapes, ribbons, fibrils, and fibers. Proc Natl Acad Sci USA 98:11857–11862

Ago H, Kanaoka AH, Irikura D et al (2007) Crystal structure of a human membrane protein involved in cysteinyl leukotriene biosynthesis. Nature 448:609–612

Ahlstrom M, Roxhed N, Nordquist L (2014) Intradermal insulin delivery: a promising future for diabetes management. J Diabetes Sci Technol 8(3):453–457. https://doi.org/10.1177/1932296814530060. Epub 2014 Apr 10. Review. PubMed PMID: 24876605; PubMed Central PMCID: PMC4455430

Ahmad S, Pecqueur L, Dreier B, Hamdane D, Aumont-Nicaise M, Plückthun A et al (2016) Destabilizing an interacting motif strengthens the association of a designed ankyrin repeat protein with tubulin. Sci Rep 6:28922. https://doi.org/10.1038/srep28922

De Alba E (2009) Structure and interdomain dynamics of apoptosis-associated speck-like protein containing a CARD (ASC). J Biol Chem 284:32932–32941

N. Kurochkina, *Protein Structure and Modeling*,
https://doi.org/10.1007/978-981-13-6601-7

Alber T, Banner DW, Bloomer AC, Petsko GA, Phillips D, Rivers PS, Wilson IA (1981) On the three-dimensional structure and catalytic mechanism of triose phosphate isomerase. Philos Trans R Soc London 293:159–171

Alderson TR, Charlier C, Torchia DA, Anfinrud P, Bax A (2017) Monitoring hydrogen exchange during protein folding by fast pressure jump NMR spectroscopy. J Am Chem Soc 139 (32):11036–11039. doi: https://doi.org/10.1021/jacs.7b06676. Epub 2017 Aug 7. PubMed PMID: 28766333; PubMed Central PMCID: PMC5586491

Allan RK, Ratajczak T (2011) Versatile TPR domains accommodate different modes of target protein recognition and function. Cell Stress and Chaperones *16*:353–367

Allard B, Fuster C (2018) When muscle Ca(2+) channels carry monovalent cations through gating pores: insights into the pathophysiology of type 1 hypokalaemic periodic paralysis. J Physiol 596(11):2019–2027. https://doi.org/10.1113/JP274955. Epub 2018 Apr 15. PubMed PMID: 29572832; PubMed Central PMCID: PMC5983135

Al Khamici H, Hossain KR, Cornell BA, Valenzuela SM (2016) Investigating sterol and redox regulation of the ion channel activity of CLIC1 using tethered bilayer membranes. Membranes (Basel). 6(4). pii: E51. PubMed PMID: 27941637; PubMed Central PMCID: PMC5192407

Alushin GM, Ramey VH, Pasqualato S, Ball DA, Grigorieff N, Musacchio A, Nogales E (2010) The Ndc80 kinetochore complex forms oligomeric arrays along microtubules. Nature 467:805–810

Alvarado JA, Betts L, Moroco JA, Smithgall TE, Yeh JI (2010) Crystal Structure of the Src Family Kinase Hck SH3-SH2 Linker Regulatory Region Supports an SH3-dominant Activation Mechanism. J Biol Chem 285(46):35455–35461

Alvarez BH, Gruber M, Ursinus A, Dunin-Horkawicz S, Lupas AN, Zeth K (2010) A transition from strong right-handed to canonical left-handed supercoiling in a conserved coiled-coil segment of trimeric autotransporter adhesins. J Struct Biol 170(2):236–245. https://doi.org/10.1016/j.jsb.2010.02.009. Epub 2010 Feb 21. PubMed PMID: 20178846

Andrade MA, Bork P (1995) HEAT repeats in the Huntington's disease protein. Nat Genet *11*:115–116

Andrade MA, Perez-Iratxeta C, Ponting CP (2001a) Protein repeats: structures, functions, and evolutions. J Struct Biol 134:117–131

Andrade MA, Petosa C, O'Donoghue SI, Müller CW, Bork P (2001b) Comparison of ARM and HEAT protein repeats. J Mol Biol *309*:1–18

Andrews SC, Smith JM, Guest JR, Harrison PM (1989) Amino acid sequence of the bacterioferritin (cytochrome b1) of Escherichia coli-K12. Biochem Biophys Res Commun 158(2):489–496. PubMed PMID: 2644932

Anstrom DM, Kallio K, Remington SJ (2003) Structure of the Escherichia coli malate synthase G: pyruvate:acetyl-coenzyme A abortive ternary complex at 1.95 Å resolution. *Protein Sci* 12:1822–1832

Arana ME, Holmes SF, Fortune JM, Moon AF, Pedersen LC, Kunkel TA (2010) Functional Residues on the Surface of the N-terminal domain of Yeast Pms1. DNA Repair 9 (4):448–457. https://doi.org/10.1016/j.dnarep.2010.01.010

Aravinda S, Shamala N, Balaram P (2008) Aib residues in peptaibiotics and synthetic sequences: analysis of nonhelical conformations. Chem Biodivers 5(7):1238–1262. https://doi.org/10.1002/cbdv.200890112

Arcy SD, Davies OR, Blundell TL, Bolanos-Garcia VM (2010) Defining the molecular basis of BubR1 kinetochore interactions and APC/C-CDC20 inhibition. J Biol Chem 285:14764–14776

Arévalo JC, Pereira DB, Yano H, Teng KK, Chao MV (2006) Identification of a switch in neurotrophin signaling by selective tyrosine phosphorylation. J Biol Chem 281:1001–1007

Arévalo JC, Wu SH, Takahashi TZH, Yu T, Yano H et al (2010) The ARMS/Kidins220 scaffold protein modulates synaptic transmission. Mol Cell Neurosci 45:92–100

Arif A (2012) Extraneuronal activities and regulatory mechanisms of the atypical cyclin-dependent kinase Cdk5. Biochem Pharmacol 84:985–993

Armstrong N, Gouaux E (2000) Mechanisms for activation and antagonism of an AMPA-sensitive glutamate receptor: crystal structures of the GluR2 ligand binding core. Neuron 28:165–181

Armstrong N, Jasti J, Beich-Frandsen M, Gouaux F (2006) Measurement of conformational changes accompanying desensitization in an ionotropic glutamate receptor. Cell 127:85–97

Arnold FH (2018) Directed evolution: bringing new chemistry to life. Angew Chem Int Ed Engl 57 (16):4143–4148. https://doi.org/10.1002/anie.201708408. Epub 2017 Nov 28. Review. PubMed PMID: 29064156; PubMed Central PMCID: PMC5901037

Arnold E, Rossmann MG (1988) The use of molecular-replacement phases for the refinement of the human rhinovirus 14 structure. Acta Crystallogr A 44(Pt 3):270–282. PubMed PMID: 2856083 Viruses (Sobhy 2014)

Aschner Y, Downey GP (2018) The importance of tyrosine phosphorylation control of cellular signaling pathways in respiratory disease: pY and pY Not. Am J Respir Cell Mol Biol. https://doi.org/10.1165/rcmb.2018-0049TR. [Epub ahead of print] PubMed PMID: 29812954

Ashburner M, Ball CA, Blake JA, Botstein D, Butler H, Cherry JM, Davis AP, Dolinski K, Dwight SS, Eppig JT, Harris MA, Hill DP, Issel-Tarver L, Kasarskis A, Lewis S, Matese JC, Richardson JE, Ringwald M, Rubin GM, Sherlock G (2000) Gene ontology: tool for the unification of biology. The gene ontology consortium. Nat Genet 25:25–29. http://www.ncbi.nlm.nih.gov/pubmed/10802651?ordinalpos=3&itool=EntrezSystem2.PEntrez.Pubmed.Pubmed_ResultsPanel.Pubmed_RVDocSum

Ataman ZA, Gakhar L, Sorensen BR, Hell JW, Shea MA (2007) The NMDA receptor NR1 C1 region bound to calmodulin: structural insights into functional differences between homologous domains. Structure 15:1603–1617

Audet-Walsh É, Dufour CR, Yee T, Zouanat FZ, Yan M, Kalloghlian G, Vernier M, Caron M, Bourque G, Scarlata E, Hamel L, Brimo F, Aprikian AG, Lapointe J, Chevalier S, Giguère V (2017) Nuclear mTOR acts as a transcriptional integrator of the androgen signaling pathway in prostate cancer. Genes Dev 31(12):1228–1242. https://doi.org/10.1101/gad.299958.117. Epub 2017 Jul 19. PubMed PMID: 28724614; PubMed Central PMCID: PMC5558925

Azzi L, Meijer L, Ostvold A-C, Lew J, Wang JH (1994) Purification of a 15-kDa cdk4- and cdk5-binding Protein. J Biol Chem 269:13279–13288

Ayala R, Shu T, Tsai LH (2007) Trecking across the brain: the journey of neuronal migration. Cell 128:29–43

Avirneni-Vadlamudi U, Galindo KA, Endicott TR, Paulson V, Cameron S, Galindo RL (2012) *Drosophila* and mammalian models uncover a role for the myoblast fusion gene *TANC1* in rhabdomyosarcoma. J Clin Invest 122:403–407

Bai Y, Luo Y, Liu S, Zhang L, Shen K, Dong Y et al (2011) PRL-1 protein promotes ERK1/2 and RhoA protein activation through a non-canonical interaction with the Src homology 3 domain of p115 Rho GTPase-activating protein. J Bioll Chem 286(49):42316–42324. https://doi.org/10.1074/jbc.M111.286302

Balaram P (2010) Hybrid polypeptides: gabapentin as a stereochemically constrained γ-amino acid residue. Biopolymers 94(6):733–741. https://doi.org/10.1002/bip.21468. PubMed PMID: 20564041

Balasco N, Smaldone G, Ruggiero A, De Simone A, Vitagliano L (2018) Local structural motifs in proteins: Detection and characterization of fragments inserted in helices. Int J Biol Macromol. pii: S0141-8130(18)32822–32828. doi: https://doi.org/10.1016/j.ijbiomac.2018.07.047. [Epub ahead of print] PubMed PMID: 30017977

Balashova TA, Shenkarev ZO, Tagaev AA, Ovchinnikova TV, Raap J, Arseniev AS (2000) NMR structure of the channel-former zervamicin IIB in isotropic solvents. FEBS Lett 466 (2-3):333–336. PubMed PMID: 10682854

Balasubramanian A, Manzano M, Teramoto T, Pilankatta R, Padmanabhan R (2016) High-throughput screening for the identification of small-molecule inhibitors of the flaviviral protease. Antiviral Res 134:6–16. https://doi.org/10.1016/j.antiviral.2016.08.014. Epub 2016 Aug 15. PubMed PMID: 27539384; PubMed Central PMCID: PMC5065773

Balchin D, Fanucchi S, Achilonu I, Adamson RJ, Burke J, Fernandes M, Gildenhuys S, Dirr HW (2010) Stability of the domain interface contributes towards the catalytic function at the H-site of class alpha glutathione transferase A1-1. Biochim Biophys Acta 1804(12):2228–2233. https://doi.org/10.1016/j.bbapap.2010.09.003. Epub 2010 Sep 15. PubMed PMID: 20833278

Banci L, Bertini I, Cefaro C, Ciofi-Baffoni S, Gallo A (2011) Functional role of two interhelical disulfide bonds in human Cox17 protein from a structural perspective. J Biol Chem 286 (39):34382–34390. https://doi.org/10.1074/jbc.M111.246223. Epub 2011 Aug 4. PubMed PMID: 21816817; PubMed Central PMCID: PMC3190761

Bankston JR, Camp SS, DiMaio F, Lewis AS, Chetkovich DM, Zagotta WN (2012) Structure and stoichiometry of an accessory subunit TRIP8b interaction with hyperpolarization-activated cyclic nucleotide-gated channels. Proc Natl Acad Sci USA 109(20):7899–7904. https://doi.org/10.1073/pnas.1201997109. Epub 2012 May 1. PubMed PMID: 22550182; PubMed Central PMCID: PMC3356637

Banner DW, Kokkinidis M, Tsernoglou D (1987) Structure of the ColE1 rop protein at 1.7 A resolution. J Mol Biol 196(3):657–675. PubMed PMID: 3681971

Barbieri (2010) 3nnn

Bariselli S, Bellone C (2016) VTA DA neuron excitatory synapses in Shank3 ?ex4-9 mouse line. Synapse. https://doi.org/10.1002/syn.21955

Baron MK, Boeckers TM, Vaida B, Faham S, Gingery M, Sawaya MR, Salyer D, Gundelfinger ED, Bowie JU (2006) An architectural framework that may lie at the core of the postsynaptic density. Science 311:531–535

Barua B (2013) Periodicities designed in the tropomyosin sequence and structure define its functions. Bioarchitecture 3(3):51–56. https://doi.org/10.4161/bioa.2561

Basu A, Chakrabarti A (2015) Hemoglobin interacting proteins and implications of spectrin hemoglobin interaction. J Proteomics 128:469–475. https://doi.org/10.1016/j.jprot.2015.06.014. Epub 2015 Jun 30. PubMed PMID: 26141508

Bauer CK, Calligari P, Radio FC, Caputo V, Dentici ML, Falah N, High F, Pantaleoni F, Barresi S, Ciolfi A, Pizzi S, Bruselles A, Person R, Richards S, Cho MT, Claps Sepulveda DJ, Pro S, Battini R, Zampino G, Digilio MC, Bocchinfuso G, Dallapiccola B, Stella L, Tartaglia M (2018) Mutations in KCNK4 that Affect Gating Sullivan R, Yau WY, O'Connor E, Houlden H. Spinocerebellar ataxia: an update. J Neurol. https://doi.org/10.1007/s00415-018-9076-4. [Epub ahead of print] PubMed PMID: 30284037

Bax JJ, Delgado V (2015) Myocardial viability as integral part of the diagnostic and therapeutic approach to ischemic heart failure. J Nucl Cardiol 22(2):229–245. https://doi.org/10.1007/s12350-015-0096-5. Epub 2015 Mar 3. Review. PubMed PMID: 25733105; PubMed Central PMCID: PMC4490177

Begley MJ, Taylor GS, Kim SA, Veine DM, Dixon JE, Stuckey JA (2003) Crystal structure of a phosphoinositide phosphatase, MTMR2: insights into myotubular myopathy and Charcot-Marie-Tooth syndrome. Mol Cell 12(6):1391–1402. PubMed PMID: 14690594

Bennett BC, Purdy MD, Baker KA, Acharya C, McIntire WE, Stevens RC, Zhang Q, Harris AL, Abagyan R, Yeager M (2016) An electrostatic mechanism for Ca(2+)-mediated regulation of gap junction channels. Nat Commun 7:8770. https://doi.org/10.1038/ncomms9770. PubMed PMID: 26753910; PubMed Central PMCID: PMC4730032

Bent CJ, Isaacs NW, Mitchell TJ, Riboldi-Tunnicliffe A (2004) Crystal structure of the response regulator 02 receiver domain, the essential YycF two-component system of *Streptococcus pneumoniae* in both complexed and native states. J Bacteriol 186(9):2872–2879. https://doi.org/10.1128/JB.186.9.2872-2879.2004

Berisio R, Vitagliano L (2012) Polyproline and triple helix motifs in host-pathogen recognition. Curr Protein Pept Sci 13(8):855–865. https://doi.org/10.2174/138920312804871157

Bernstein FC, Koetzle TF, Williams GJB, Meyer EF Jr, Brice MD, Rodgers JR, Kennard O, Shimanouchi T, Tasumi M (1977) The protein data bank: a computer-based archival file for macromolecular structures. J Mol Biol 112:535

Biederer T, Südhof TC (2000) Mints as adaptors. Direct binding to neurexins and recruitment of munc18. J Biol Chem 275(51):39803–39806

Binukumar BK, Shukla V, Amin ND, Reddy P, Skuntz S, Grant P, Pant HC (2013) Topographic regulation of neuronal intermediate filaments by phosphorylation, role of peptidyl-prolyl isomerase 1: significance in neurodegeneration. Histochem Cell Biol 140(1):23–32. https://doi.org/10.1007/s00418-013-1108-7

Binz HK, Stumpp MT, Forrer P, Amstutz P, Pluckthun A (2003) Designing Repeat Proteins: Well-expressed, Soluble and Stable Proteins from Combinatorial Libraries of Consensus Ankyrin Repeat Proteins. J Mol Biol 332:489–503

Bhyravbhatla B, Watowich SJ, Caspar DL (1998) Refined atomic model of the four-layer aggregate of the tobacco mosaic virus coat protein at 2.4-Å resolution. Biophys J 74:604–615

Blatch GL, Lassle M (1999) The tetratricopeptide repeat: a structural motif mediating protein-protein interactions. Bioassays 21:932–939

Bochtler M (2012) Structural basis of the TAL effector–DNA interaction. Biol Chem 393:1055–1066

Board PG, Coggan M, Watson S, Gage PW, Dulhunty AF (2004) CLIC-2 modulates cardiac ryanodine receptor Ca2+ release channels. Int J Biochem Cell Biol 36(8):1599–1612. PubMed PMID: 15147738

Bodey AJ, Kikkawa M, Moores CA (2009) 9-Angström structure of a microtubule-bound mitotic motor. J Mol Biol 388(2):218–224. https://doi.org/10.1016/j.jmb.2009.03.008. Epub 2009 Mar 10. PubMed PMID: 19285086

Bodra N, Young D, Astolfi Rosado L, Pallo A, Wahni K, De Proft F, Huang J, Van Breusegem F, Messens J (2017) Arabidopsis thaliana dehydroascorbate reductase 2: Conformational flexibility during catalysis. Sci Rep 7:42494. doi: 10.1038/srep42494. Erratum in: Sci Rep 7:46896. PubMed PMID: 28195196; PubMed Central PMCID: PMC5307357

Borgdorff AJ, Choquet D (2002) Regulation of AMPA receptor lateral movements. Nature 417:649–653

Born D, Reuter L, Mersdorf U, Mueller M, Fischer MG, Meinhart A, Reinstein J (2018) Capsid protein structure, self-assembly, and processing reveal morphogenesis of the marine virophage mavirus. Proc Natl Acad Sci USA 115(28):7332–7337. https://doi.org/10.1073/pnas.1805376115. Epub 2018 Jun 25. PubMed PMID: 29941605; PubMed Central PMCID: PMC6048507

Boscutti G, Nardon C, Marchiò L, Crisma M, Biondi B, Dalzoppo D, Dalla Via L, Formaggio F, Casini A, Fregona D (2018) Anticancer gold(III) peptidomimetics: from synthesis to in vitro and ex vivo biological evaluation. ChemMedChem. https://doi.org/10.1002/cmdc.201800098. [Epub ahead of print] PubMed PMID: 29570944

Boucher LE, Bosch J (2015) The apicomplexan glideosome and adhesins -- structures and function. J Struct Biol 190(2):93–114. https://doi.org/10.1016/j.jsb.2015.02.008

Boudker O, Ryan RM, Yernool D, Shimamoto K, Gouaux E (2007) Coupling substrate and ion binding to extracellular gate of a sodium-dependent aspartate transporter. Nature 445 (7126):387–393. Epub 2007 Jan 17. PubMed PMID: 17230192

Boudko SP, Bächinger HP (2016) Structural insight for chain selection and stagger control in collagen. Sci Rep 6:37831. https://doi.org/10.1038/srep37831

Boumil EF, Vohnoutka R, Lee S, Pant H, Shea TB (2018) Assembly and turnover of neurofilaments in growing axonal neurites. Biol Open 7(1). pii: bio028795). https://doi.org/10.1242/bio.028795

Bouzidi M, Tricaud N, Giraud P, Kordeli E, Caillol G, Deleuze C, Couraud F, Alcaraz G (2002) Interaction of the Nav1.2a subunit of the voltage-dependent sodium channel with nodal ankyrinG. In vitro mapping of the interacting domains and association in synaptosomes. J Biol Chem 277:28996–28900

Branden, C., Tooze, J (1991) Introduction to protein structure.

Brennan SO (2015) Variation of fibrinogen oligosaccharide structure in the acute phase response: possible haemorrhagic implications. BBA Clinical 3:221–226. https://doi.org/10.1016/j.bbacli.2015.02.007

Brinkkoetter PT, Olivier P, Wu JS, Henderson S, Krofft RD, Pippin JW, Hockenbery D, Roberts JM, Shankland SJ (2009) J Clin Invest 119:3089

Brinkmann U, Di Carlo A, Vasmatzis G, Kurochkina N, Beers R, Lee B, Pastan I (1997) Stabilization of a recombinant Fv fragment by base loop interconnection and VH-VL permutation. J Mol Biol 268:107–117

Brock J, Board PG, Oakley AJ (2013) Structural insights into omega-class glutathione transferases: a snapshot of enzyme reduction and identification of a non-catalytic ligandin site. PLoS One 8 (4):e60324. https://doi.org/10.1371/journal.pone.0060324. Print 2013. PubMed PMID: 23593192; PubMed Central PMCID: PMC3621891

Brown NR, Noble MEM, Endicott JE, Johnson LN (1999) The structural basis for specificity of substrate and recruitment peptides for cyclin-dependent kinases. Cell biology 1:438–443

Brown PN, Hill CP, Blair DF (2002) Crystal structure of the middle and C-terminal domains of the flagellar rotor protein FliG. EMBO J 21(13):3225–3234. PubMed PMID: 12093724; PubMed Central PMCID: PMC126082

Brunner JD, Lim NK, Schenck S, Duerst A, Dutzler R (2014) X-ray structure of a calcium-activated TMEM16 lipid scramblase. Nature 516(7530):207–212. https://doi.org/10.1038/nature13984. Epub 2014 Nov 12. PubMed PMID: 25383531

Bryda EC, Ling H, Rathbun DE, Burmeister M, Flaherty L (1996) Fine genetic map of mouse chromosome 10 around the polycystic kidney disease gene, jcpk, and ankyrin 3. Genomics 35 (3):425–430. PubMed PMID: 8812475

Buescher JL, Phiel CJ (2010) A noncatalytic domain of glycogen synthase kinase-3 (GSK-3) is essential for activity. J Biol Chem 285(11):7957–7963. https://doi.org/10.1074/jbc.M109.091603. Epub 2010 Jan 15. PubMed PMID: 20080974; PubMed Central PMCID: PMC2832946

Bunick CG, Milstone LM (2017) The X-ray crystal structure of the keratin 1-keratin 10 helix 2B heterodimer reveals molecular surface properties and biochemical insights into human skin disease. J Invest Dermatol 137(1):142–150. https://doi.org/10.1016/j.jid.2016.08.018. Epub 2016 Sep 3.

Burgess AW, Leach SJ (1973) An obligatory alpha-helical amino acid residue. Biopolymers 12 (11):2599–2605. PubMed PMID: 4780721

Buxbaum JD, Daly MJ, Devlin B, Lehner T, Roeder K, State MW, The Autism Sequencing Consortium (2012) The autism sequencing consortium: large scale, high throughput sequencing in autism spectrum disorders. Neuron 76:1052–1056

Cabrele C, Langer M, Bader R, Wieland HA, Doods HN, Zerbe O, Beck-Sickinger AG (2000) The first selective agonist for the neuropeptide YY5 receptor increases food intake in rats. J Biol Chem 275(46):36043–36048

Caffrey M, Cai M, Kaufman J, Stahl SJ, Wingfield PT, Covell DG, Gronenborn AM, Clore GM (1998) Three-dimensional solution structure of the 44 kDa ectodomain of SIV gp41. Embo J 17:4572–4584

Caing-Carlsson R, Goyal P, Sharma A, Ghosh S, Setty TG, North RA et al (2017) Crystal structure of N-acetylmannosamine kinase from Fusobacterium nucleatum. Acta Crystallogr Sect F Struct Biol Cryst Commun 73(Pt 6):356–362. https://doi.org/10.1107/S2053230X17007439

Caldieri G, Malabarba MG, Di Fiore PP, Sigismund S (2018) EGFR trafficking in physiology and cancer. Prog Mol Subcell Biol 57:235–272. https://doi.org/10.1007/978-3-319-96704-2_9. PubMed PMID: 30097778

Cantero-Recasens G, Butnaru CM, Valverde MA, Naranjo JR, Brouwers N, Malhotra V (2018) KChIP3 coupled to Ca(2+) oscillations exerts a tonic brake on baseline mucin release in the colon. Elife. 7. pii: e39729 Phillips L, Trivedi JR. Skeletal Muscle Channelopathies. Neurotherapeutics doi: https://doi.org/10.1007/s13311-018-00678-0. [Epub ahead of print] Review. PubMed PMID: 30341599

Cao DS, Zhong L, Hsieh TH, Abooj M, Bishnoi M, Hughes L, Premkumar LS (2012) Expression of transient receptor potential ankyrin 1 (TRPA1) and its role in insulin release from rat pancreatic beta cells. PLoS One 7(5):e38005. https://doi.org/10.1371/journal.pone.0038005. Epub 2012 May 31. PubMed PMID: 22701540; PubMed Central PMCID: PMC3365106

Carlacci L (1991) New development on energetic approach to the packing in proteins. J Comput Chem 12:410–415

Cervantes-Madrid D, Dominguez-Gomez G, Gonzalez-Fierro A, Perez-Cardenas E, Taja-Chayeb L, Trejo-Becerril C, Duenas-Gonzalez A (2017) Feasibility and antitumor efficacy in vivo, of simultaneously targeting glycolysis, glutaminolysis and fatty acid synthesis using lonidamine, 6-diazo-5-oxo-L-norleucine and orlistat in colon cancer. Oncology Letters 13(3):1905–1910. https://doi.org/10.3892/ol.2017.5615

Chai G, Brewer J, Lovelace L, Aoki T, Minor W, Lebioda L (2004) Expression, purification and the 1.8 Å resolution crystal structure of human neuron specific enolase. J Mol Biol 341:1015–1021

Chan W, Sit ST, Manser E (2011a) The Cdc42-associated kinase ACK1 is not autoinhibited but requires Src for Activation. Biochem J. (2011) 435:355–364

Chan PW, Yakunin AF, Edwards EA, Pai EF (2011b) Mapping the reaction coordinates of enzymatic defluorination. J Am Chem Soc 133(19):7461–7468. https://doi.org/10.1021/ja200277d. Epub 2011 Apr 21. PubMed PMID: 21510690; PubMed Central PMCID: PMC3101105

Charlier C, Alderson TR, Courtney JM, Ying J, Anfinrud P, Bax A (2018) Study of protein folding under native conditions by rapidly switching the hydrostatic pressure inside an NMR sample cell. Proc Natl Acad Sci USA 115(18):E4169–E4178. https://doi.org/10.1073/pnas.1803642115. Epub 2018 Apr 16. PubMed PMID: 29666248; PubMed Central PMCID: PMC5939115

Chavez-Calvillo G, Martin S, Hamm C, Sztuba-Solinska J (2018) The structure-to-function relationships of Gammaherpesvirus-encoded long non-coding RNAs and their contributions to viral pathogenesis. Noncoding RNA 4(4). pii: E24. doi: 10.3390/ncrna4040024. Review. PubMed PMID: 30261651

Chen J, Dou Y, Zheng X, Leng T, Lu X, Ouyang Y, Sun H, Xing F, Mai J, Gu J, Lu B, Yan G, Lin J, Zhu W (2016a) TRPM7 channel inhibition mediates midazolam-induced proliferation loss in human malignant glioma. Tumour Biol 37(11):14721–14731. Epub 2016 Sep 14. PubMed PMID: 27629139

Chen J, Zhang W, Lin J, Wang F, Wu M, Chen C, Zheng Y, Peng X, Li J, Yuan Z (2014a) An efficient antiviral strategy for targeting Hepatitis B virus genome Using transcription activator-like effector nucleases. Molecular Therapy 22:303–311

Chen J, Yu S, Fu Y, Li X (2014b) Synaptic proteins and receptors defects in autism spectrum disorders. Front Cell Neurosci 8:276

Chen Q, Wells MM, Arjunan P, Tillman TS, Cohen AE, Xu Y, Tang P (2018) Structural basis of neurosteroid anesthetic action on GABA(A) receptors. Nat Commun 9(1):3972. https://doi.org/10.1038/s41467-018-06361-4. PubMed PMID: 30266951; PubMed Central PMCID: PMC6162318

Chen PE, Wyllie DJA (2006) Pharmacological insights obtained from structure–function studies of ionotropic glutamate receptors. BJ Pharm 147:839–853

Chen SC, Huang CH, Lai SJ, Yang CS, Hsiao TH, Lin CH, Fu PK, Ko TP, Chen Y (2016b) Mechanism and inhibition of human UDP-GlcNAc 2-epimerase, the key enzyme in sialic acid biosynthesis. Sci Rep 6:23274. https://doi.org/10.1038/srep23274. PubMed PMID: 26980148; PubMed Central PMCID: PMC4793188

Cheng TC, Akey IV, Yuan S, Yu Z, Ludtke SJ, Akey CW (2017) A near-atomic structure of the dark apoptosome provides insight into assembly and activation. Structure 25(1):40–52. https://doi.org/10.1016/j.str.2016.11.002. Epub 2016 Dec 1. PubMed PMID: 27916517; PubMed Central PMCID: PMC5214966

Chin L, Gray JW (2008) Translating insights from the cancer genome into clinical practice. Nature 452:553–563

Chingle R, Proulx C, Lubell WD (2017) Azapeptide synthesis methods for expanding side-chain diversity for biomedical applications. Acc Chem Res 50(7):1541–1556. https://doi.org/10.1021/acs.accounts.7b00114. Epub 2017 Jun 9

Cho HY, Maeng SJ, Cho HJ, Choi YS, Chung JM, Lee S, Kim HK, Kim JH, Eom CY, Kim YG, Guo M, Jung HS, Kang BS, Kim S (2015) Assembly of multi-tRNA Synthetase complex via heterotetrameric glutathione transferase-homology domains. J Biol Chem 290 (49):29313–29328. https://doi.org/10.1074/jbc.M115.690867. Epub 2015 Oct 15. PubMed PMID: 26472928; PubMed Central PMCID: PMC4705937

Choquet D, Triller A (2003) The role of receptor diffusion in the organization of the postsynaptic membrane. Nat Rev Neurosci 4:251–265

Chothia C (1973) Conformation of twisted beta-pleated sheets in proteins. J Mol Biol 75:295–302

Chothia C (1975) Structural invariants in protein folding. Nature 254:304–308

Chothia C, Levitt M, Richardson D (1981) Helix to helix packing in proteins. J Mol Biol 145:215–250

Chothia C, Janin J (1982) Orthogonal packing of beta-pleated sheets in proteins. Biochemistry 21 (17):3955–3965. PubMed PMID: 6751382

Chothia C, Janin J (1981) Relative orientation of close-packed beta-pleated sheets in proteins. Proc Natl Acad Sci USA 78(7):4146–4150. PubMed PMID: 16593054; PubMed Central PMCID: PMC319745

Chou KC (1988) Review: Low-frequency collective motion in biomacromolecules and its biological functions. Biophysical Chemistry 30:3–48

Chou KC (1992) Energy-optimized structure of antifreeze protein and its binding mechanism. J Mol Biol 223:509–517

Chou KC, Nemethy G, Scheraga HA (1990) Review: Energetics of interactions of regular structural elements in proteins. Acc Chem Res 23:134–141

Chou KC (2004) Review: Structural bioinformatics and its impact to biomedical science. Curr Med Chem 11:2105–2134

Chou KC, Carlacci L (1991) Energetic approach to the folding of alpha/beta barrels. Proteins: Struct., Funct., Genet 9:280–295

Chou KC, Maggiora GM, Nemethy G, Scheraga HA (1988) Energetics of the structure of the four-alpha-helix bundle in proteins. Proc Natl Acad Sci USA 85:4295–4299

Chou KC, Maggiora GM, Scheraga HA (1992) The role of loop-helix interactions in stabilizing four-helix bundle proteins. Proc Natl Acad Sci USA 89:7315–7319

Chua TK, Bujnicki JM, Tan T-C, Huynh F, Patel BKC, Sivaraman J (2008) The structure of sucrose phosphate synthase from halothermothrix orenii reveals its mechanism of action and binding mode. Plant Cell 20:1059–1072

Chun J, Yang HE, Kim DH (2018) Identification and molecular characterization of a novel partitivirus from trichoderma atroviride NFCF394. Viruses 10(11). pii: E578.: 10.3390/v10110578. PubMed PMID: 30360465

Ciferri C, Pasqualato S, Screpanti E, Varetti G, Santaguida S, Reis GD et al (2008) Implications for kinetochore-microtubule attachment from the structure of an engineered Ndc80 complex. Cell 133(3):427–439. https://doi.org/10.1016/j.cell.2008.03.020

Clapham DE, Miller C (2011) A thermodynamic framework for understanding temperature-sensing by TRP channels. Proc Natl Acad Sci USA 108:19492–19497

Constals A, Penn AC, Compans B, Toulmé E, Phillipat A, Marais S, Retailleau N, Hafner AS, Coussen F, Hosy E, Choquet D (2015) Glutamate-induced AMPA receptor desensitization increases their mobility and modulates short-term plasticity through unbinding from Stargazin. Neuron 85:787–803

Collins FS, Gottlieb S (2018) The next phase of human gene-therapy oversight. N Engl J Med doi: 10.1056/NEJMp1810628. [Epub ahead of print] PubMed PMID: 30110242

Comba FN, Romero MR, Garay FS, Baruzzi AM (2018) Mucin and carbon nanotube-based biosensor for detection of glucose in human plasma. Anal Biochem 550:34–40. doi: 10.1016/j.ab.2018.04.006. [Epub ahead of print] PubMed PMID: 29649474

Contreras-Vallejos E, Utreras E, Gonzalez-Billault C (2012) Going out of the brain: Non-nervous system physiological and pathological functions of Cdk5. Cellular Signalling 24(2012):44–52

Cook JD, Soto-Montoya H, Korpela MK, Lee JE (2015) Electrostatic architecture of the Infectious Salmon Anemia Virus (ISAV) Core fusion protein illustrates a Carboxyl-Carboxylate pH Sensor. J Biol Chem 290(30):18495–18504. https://doi.org/10.1074/jbc.M115.644781. Epub 2015 Jun 16. PubMed PMID: 26082488; PubMed Central PMCID: PMC4513110

Coquille S, Filipovska A, Chia T, Rajappa L, Lingford JP, Razif MF, Thore S, Rackham O (2014) An artificial PPR scaffold for programmable RNA recognition. Nat Commun 5:5729: https://doi.org/10.1038/ncomms6729. PubMed PMID: 25517350

Cordero-Morales JF, Gracheva EO, Julius D (2011) Cytoplasmic ankyrin repeats of transient receptor potential A1 (TRPA1) dictate sensitivity to thermal and chemical stimuli. Proc Natl Acad Sci 108:E1184–E1191

Cordopatis P, Manessi-Zoupa E, Theodoropoulos D, Bossé R, Bouley R, Gagnon S, Escher E (1994) Methylation in positions 1 and 7 of angiotensin II. A structure-activity relationship study. Int J Pept Protein Res 44(4):320–324. PubMed PMID: 7875933

Cortés RY, Arévalo JC, Magby JP, Chao MV, Plummer MR (2007) Developmental and activity-dependent regulation of ARMS/Kidins220 in cultured rat hippocampal neurons. Dev Neurobiol 67:1687–1698

Costales JL, Kolevzon A (2015) Phelan–McDermid Syndrome and SHANK3: Implications for Treatment. Neurotherapeutics 12:620–630

Costil R, Fernández-Nieto F, Atkinson RC, Clayden J (2018) α-Methyl phenylglycines by asymmetric α-arylation of alanine and their effect on the conformational preference of helical Aib foldamers. Org Biomol Chem doi: 10.1039/c8ob00551f. [Epub ahead of print] PubMed PMID: 29595846

Cox K, Combs B, Abdelmesih B, Morfini G, Brady ST, Kanaan NM (2016) Analysis of isoform-specific tau aggregates suggests a common toxic mechanism involving similar pathological conformations and axonal transport inhibition. Neurobiol Aging 47:113–126. https://doi.org/10.1016/j.neurobiolaging.2016.07.015

Crick F (1953) Acta Crystallogr 6:689

Cunha ES, Hatem CL, Barrick D (2016) Synergistic enhancement of cellulase pairs linked by consensus ankyrin repeats: determination of the roles of spacing, orientation and enzyme identity. Proteins 84(8):1043–1054. https://doi.org/10.1002/prot.25047

Dadheech T, Shah R, Pandit R, Hinsu A, Chauhan PS, Jakhesara S, Kunjadiya A, Rank D, Joshi C (2018) Cloning, molecular modeling and characterization of acidic cellulase from buffalo rumen and its applicability in saccharification of lignocellulosic biomass. Int J Biol Macromol 113:73–81. https://doi.org/10.1016/j.ijbiomac.2018.02.100. Epub 2018 Feb 15. PubMed PMID: 29454942

D'Andrea LD, Regan L (2003) TPR proteins: the versatile helix. Trends Biochem Sci 28 (12):655–662. Review. PubMed PMID: 14659697

Dao EH, Poitevin F, Sierra RG, Gati C, Rao Y, Ciftci HI, Aksit F, McGurk A, Obrinski T, Mgbam P, Hayes B, DE Lichtenberg C, Pardo-Avila F, Corsepius N, Zhang L, Seaberg MH, Hunter MS, Liang M, Koglin JE, Wakatsuki S, Demirci H (2018) Structure of the 30S ribosomal decoding complex at ambient temperature. RNA. pii: rna.067660.118. doi: 10.1261/rna.067660.118. [Epub ahead of print] PubMed PMID: 30139800

Das T, Häring M, Haldar D, Díaz DD (2017) Phenylalanine and derivatives as versatile low-molecular-weight gelators: design, structure and tailored function. Biomater Sci 6(1):38–59. https://doi.org/10.1039/c7bm00882a. Review. PubMed PMID: 29164186

Da Silva Giotto MT, Hannaert V, Vertommen D, Navarro M, Rider MH, Michels AM, Garratt RC, Rigden DJ (2003) The crystal structure of trypanosome brucei enolase: visualization of the inhibitory metal binding site III and potential as target for selective, irreversible inhibition. J Mol Biol 331:653–665

Davenport RC, Bash PA, Seaton BA, Karplus M, Petsko GA, Ringe D (1991) Structure of triose phosphate isomerase phosphoglycolohydroxamate complex: an analogue of the intermediate on the reaction pathway. Biochemistry 30:5821

Davies TG, Hubbard RE, Tame JR (1999) Relating structure to thermodynamics: the crystal structures and binding affinity of eight OppA-peptide complexes. Protein Sci 8(7):1432–1444

Dautant A, Velours J, Giraud MF (2010) Crystal structure of the Mg·ADP-inhibited state of the yeast F1c10-ATP synthase. J Biol Chem 285(38):29502–29510. https://doi.org/10.1074/jbc.M110.124529. Epub 2010 Jul 7. PubMed PMID: 20610387; PubMed Central PMCID: PMC2937982

Dayam RM, Saric A, Shilliday RE, Botelho RJ (2015) The Phosphoinositide-Gated Lysosomal Ca (2+) channel, TRPML1, Is required for Phagosome maturation. Traffic 16(9):1010–1026. https://doi.org/10.1111/tra.12303. Epub 2015 Jun 18. PubMed PMID: 26010303

Degenkolb T, Brückner H (2008) Peptaibiomics: towards a myriad of bioactive peptides containing C(alpha)-dialkylamino acids? Chem Biodivers 5(9):1817–1843. https://doi.org/10.1002/cbdv.200890171. Review. PubMed PMID: 18816513

Delmar M (2012) Connexin43 regulates sodium current; ankyrin-G modulates gap junctions: the intercalated disc exchanger. Cardiovasc Res 93(2):220–222. https://doi.org/10.1093/cvr/cvr343. Epub 2011 Dec 16. PubMed PMID: 22180603

Demelash A, Rudrabhatla P, Pant HC, Wang X, Amin ND, McWhite CD, Xu N, Ilona Linnoila R (2012) Achaete-scute homologue-1 (ASH1) stimulates migration of lung cancer cells through Cdk5/p35 pathway. Mol Biol Cell 23:2856–2866

Deng L, Guindon J, Vemuri VK, Thakur GA, White FA, Makriyannis A, Hohmann AG (2012) The maintenance of cisplatin- and paclitaxel-induced mechanical and cold allodynia is suppressed by cannabinoid CB2 receptor activation and independent of CXCR4 signaling in models of chemotherapy-induced peripheral neuropathy. Mol Pain 8:71–83

De Rubeis S, He X, Goldberg AP, Poultney CS, Samocha K, Cicek AE et al (2014) Synaptic, transcriptional, and chromatin genes disrupted in autism. Nature 515:209–215

De Simone G, Monti SM, Alterio V, Buonanno M, De Luca V, Rossi M, Carginale V, Supuran CT, Capasso C, Di Fiore A (2015) Crystal structure of the most catalytically effective carbonic anhydrase enzyme known, SazCA from the thermophilic bacterium Sulfurihydrogenibium azorense. Bioorg Med Chem Lett 25(9):2002–2006. https://doi.org/10.1016/j.bmcl.2015.02.068. Epub 2015 Mar 6. PubMed PMID: 25817590

Devaux JJ (2010) The C-terminal domain of ßIV-spectrin is crucial for KCNQ2 aggregation and excitability at nodes of Ranvier. J Physiol. 588(Pt 23):4719–4730. https://doi.org/10.1113/jphysiol.2010.196022. Epub 2010 Oct 20. PubMed PMID: 20962009; PubMed Central PMCID: PMC3010141

Di Fiore PP, Scita G (2002) Eps8 in the midst of GTPases. Int J Biochem Cell Biol 34:1178–1183

Ding J, Smith AD, Geisler SC, Ma X, Arnold GF, Arnold E (2002) Crystal structure of a human rhinovirus that displays part of the HIV-1 V3 loop and induces neutralizing antibodies against HIV-1. Structure 10(7):999–1011. PubMed PMID: 12121655

Dluzewski AR, Fryer PR, Griffiths S, Wilson RJ, Gratzer WB (1989) Red cell membrane protein distribution during malarial invasion. J Cell Sci 92(Pt4):691–699. PubMed PMID: 2532219

Dodonova SO, Diestelkoetter-Bachert P, von Appen A, Hagen WJ, Beck R, Beck M, Wieland F, Briggs JA (2015) Vesicular transport. A structure of the COPI coat and the role of coat proteins in membrane vesicle assembly. Science 349(6244):195–198. https://doi.org/10.1126/science.aab1121. PubMed PMID: 26160949

Dong Y, Liu Y, Jiang W, Smith TJ, Xu Z, Rossmann MG (2017) Antibody-induced uncoating of human rhinovirus B14. Proc Natl Acad Sci USA 114(30):8017–8022. https://doi.org/10.1073/pnas.1707369114. Epub 2017 Jul 10. PubMed PMID: 28696310; PubMed Central PMCID: PMC5544331

Dorstyn L, Akey CW, Kumar S (2018) New insights into apoptosome structure and function. Cell Death Differ 25(7):1194–1208. https://doi.org/10.1038/s41418-017-0025-z. Epub 2018 May 15. Review. PubMed PMID: 29765111; PubMed Central PMCID: PMC6030056

Dosemeci A, Toy D, Reese TS, Tao-Cheng J-H (2015) AIDA-1 moves out of the postsynaptic density core under excitatory conditions. PLoS ONE 10:e0137216

Dosemeci A, Weinberg RJ, Reese TS, Tao-Cheng J-H (2016) The Postsynaptic Density: there is more than meets the eye. Front Synaptic Neurosci 8:23

Dou Y, Li Y, Chen J, Wu S, Xiao X, Xie S, Tang L, Yan M, Wang Y, Lin J, Zhu W, Yan G (2013) Inhibition of cancer cell proliferation by midazolam by targeting transient receptor potential melastatin 7. Oncol Lett 5(3):1010–1016. Epub 2013 Jan 11. PubMed PMID: 23426784; PubMed Central PMCID: PMC3576402

Dowling DP, Croft AK, Drennan CL (2012) Radical use of Rossmann and TIM barrel architectures for controlling coenzyme B12 chemistry. Annu Rev Biophys 41:403–427. https://doi.org/10.1146/annurev-biophys-050511-102225. Review. PubMed PMID: 22577824

Duclohier H (2010) Antimicrobial peptides and peptaibols, substitutes for conventional antibiotics. Curr Pharm Des 16(28):3212–3223. Review. PubMed PMID: 2068788

Dufour A, Bellac CL, Eckhard U, Solis N, Klein T, Kappelhoff R, Fortelny N, Jobin P, Rozmus J, Mark J, Pavlidis P, Dive V, Barbour SJ, Overall CM (2018) C-terminal truncation of IFN-? inhibits proinflammatory macrophage responses and is deficient in autoimmune disease. Nat Commun 9(1):2416. https://doi.org/10.1038/s41467-018-04717-4. PubMed PMID: 29925830; PubMed Central PMCID: PMC6010466

Dulhunty A, Gage P, Curtis S et al (2001) The glutathione transferase structural family includes a nuclear chloride channel and a ryanodine receptor calcium release channel modulator. J Biol Chem 276:3319–3323

Dunbrack RL Jr, Karplus M (1993) Backbone-dependent rotamer library for proteins. Application to side-chain prediction. J Mol Biol 230(2):543–574

Durak O, de Anda FC, Singh KK, Leussis MP, Petryshen TL, Sklar P, Tsai L-H (2015) Ankyrin-G regulates neurogenesis and Wnt signaling by altering the subcellular localization of β-catenin. Mol Psychiatry 20:388–397

Durand CM, Perroy J, Loll F, Perrais D, Fagni L, Bourgeron T et al (2012) *SHANK3* mutations identified in autism lead to modification of dendritic spine morphology via an actin-dependent mechanism. Mol Psychiatry 17:71–84

Edwards RA, Lee MS, Tsutakawa SE, Williams RS, Tainer JA, Glover JNM (2008) The BARD1 C-terminal domain structure and interactions with polyadenylation factor CstF-50. Biochemistry *47*:11446–11456

Effertz K, Hinderlich S, Reutter W (1999) Selective Loss of either the Epimerase or Kinase Activity of UDP-*N*-acetylglucosamine 2-Epimerase/*N*-Acetylmannosamine Kinase due to Site-directed Mutagenesis Based on Sequence Alignments. J Biol Chem 274:28771

Efimov AV (1979) Packing of alpha-helices in globular proteins. Layer-structure of globin hydrophobic cores. J Mol Biol 134(1):23–40

Eisenberg I, Avidan N, Potikha T, Hochner H, Chen M, Olender T, Barash M, Shemesh M, Sadeh M, Grabov-Nardini G, Shmilevich I, Friedmann A, Karpati G, Bradley WG, Baumbach L, Lancet D, Asher EB, Beckmann JS, Argov Z, Mitrani-Rosenbaum S (2001) The UDP-N-acetylglucosamine 2-epimerase/N-acetylmannosamine kinase gene is mutated in recessive hereditary inclusion body myopathy. Nat Genet 29:83

Elkin I, Maris T, Melkoumov A, Hildgen P, Banquy X, Leclair G, Barrett C (2018) Crystal structure of 2-oxopyrrolidin-3-yl 4-(2-phenyl-diazen-1-yl)benzoate. Acta Crystallogr E Crystallogr Commun 6;74(Pt 4):458–460. doi: 10.1107/S205698901800333X. eCollection 2018 Apr 1. PubMed PMID: 29765745; PubMed Central PMCID: PMC5946967

Eriksson M, Hassan S, Larsson R, Linder S, Ramqvist T, Lövborg H, Vikinge T, Figgemeier E, Müller J, Stetefeld J, Dalianis T, Ozbek S (2009) Utilization of a right-handed coiled-coil protein from archaebacterium Staphylothermus marinus as a carrier for cisplatin. Anticancer Res 29(1):11–18. PubMed PMID: 19331128

Erler I, Hirnet D, Wissenbach U, Flockerzi V, Niemeyer BA (2004) J Biol Chem 279(34456)

Erreger K, Chen PE, Wyllie DJ, Traynellis SF (2004) Glutamate receptor gating. Crit Rev Neurobiol 16:187–124

Eydt K, Davies KM, Behrendt C, Wittig I, Reichert AS (2017) Cristae architecture is determined by an interplay of the MICOS complex and the F(1)F(O) ATP synthase via Mic27 and Mic10. Microb Cell 4(8):259–272. https://doi.org/10.15698/mic2017.08.585. PubMed PMID: 28845423; PubMed Central PMCID: PMC5568431

Faurobert E, Albiges-Rizo C (2010) Recent insights into cerebral cavernous malformations: a complex jigsaw puzzle under construction. The Febs Journal 277:1084–1096

Fenwick MK, Oswald RE (2008) NMR spectroscopy of the ligand-binding core of ionotropic glutamate receptor 2 bound to 5-substituted willardiine partial agonists. J Mol Biol. 378:673–685

Ferguson AD, Mckeever BM, Wisniewski D et al (2007) Crystal structure of inhibitor-bound human 5-lipoxygenase-activating protein. Science 317:510–512

Fermi G, Perutz MF, Shaanan B, Fourme R (1984) The crystal structure of human deoxyhaemoglobin at 1.74 A resolution. J Mol Biol 175(2):159–174. PubMed PMID: 6726807

Ferner J, Suhartono M, Breitung S, Jonker HR, Hennig M, Wöhnert J, Göbel M, Schwalbe H (2009) Structures of HIV TAR RNA-ligand complexes reveal higher binding stoichiometries. Chembiochem 10(9):1490–1494. https://doi.org/10.1002/cbic.200900220

Ferre-D'Amare AR, Prendergast GC, Ziff EB, Burley SK (1993) Recognition by MAX of its cognate DNA through a dimeric B/HLH/Z domain. Nature 363:38

Ferreiro DU, Cervantes CF, Truhlar Stephanie ME, Cho SS, Wolynes PG, Komives EA (2007) Stabilizing IκBα by 'consensus' design. J Mol Biol 365:1201–1216

Ferrone FA (2016) Sickle cell disease: Its molecular mechanism and the one drug that treats it. Int J Biol Macromol 93(Pt A):1168–1173. https://doi.org/10.1016/j.ijbiomac.2016.09.073. Epub 2016 Sep 22

Ferofontov A, Strulovich R, Marom M, Giladi M, Haitin Y (2018) Inherent flexibility of CLIC6 revealed by crystallographic and solution studies. Sci Rep 8(1):6882. https://doi.org/10.1038/s41598-018-25231-z. PubMed PMID: 29720717; PubMed Central PMCID: PMC5931990

Filippakopoulos P, Müller S, Knapp S (2009) SH2 domains: modulators of nonreceptor tyrosine kinase activity. Curr Opin Struct Biol 19(6):643–649. https://doi.org/10.1016/j.sbi.2009.10.001. Epub 2009 Nov 18. Review. PubMed PMID: 19926274; PubMed Central PMCID: PMC2791838

Fisher E (1901) Z Physiol Chem 33:151

Fisher E (1902) Chem Ber 35:2660

Formaggio F, Crisma M, Bonora GM, Pantano M, Valle G, Toniolo C, Aubry A, Bayeul D, Kamphuis J (1995) (R)-isovaline homo-peptides adopt the left-handed 3(10)-helical structure. Pept Res 8(1):6–15. PubMed PMID: 7756755

Forwood JK, Lange A, Zachariae U, Marfori M, Preast C, Grubmüller H, Stewart M, Corbett AH, Kobe B (2010) Quantitative structural analysis of importin-β flexibility: paradigm for solenoid protein structures. Structure *18*:1171–1183

Fredslund F, Borchert MS, Poulsen JN, Mortensen SB, Perner M, Streit WR, Lo LL (2018) Structure of a hyperthermostable carbonic anhydrase identified from an active hydrothermal vent chimney. Enzyme Microb Technol 114:48–54. https://doi.org/10.1016/j.enzmictec.2018.03.009. Epub 2018 Mar 30. PubMed PMID: 29685353

Fritz-Wolf K, Becker K, Rahlfs S, Harwaldt P, Schirmer RH, Kabsch W, Becker K (2003) X-ray structure of glutathione S-transferase from the malarial parasite Plasmodium falciparum. 100 (24):13821–13826

Fu Q, Shaik MM, Cai Y, Ghantous F, Piai A, Peng H, Rits-Volloch S, Liu Z, Harrison SC, Seaman MS, Chen B, Chou JJ (2018) Structure of the membrane proximal external region of HIV-1 envelope glycoprotein. Proc Natl Acad Sci USA 115(38):E8892–E8899. doi: https://doi.org/10.1073/pnas.1807259115. Epub 2018 Sep 5. PubMed PMID: 30185554; PubMed Central PMCID: PMC6156635

Fujita-Jimbo E, Tanabe Y, Yu Z, Kojima K, Mori M, Li H et al (2015) The association of GPR85 with PSD-95-neuroligin complex and autism spectrum disorder: a molecular analysis. Molecular Autism 6: 17 *Proc*. Natl Acad Sci *100*:13821–13826

Fung T, Asiri YI, Wall R, Schwarz SKW, Puil E, MacLeod BA (2017) Variations of isovaline structure related to activity in the formalin foot assay in mice. Amino Acids 49(7):1203–1213. https://doi.org/10.1007/s00726-017-2421-6. Epub 2017 Apr 21. PubMed PMID: 28432424

Furukawa H, Gouaux E (2003) Mechanisms of activation, inhibition and specificity: crystal structures of the NMDA receptor NR1 ligand-binding core. Embo J 22:2873–2885

Furukawa H, Singh SK, Mancusso R, Gouaux E (2005) Subunit arrangement and function in NMDA receptors. Nature 438:185–192

Gamblin SJ, Davies GJ, Grimes JM, Jackson RM, Littlechild JA, Watson HC (1991) Activity and specificity of human aldolases. J Mol Biol 219:573–576

Gamir-Morralla A, López-Menéndez C, Ayuso-Dolado S, Tejeda GS, Montaner J, Rosell A et al (2015) Development of a neuroprotective peptide that preserves survival pathways by preventing Kidins220/ARMS calpain processing induced by excitotoxicity. Cell Death & Disease 6:e1939. https://doi.org/10.1038/cddis.2015.307

Gamlin CR, Yu WQ, Wong ROL, Hoon M (2018) Assembly and maintenance of GABAergic and Glycinergic circuits in the mammalian nervous system. Neural Dev 13(1):12. https://doi.org/10.1186/s13064-018-0109-6. Review. PubMed PMID: 29875009; PubMed Central PMCID: PMC5991458

Gan Z, Xu H (2017) Photoluminescence of diphenylalanine peptide nano/microstructures: from mechanisms to applications. Macromol Rapid Commun 38(22). https://doi.org/10.1002/marc.201700370. Epub 2017 Sep 13. Review. PubMed PMID: 28902961

Gardoni F, Di Luca M (2006) New targets for pharmacological intervention in the glutamatergic synapse. Eur J Pharmacol 545:2–10

Gaubitz C, Prouteau M, Kusmider B, Loewith R (2016) TORC2 structure and function. Trends Biochem Sci 41(6):532–545: https://doi.org/10.1016/j.tibs.2016.04.001. Epub 2016 May 5. Review. PubMed PMID: 27161823

Gaudet R (2008) A primer on ankyrin repeat function in TRP channels and beyond. Mol Biosyst 4:372–379

Gernert KM, Surles MC, Labean TH, Richardson JS, Richardson DC (1995) Alacoil: a very tight, antiparallel coiled-coil of helices. Protein Science 4:2252–2260

Gerritsen M, Nemethy G, Scheraga HA (1985) Energetics of multi-helix interactions in protein folding: Application to myoglobin. Biopolymers 24:1271–1293

Gessmann R, Brückner H, Petratos K (2016) The crystal structure of Z-(Aib)10-OH at 0.65 Å resolution: three complete turns of 310-helix. J Pept Sci 22(2):76–81. https://doi.org/10.1002/psc.2842. Epub 2015 Dec 18

Ghaderi D, Strauss HM, Reinke S, Cirak S, Reutter W, Lucka L, Hinderlich S (2007) Evidence for dynamic interplay of different oligomeric states of UDP-N-acetylglucosamine 2-epimerase/N-acetylmannosamine kinase by biophysical methods. J Mol Biol 369:746–758

Ghazawi FM, Faller EM, Parmar P, El-Salfiti A, MacPherson PA (2016) Suppressor of cytokine signaling (SOCS) proteins are induced by IL7 and target surface CD127 protein for degradation in human CD8 T cells. Cell Immunol 306-307:41–52. https://doi.org/10.1016/j.cellimm.2016.07.002

Ghersi E, Noviello C, D'Adamio L (2004) Amyloid-beta protein precursor (AbetaPP) intracellular domain-associated protein-1 proteins bind to AbetaPP and modulate its processing in an isoform-specific manner. J Biol Chem 279:49105–49112

Glondu-Lassis M, Dromard M, Chavey C, Puech C, Fajas L, Hendriks W, Freiss G (2009) Downregulation of protein tyrosine phosphatase PTP-BL represses adipogenesis. Int J Biochem Cell Biol 41:2173–2180

Gmeiner WH, Horita DA (2001) Implications of SH3 domain structure and dynamics for protein regulation and drug design. Cell Biochem Biophys 35(2):127–140. Review. PubMed PMID: 11892788

Goh KL, Holmes DF (2017) Collagenous extracellular matrix biomaterials for tissue engineering: lessons from the common sea urchin tissue. Int J Mol Sci 18(5):901. https://doi.org/10.3390/ijms18050901

Kapitein LC, Hoogenraad CC (2011) Which way to go? Cytoskeletal organization and polarized transport in neurons. Mol Cell Neurosci 46(1):9–20. https://doi.org/10.1016/j.mcn.2010.08.015. Epub 2010 Sep 9

Gonzalez-Reyes LE, Ladas TP, Chiang CC, Durand DM (2013) TRPV1 antagonist capsazepine suppresses 4-AP-induced epileptiform activity in vitro and electrographic seizures in vivo. Exp Neurol 250:321–332. doi: https://doi.org/10.1016/j.expneurol.2013.10.010. Epub 2013 Oct 18. PubMed PMID: 24145133; PubMed Central PMCID: PMC4104534

Görbitz CH, Karen P, Dušek M, Petrícek V (2016) An exceptional series of phase transitions in hydrophobic amino acids with linear side chains. IUCrJ 3(Pt 5):341–353. https://doi.org/10.1107/S2052252516010472

Gord JR, Hewett DM, Hernandez-Castillo AO, Blodgett KN, Rotondaro MC, Varuolo A, Kubasik MA, Zwier TS (2016) Conformation-specific spectroscopy of capped, gas-phase Aib oligomers: tests of the Aib residue as a 310-helix former. Phys Chem Chem Phys 18(36):25512–25527

Goto M, Miyahara I, Hayashi H, Kagamiyama H, Hirotsu K (2003) Crystal structures of branched-chain amino acid aminotransferase complexed with glutamate and glutarate: true reaction intermediate and double substrate recognition of the enzyme. Biochemistry 42 (13):3725–3733. PubMed PMID: 12667063

Granot-Hershkovitz E, Raas-Rothschild A, Frumkin A, Granot D, Silverstein S, Abeliovich D (2011) Complex chromosomal rearrangement in a girl with psychomotor-retardation and a de novo inversion: inv(2)(p15;q24.2). Am J Med Genet A 155A:1825–1832

Grauer AA, Cabrele C, Zabel M, König B (2009) Stable right- and left-handed peptide helices containing C(alpha)-tetrasubstituted alpha-amino acids. J Org Chem 74(10):3718–3726. doi: https://doi.org/10.1021/jo900222g. PubMed PMID: 19354242

Greber BJ, Bieri P, Leibundgut M, Leitner A, Aebersold R, Boehringer D, Ban N (2015) Ribosome. The complete structure of the 55S mammalian mitochondrial ribosome. Science 348 (6232):303–308. doi: https://doi.org/10.1126/science.aaa3872. Epub 2015 Apr 2. PubMed PMID: 25837512

Greenberg KP, Pham A, Werblin FS (2011) Differential targeting of optical neuromodulators to ganglion cell soma and dendrites allows dynamic control of center-surround antagonism. Neuron 69:713–720

Gregersen JL, Mattle D, Fedosova NU, Nissen P, Reinhard L (2016) Isolation, crystallization and crystal structure determination of bovine kidney Na(+),K(+)-ATPase. Acta Crystallogr F Struct Biol Commun 72(Pt4):282–287. doi: https://doi.org/10.1107/S2053230X1600279X. Epub 2016 Mar 16. PubMed PMID: 27050261; PubMed Central PMCID: PMC4822984

Groothuizen FS, Winkler I, Cristóvão M, Fish A, Winterwerp HH, Reumer A et al (2015) MutS/MutL crystal structure reveals that the MutS sliding clamp loads MutL onto DNA. eLife 4: e06744. https://doi.org/10.7554/eLife.06744

Groves MR, Hanlon N, Turowski P, Hemmings BA, Barford D (1999) The structure of the PROTEIN Phosphatase 2A PR65/A Subunit Reveals the Conformation of Its 15 Tandemly Repeated HEAT Motifs. Cell *96*:99–110

Grünwald M, Bono F (2011) Structure of Importin13-Ubc9 complex: nuclear import and release of a key regulator of sumoylation. EMBO J 30(2):427–438. https://doi.org/10.1038/emboj.2010.320. Epub 2010 Dec 7. PubMed PMID: 21139563; PubMed Central PMCID: PMC3025465

Gu L, Hong Y, McCulloch S, Watanabe H, Li GM (1998) ATP-dependent interaction of human mismatch repair proteins and dual role of PCNA in mismatch repair. Nucleic Acids Res 26 (5):1173–1178. PubMed PMID: 9469823; PubMed Central PMCID: PMC147380

Guarné A, Junop MS, Yang W (2001) Structure and function of the N-terminal 40 kDa fragment of human PMS2: a monomeric GHL ATPase. The EMBO Journal 20(19):5521–5531. https://doi.org/10.1093/emboj/20.19.5521

Guerra ME, Fadel V, Maltarollo VG, Baldissera G, Honorio KM, Ruggiero JR, Dos Santos Cabrera MP (2017) MD simulations and multivariate studies for modeling the antileishmanial activity of peptides. Chem Biol Drug Des 2017 Mar 7. doi: https://doi.org/10.1111/cbdd.12970. [Epub ahead of print]

Gunning PW, Hardeman EC, Lappalainen P, Mulvihill DP (2015) Tropomyosin - master regulator of actin filament function in the cytoskeleton. J Cell Sci 128(16):2965–2974. https://doi.org/10.1242/jcs.172502. Epub 2015 Aug 3

Haa BH, Matthew J, Davisa B, Chena C, Loua HJ, Jia Gaoa C, Zhanga R, Krauthammerd M, Halabane R, Schlessingera J, Turka BE, Boggon TJ (2012) Type II p21-activated kinases (PAKs) are regulated by an autoinhibitory pseudosubstrate. Proc Natl Acad Sci 109:16107–16112

Hafner AS, Penn AC, Grillo-Bosch D, Retailleau N, Poujol C, Philippat A, Coussen F, Sainlos M, Opazo P, Choquet D (2015) Lengthening of the Stargazin Cytoplasmic Tail Increases Synaptic Transmission by Promoting Interaction to Deeper Domains of PSD-95. Neuron 86:475–489

Halazonetis TD, Gorgoulis VG, Bartek J (2008) An oncogene-induced DNA damage model for cancer development. Science 319:1352–1355

Hald H, Naur P, Pickering DS, Sprogoe D, Madsen U, Timmermann DB, Ahring PK, Liljefors T, Schousboe A, Edebjerg J, Gajhede M, Kastrup JS (2007) Partial agonism and antagonism of the Ionotropic Glutamate Receptor iGLuR5. J Biol Chem 282:25726–25736

Hales EC, Taub JW, Matherly LH (2014) New insights into Notch1 regulation of the PI3K-AKT-mTOR1 signaling axis: targeted therapy of ?-secretase inhibitor resistant T-cell acute lymphoblastic leukemia. Cell Signal 26(1):149–161. https://doi.org/10.1016/j.cellsig.2013.09.021. Epub 2013 Oct 16. Review. PubMed PMID: 24140475

Hall BE, Prochazkova M, Sapio MR, Minetos P, Kurochkina N, Binukumar BK, Amin ND, Terse A, Joseph J, Raithel SJ, Mannes AJ, Pant HC, Chung MK, Iadarola MJ, Kulkarni AB (2018) Phosphorylation of the Transient Receptor Potential Ankyrin 1 by Cyclin-dependent Kinase 5 affects Chemo-nociception. Sci Rep 8(1):1177. https://doi.org/10.1038/s41598-018-19532-6. PubMed PMID: 29352128; PubMed Central PMCID: PMC5775258

Hall TM (2016) De-coding and re-coding RNA recognition by PUF and PPR repeat proteins. Curr Opin Struct Biol 36:116–121. https://doi.org/10.1016/j.sbi.2016.01.010. Epub 2016 Feb 11. Review. PubMed PMID: 26874972; PubMed Central PMCID: PMC4757904

Han D, Kim K, Oh J, Park J, Kim Y (2007) TPR domain of NrfG mediates complex formation between heme lyase and formate-dependent nitrite reductase in Escherichia coli O157:H7. Proteins *70*:900–914

Han S, Nam J, Li Y, Kim S, Cho SH, Cho YS, Choi SY, Choi J, Han K, Kim Y, Na M, Kim H, Bae YC, Choi SY, Kim E (2010) Regulation of dendritic spines, spatial memory, and embryonic development by the TANC family of PSD-95-interacting proteins. J Neurosci 30:15102–15112

Harbury PB, Zhang T, Kim PS, Alber T (1993) A switch between two-, three-. and four-stranded coiled coils in GCN4 leucine zipper mutants. Science 262(5138):1401–1407

Harbury PB, Plecs JJ, Tidor B, Alber T, Kim PS (1998) High-resolution protein design with backbone freedom. Science 282(5393):1462–1467. PubMed PMID: 9822371

Harris NL, Presnell SR, Cohen FE (1994) Four helix bundle diversity in proteins. J Mol Biol 236:1356–1368

Harrop SJ, DeMaere MZ, Fairlie WD et al (2001) Crystal structure of a soluble form of the intracellular chloride ion channel CLIC1 (NCC27) at 1.4-A resolution. J Biol Chem 276:44993–45000

Harsini FM, Chebrolu S, Fuson KL, White MA, Rice AM, Sutton RB (2018) FerA is a membrane-associating four-helix bundle domain in the Ferlin family of membrane-fusion proteins. Sci Rep 8(1):10949. doi: https://doi.org/10.1038/s41598-018-29184-1. PubMed PMID: 30026467; PubMed Central PMCID: PMC6053371.

Hausmann J, Kamtekar S, Christodoulou E, Day JE, Wu T, Fulkerson Z et al (2011) Structural basis for substrate discrimination and integrin binding by autotaxin. Nat Struct Mol Biol 18 (2):198–204. https://doi.org/10.1038/nsmb.1980

Hayashi MK (2018) Structure-Function Relationship of Transporters in the Glutamate-Glutamine Cycle of the Central Nervous System. Int J Mol Sci 19(4):pii: E1177. https://doi.org/10.3390/ijms19041177. Review. PubMed PMID: 29649168; PubMed Central PMCID: PMC5979278

Hayashi MK, Tang C, Verpelli C, Narayanan R, Stearns MH, Xu R-M et al (2009) The postsynaptic density proteins Homer and Shank form a polymeric network structure. Cell 137:159–171

Heine M, Groc L, Frischknecht R, Béïque J-C, Lounis B, Rumbaugh G et al (2008) Surface Mobility of Postsynaptic AMPARs Tunes Synaptic Transmission. Science 320:201–205

Heldwein EE, Macia E, Wang J, Yin HL, Kirchhausen T, Harrison SC (2004) Crystal structure of the clathrin adaptor protein 1 core. Proc Natl Acad Sci U S A 101(39):14108–14113. Epub 2004 Sep 17. PubMed PMID: 15377783; PubMed Central PMCID: PMC521094

Hendrickson WA, Ward KB (1977) Pseudosymmetry in the structure of myohemerythrin. *J Biol Chem 252*:3012–3018

Herberich SE, Klose R, Moll I, Yang W-J, Wüstehube-Lausch J, Fischer A (2015) ANKS1B interacts with the cerebral cavernous malformation Protein-1 and controls endothelial permeability but Not sprouting angiogenesis. PLoS ONE 10:e0145304

Hester G, Brenner-Holzach O, Rossi FA, Struck-Donatz M, WinterHalter KH, Smit J, Piontek K (1991) The crystal structure of fructose-1,6-biphosphate aldolase from *Drosophila Melanogaster* at 2.5 Å resolution. Febs Letters 292:237–242

Hiller S, Kohl A, Fiorito F, Herrmann T, Wider G, Tschopp J, Grutter MG, Wuthrich K (2003) NMR structure of the apoptosis- and inflammation-related NALP1 Pyrin domain. structure 11:1199–1205

Hinderlich S, Stasche R, Zeitler R, Reitter W (1997) A bifunctional enzyme catalyzes the first two steps in N-acetylneuraminic acid biosynthesis of rat liver. Purification and characterization of UDP-N-acetylglucosamine 2-epimerase/N-acetylmannosamine kinase. J Biol Chem 272:24313–24318

Hinderlich S, Weidemann W, Yardeni T, Horstkorte R, Huizing M (2015) UDP-GlcNAc 2-epimerase/ManNAc kinase (GNE), a master regulator of sialic acid synthesis. Topics in Current Chemistry 366:97–137. https://doi.org/10.1007/128_2013_464

Hinds TD Jr, Sánchez ER (2008) Protein Phosphatase 5. Int J Biochem Cell Biol *40*:2358–2362

Ho KL, Gabrielsen M, Beh PL, Kueh CL, Thong QX, Streetley J, Tan WS, Bhella D (2018) Structure of the Macrobrachium rosenbergii nodavirus: A new genus within the Nodaviridae?, e3000038. PLoS Biol 16(10). https://doi.org/10.1371/journal.pbio.3000038. [Epub ahead of print] PubMed PMID: 30346944

Hodges RS, Sodek J, Smillie LB et al (1972) Amino-acid sequence of rabbit skeletal tropomyosin and its coiled-coil structure. Proc Natl Acad Sci 69:3800–3804

Hodkinson PS, Elliott PA, Lad Y, McHugh BJ, MacKinnon AC, Haslett C, Sethi T (2007) Mammalian NOTCH-1 activates beta1 integrins via the small GTPase R-Ras. J Biol Chem 282:28991–29001

Hoffmann E, Machelart A, Song OR, Brodin P (2018) Proteomics of mycobacterium infection: moving towards a better understanding of pathogen-driven immunomodulation. Front Immunol 9:86. https://doi.org/10.3389/fimmu.2018.00086. eCollection 2018. Review. PubMed PMID: 29441067; PubMed Central PMCID: PMC5797607

Hofmann A, Proust J, Dorowski A et al (2000) Annexin 24 from *Capsicum annuum*. J Biol Chem 275:8072–8082

Hollman AL, Tchounwou PB, Huang HC (2016) The Association between gene-environment interactions and diseases involving the human GST superfamily with SNP variants. Int J Environ Res Public Health 13(4):379. https://doi.org/10.3390/ijerph13040379. Review. PubMed PMID: 27043589; PubMed Central PMCID: PMC4847041

Holm PJ, Bhakat P, Jegerschold C, Gyobu N, Mitsuoka K, Fujiyoshi Y, Morgenstern R, Hebert H (2006) Structural basis for detoxification and oxidative stress protection in membranes. J Mol Biol *360*:934–945

Holm L, Sander C (1993) Structural alignment of globins, phycocyanins and colicin A. FEBS letters 315:301–306

Holmes TC (2002) Novel peptide-based biomaterial scaffolds for tissue engineering. Trends Biotechnol 20(1):16–21. Review. PubMed PMID: 11742673

Holton A, Alber J (2004) Automated protein crystal structure determination using ELVES. Proc Natl Acad Sci 101:1537–1542

Holzer S, Ban N, Klinge S (2013) Crystal structure of the yeast ribosomal protein rpS3 in complex with its chaperone Yar1. J Mol Biol 425(22):4154–4160. https://doi.org/10.1016/j.jmb.2013.08.022. Epub 2013 Sep 7. PubMed PMID: 24021814

Holzer P, Izzo AA (2014) The pharmacology of TPR channels. British J of Pharmacology 171:2469–2473

Hong TJ, Park K, Choi EW, Hahn JS (2017) Ro 90-7501 inhibits PP5 through a novel. TPR-dependent mechanism. Biochem Biophys Res Commun. 482(2):215–220. https://doi.org/10.1016/j.bbrc.2016.11.043. Epub 2016 Nov 10. PubMed PMID: 27840051

Hou Q, Bourgeas R, Pucci F, Rooman M (2018) Computational analysis of the amino acid interactions that promote or decrease protein solubility. Sci Rep 8(1):14661. https://doi.org/10.1038/s41598-018-32988-w. PubMed PMID: 30279585; PubMed Central PMCID: PMC6168528

Huang L, Hsiao JP, Powierza C, Taylor RM 2nd, Lord ST (2014) Does topology drive fiber polymerization? Biochemistry 53(49):7824–7834. https://doi.org/10.1021/bi500986z. Epub 2014 Dec 5. PubMed PMID: 25419972; PubMed Central PMCID: PMC4270379

Huber R, Romish J, Paques EP (1990) The crystal and molecular structure of human annexin V, an anticoagulant protein that binds to calcium and membranes. Embo J 9:3867–3874

Huber AH, Nelson WJ, Weis WI (1997) Three-Dimensional Structure of the Armadillo Repeat Region of *b*-Catenin. Cell *90*:871–882

Huggins DJ, Grant GH (2005) The function of the amino-terminal domain in NMDA receptor modulation. J Mol Graph Model 23:381–388

Huizing M (2005) Disease mechanisms associated with mutations of the GNE gene. Drug discovery today: Disease mechanisms 2:519–527

Huizing M, Helip-Wooley A, Westbroek W, Gunay-Aygun M, Gahl WA (2008) Disorders of lysosome-related organelle biogenesis: clinical and molecular genetics. Annu Rev Genomics Hum Genet. 9:359–386. https://doi.org/10.1146/annurev.genom.9.081307.164303. Review. PubMed PMID: 18544035; PubMed Central PMCID: PMC2755194

Huizing M, Krasnewich DM (2009) Hereditary inclusion body myopathy: a decade of progress. Biochim Biophys Acta 1792:881

Huizing M, Rakocevic G, Sparks SE, Mamali I, Shatunov A, Goldfarb L, Krasnewich D, Gahl WA, Dalakas MC (2004) Hypoglycosylation of alpha-dystroglycan in patients with hereditary IBM due to GNE mutations. Mol Genet Metab 81:196

Hutchins BI, Wray S (2014) Capture of microtubule plus-ends at the actin cortex promotes axophilic neuronal migration by enhancing microtubule tension in the leading process. Front Cell Neurosci 8:400. https://doi.org/10.3389/fncel.2014.00400

Iadarola MJ, Sapio MR, Raithel SJ, Mannes AJ, Brown DC (2018) Long-term pain relief in canine osteoarthritis by a single intra-articular injection of resiniferatoxin, a potent TRPV1 agonist. Pain 159(10):2105–2114. https://doi.org/10.1097/j.pain.0000000000001314. PubMed PMID: 30015705

Ida M, Sato A, Matsumoto I et al (2004) Human annexin V binds to sulfatide: contribution to regulation of blood coagulation. J Mol Biol 135:583–588

Igarashi M, Nozumi M, Wu LG, Cella Zanacchi F, Katona I, Barna L, Xu P, Zhang M, Xue F, Boyden E (2018) New observations in neuroscience using superresolution microscopy. J Neurosci 38(44):9459–9467. https://doi.org/10.1523/JNEUROSCI.1678-18.2018. PubMed PMID: 30381437

Ikeda K, Yamamoto R, Wirschell M, Yagi T, Bower R, Porter ME, Sale WS, Kamiya R (2009) A novel ankyrin-repeat protein interacts with the regulatory proteins of inner arm dynein f (I1) of Chlamydomonas reinhardtii. Cell Motil Cytoskeleton 66(8):448–456. https://doi.org/10.1002/cm.20324. PubMed PMID: 19021242; PubMed Central PMCID: PMC3102495

Ingles-Prieto A, Ibarra-Molero B, Delgado-Delgado A, Perez-Jimenez R, Fernandez JM, Gaucher EA, Sanchez-Ruiz JM, Gavira JA (2013) Conservation of protein structure over four billion years. Structure. 21(9):1690–1697. https://doi.org/10.1016/j.str.2013.06.020. Epub 2013 Aug 8. PubMed PMID: 23932589; PubMed Central PMCID: PMC3774310

Isaka Y, Takabatake Y, Takahashi A, Saitoh T, Yoshimori T (2016) Hyperuricemia-induced inflammasome and kidney diseases. Nephrol Dial Transplant 31(6):890–896. https://doi.org/10.1093/ndt/gfv024. Epub 2015 Mar 31. Review. PubMed PMID: 25829326

Ishitsuka R, Kojima K, Utsumi H et al (1998) Glycosaminoglycan Binding Properties of Annexin IV, V, and VI. J Biol Chem 273:9935–9941

Izard T, Sygusch J (2004) Induced Fit Movements and Metal Cofactor Selectivity of Class II Aldolases. Structure of Thermus aquaticus fructose-1,6-biphospate aldolase. J Biol Chem 279:11825–11833

IUPAC-IUB Commission on biochemical nomenclature (1970) Abbreviations and symbols for the description of the conformation of polypeptide chains. J Mol Biol 52:1–17

Lai H et al (2012) Characterization of 8-hydroxyquinoline derivatives containing aminobenzothiazole as inhibitors of dengue virus type 2 protease in vitro. Antiviral Res

Jacob AL, Jordan BA, Weinberg R (2010) The organization of amyloid-β protein precursor intracellular domain-associated protein-1 in the rat forebrain. J Comp Neurol 518:3221–3236

Jacobs MD, Harrison SC (1998) Structure of an IkappaBalpha/NF-kappaB complex. Cell 95 (6):749–758. PubMed PMID: 9865693

Jacobsson PJ, Mansini JA, Riendeau D et al (1997) Identification and characterization of a novel microsomal enzyme with glutathione-dependent transferase and peroxidase activities. J Biol Chem 272:22934–22939

Jaffe EK (2017) New protein structures provide an updated understanding of phenylketonuria. Mol Genet Metab 121(4):289–296. https://doi.org/10.1016/j.ymgme.2017.06.005. Epub 2017 Jun 15. Review. PubMed PMID: 28645531; PubMed Central PMCID: PMC5549558

Janus BM, van Dyk N, Zhao X, Howell KA, Soto C, Aman MJ, Li Y, Fuerst TR, Ofek G (2018) Structural basis for broad neutralization of ebolaviruses by an antibody targeting the glycoprotein fusion loop. Nat Commun 9(1):3934. https://doi.org/10.1038/s41467-018-06113-4. PubMed PMID: 30258051; PubMed Central PMCID: PMC6158212

Jeffrey PD, Russo AA, Polyak K, Gibbs E, Hurwltz J, Massague J, Pavletich NP (1995) Mechanism of CDK activationrevealed by the structure of a cyclinA-CDK2 complex. Nature 376:313–320

Jenkins HT, Baker-Wilding R, Edwards TA (2009) Structure and RNA binding of the mouse Pumilio-2 Puf domain. J Struct Biol. 167(3):271–276. https://doi.org/10.1016/j.jsb.2009.06.007. Epub 2009 Jun 18. PubMed PMID: 19540345

Jenkins PM, Kim N, Jones SL, Tseng WC, Svitkina TM, Yin HH, Bennett (2015) Giant ankyrin-G: A critical innovation in vertebrate evolution of fast and integrated neuronal signaling. Proc Natl Acad Sci USA 112:957–964

Jeppesen MG, Ortiz P, Shepard W et al (2003) The Crystal Structure of the Glutathione *S*-Transferase-like Domain of Elongation Factor 1B from *Saccharomyces cerevisiae*. J Biol Chem 278:47190–47198

Jeyaprakash AA, Santamaria A, Jayachandran U, Chan YW, Benda C, Nigg EA, Conti E (2012) Structural and functional organization of the Ska complex, a key component of the kinetochore-microtubule interface. Mol Cell 46(3):274–286. https://doi.org/10.1016/j.molcel.2012.03.005. Epub 2012 Apr 5. PubMed PMID: 22483620

Jia Y, Yun CH, Park E, Ercan D, Manuia M, Juarez J, Xu C, Rhee K, Chen T, Zhang H, Palakurthi S, Jang J, Lelais G, DiDonato M, Bursulaya B, Michellys PY, Epple R, Marsilje TH, McNeill M, Lu W, Harris J, Bender S, Wong KK, Jänne PA, Eck MJ (2016) Overcoming EGFR(T790M) and EGFR(C797S) resistance with mutant-selective allosteric inhibitors. Nature 534(7605):129–132. doi: https://doi.org/10.1038/nature17960. Epub 2016 May 25. PubMed PMID: 27251290; PubMed Central PMCID: PMC4929832

Jiang Y, Lee A, Chen J, Ruta V, Cadene M, Chait BT, MacKinnon R (2003) X-ray structure of a voltage-dependent k+ channel. Nature 423:33–41

Jiang YY, Hou HT, Yang Q, Liu XC, He GW (2017) Chloride channels are involved in the development of atrial fibrillation–A transcriptomic and proteomic study. Sci Rep 7(1):10215. doi: 10.1038

Jiang Q, Li K, Lu WJ, Li S, Chen X, Liu XJ, Yuan J, Ding Q, Lan F, Cai SQ (2018) Identification of small-molecule ion channel modulators in C. elegans channelopathy models. Nat Commun 9 (1):3941. https://doi.org/10.1038/s41467-018-06514-5. PubMed PMID: 30258187; PubMed Central PMCID: PMC6158242

Jin P, Bulkley D, Guo Y, Zhang W, Guo Z, Huynh W, Wu S, Meltzer S, Cheng T, Jan LY, Jan YN, Cheng Y (2017) Electron cryo-microscopy structure of the mechanotransduction channel NOMPC. Nature. 547(7661):118-122. https://doi.org/10.1038/nature22981. Epub 2017 Jun 26. PubMed PMID: 28658211; PubMed Central PMCID: PMC5669069.

Jin AJ, Lafer EM, Peng JQ, Smith PD, Nossal R (2013) Unraveling protein–protein interactions in clathrin assemblies via atomic force spectroscopy. Methods *59*:316–327

Jínek M, Rehwinkel J, Lazarus BD, Izaurralde E, Hanover JA, Conti E (2004) The superhelical TPR-repeat domain of O-linked GlcNAc transferase exhibits structural similarities to importin alpha. Nat Struct Mol Biol 11(10):1001–1007. Epub 2004 Sep 12. PubMed PMID: 15361863

Johansen TN, Greenwood JR, Frydenvang K, Madsen U, Kroogsgaard-Larsen P (2003) Stereostructure-activity studies on agonists of AMPA and kainate subtypes of ionotropic glutamate receptors. Chirality 15:167–179

John Peter AT, Lachmann J, Rana M, Bunge M, Cabrera M, Ungermann C (2013) The BLOC-1 complex promotes endosomal maturation by recruiting the Rab5 GTPase-activating protein Msb3. J Cell Biol 201(1):97-111. doi: https://doi.org/10.1083/jcb.201210038. PubMed PMID: 23547030; PubMed Central PMCID: PMC3613695

Joe PA, Banerjee A, Ludueña RF (2009) Roles of beta-tubulin residues Ala428 and Thr429 in microtubule formation in vivo. J Biol Chem 284(7):4283–4291. https://doi.org/10.1074/jbc.M807491200. Epub 2008 Dec 13. PubMed PMID: 19074767; PubMed Central PMCID: PMC3837401

Johansen TN, Clausen RP (2007) A tetrazolyl-substituted subtype-selective AMPA receptor agonist. J Med Chem 17:2408–2414

Johansson P, Wiltschi B, Kumari P, Kessler B, Vonrhein C, Vonck J, Oesterhelt D, Grininger M (2008) Inhibition of the fungal fatty acid synthase type I multienzyme complex. Proc Natl Acad Sci U S A. 105(35):12803–12808. https://doi.org/10.1073/pnas.0805827105. Epub 2008 Aug 25. PubMed PMID: 18725634; PubMed Central PMCID: PMC2529065

Jogl J, Rozovsky S, MsDermott AE, Tong L (2003) Optimal alignment for enzymatic proton transfer: Structure of the Michaelis complex of triosephosphate isomerase at 1.2-Å resolution. PNAS 100:50–55

Jonas-Straube E, Hutin C, Hoffman NE, Schünemann D (2001) Functional analysis of the protein-interacting domains of chloroplast SRP43. J Biol Chem.276(27):24654-24660. Epub 2001 Apr 16. PubMed PMID: 11306572

Juárez-Martínez AB, Sotelo-Mundo RR, Rudiño-Piñera E (2017) Crystal structure of a class-mu glutathione S-transferase from whiteleg shrimp Litopenaeus vannamei: structural changes in the xenobiotic binding H-site may alter the spectra of molecules bound. J Biochem Mol Toxicol 31 (2). https://doi.org/10.1002/jbt.21838. Epub 2016 Sep 22. PubMed PMID: 27717103

Juncosa JI, Takaya K, Le HV, Moschitto MJ, Weerawarna PM, Mascarenhas R, Liu D, Dewey SL, Silverman RB (2018) Design and Mechanism of (S)-3-Amino-4-(difluoromethylenyl) cyclopent-1-ene-1-carboxylic Acid, a Highly Potent ?-Aminobutyric Acid Aminotransferase Inactivator for the Treatment of Addiction. J Am Chem Soc 140(6):2151–2164. https://doi.org/10.1021/jacs.7b10965. Epub 2018 Jan 30. PubMed PMID: 29381352; PubMed Central PMCID: PMC5812813

Juneja M, Burns J, Saporta MA, Timmerman V (2018) Challenges in modelling the Charcot-Marie-Tooth neuropathies for therapy development. J Neurol Neurosurg Psychiatry. pii: jnnp-2018-318834. doi: 10.1136/jnnp-2018-318834. [Epub ahead of print] PubMed PMID: 30018047

Jung H, Shin J-H, Park Y-S, Chang M-S (2014) Ankyrin repeat-rich membrane spanning (ARMS)/ Kidins220 scaffold protein regulates neuroblastoma cell proliferation through p21. Mol Cells 37:881–887

Kadir M, Wang X, Zhu B, Liu J, Harland D, Popescu C (2017) The structure of the "amorphous" matrix of keratins. J Struct Biol 198(2):116–123. doi: 10.1016/j.jsb.2017.04.001. Epub 2017 Apr 5

Kajander T, Cortajarena AL, Mochrieb S, Regan L (2007) Structure and stability of designed TPR protein superhelices: unusual crystal packing and implications for natural TPR proteins. Acta Cryst *D63*:800–811

Kajander T, Cortajarena AL, Regan L (2006) Consensus design as a tool for engineering repeat proteins. Methods Mol Biol *340*:151–170

Kajava AV (1998) Structural diversity of leucine-rich repeat proteins. J Mol Biol *277*:519–527

Kenji K, Mitsuya M, Nishimura T, Eiki J-i, Nagata Y (2004) Structural basis for allosteric regulation of the monomeric allosteric enzyme human glucokinase. Structure 12:429–438,. March, 2004, 2004 Elsevier Science Ltd. All rights reserved. https://doi.org/10.1016/j.str.2004.02.005

Kamei H, Saito T, Ozawa M, Fujita‡ Y, Asada‡ A, Bibb§ JA, Takaomi C, Saido HS, Hisanaga S-i (2007) Cleavage of the CDK5 activator p35 to p25 by site-specific phosphorylation. J Biol Chem *282*:1687–1694

Kantharaju RS, Aravinda S, Shamala N, Balaram P (2010) Helical conformations of hexapeptides containing N-terminus diproline segments. Biopolymers 94(3):360–370. https://doi.org/10.1002/bip.21395

Kapinos LE, Burkhard P, Herrmann H, Aebi U, Strelkov SV (2011) Simultaneous formation of right- and left-handed anti-parallel coiled-coil interfaces by a coil2 fragment of human lamin A. J Mol Biol 408(1):135–146. https://doi.org/10.1016/j.jmb.2011.02.037. Epub 2011 Feb 24. PubMed PMID: 21354179

Kaplan E, Zubedat S, Radzishevsky I, Valenta AC, Rechnitz O, Sason H, Sajrawi C, Bodner O, Konno K, Esaki K, Derdikman D, Yoshikawa T, Watanabe M, Kennedy RT, Billard JM, Avital A, Wolosker H (2018) ASCT1 (Slc1a4) transporter is a physiologic regulator of brain d-serine and neurodevelopment. Proc Natl Acad Sci U S A. 115(38):9628–9633. https://doi.org/10.1073/pnas.1722677115. Epub 2018 Sep 5. PubMed PMID: 30185558; PubMed Central PMCID: PMC6156681

Karasik A, Shanmuganathan A, Howard MJ, Fierke CA, Koutmos M (2016) Nuclear protein-only ribonuclease P2 structure and biochemical characterization provide insight into the conserved properties of tRNA 5′ end processing enzymes. J Mol Biol 428(1):26–40. https://doi.org/10.1016/j.jmb.2015.11.025. Epub 2015 Dec 3. PubMed PMID: 26655022; PubMed Central PMCID: PMC4738078

Karpusas M, Axarli I, Chiniadis L, Papakyriakou A, Bethanis K, Scopelitou K, Clonis YD, Labrou NE (2013) The interaction of the chemotherapeutic drug chlorambucil with human glutathione transferase A1-1: kinetic and structural analysis. PLoS One 8(2):e56337. https://doi.org/10.1371/journal.pone.0056337. Epub 2013 Feb 27. PubMed PMID: 23460799; PubMed Central PMCID: PMC3584069

Kasper C, Pickering OS, Mirza O, Olsen L, Kristensen AS, Greenwood JR, Liljefors T, Schousboe A, Watjen F, Gajhede M, Siqurskjold BW, Kastrup JS (2006) The structure of a mixed GluR2 ligand-binding core dimer in complex with (S)-glutamate and the antagonist (S)-NS1209. J Mol Biol 357:1184–1201

Kaszas K, Keller JM, Coddou C, Mishra SK, Hoon MA, Stojilkovic S, Jacobson KA, Iadarola MJ (2012) Small molecule positive allosteric modulation of TRPV1 activation by Vanilloids and Acidic pH. *J. Pharm. Exp. Therapeutics 340*:152–160

Kasznel AJ, Zhang Y, Hai Y, Chenoweth DM (2017) Structural basis for aza-glycine stabilization of collagen. J Am Chem Soc 2017 Jul 19;139(28):9427–9430. https://doi.org/10.1021/jacs.7b03398. Epub 2017 Jul 6.

Kato K, Nishimasu H, Okudaira S, Mihara E, Ishitani R, Takagi J et al (2012) Crystal structure of Enpp1, an extracellular glycoprotein involved in bone mineralization and insulin signaling. Proc Natl Acad Sciences U S A *109*(42):16876–16881. https://doi.org/10.1073/pnas.1208017109

Kauzmann W (1959) Some factors in the interpretation of protein denaturation. Adv Protein Chem 14:1–63

Ke J, Chen RZ, Ban T, Zhou XE, Gu X, Tan MH, Chen C, Kang Y, Brunzelle JS, Zhu JK, Melcher K, Xu HE (2013) Structural basis for RNA recognition by a dimeric PPR-protein complex. Nat Struct Mol Biol 20(12):1377–1382. doi: 10.1038/nsmb.2710. Epub 2013 Nov 3. PubMed PMID: 24186060

Kebache S, Ash J, Annis MG, Hagan J, Huber M, Hassard J, Stewart CL, Whiteway M, Nantel A (2007) Grb10 and active Raf-1 kinase promote Bad-dependent cell survival. J Biol Chem 282 (30):21873–21883. Epub 2007 May 29. PubMed PMID: 17535812

Kedei N, Szabo T, Lile JD, Treanor JJ, Olah Z, Iadarola MJ, Blumberg PM (2001) Analysis of the native quaternary structure of vanilloid receptor 1. J Biol Chem 276:28613–28619. Epub 2001 May 17

Kefalas P, Brown TR, Brickell PM (1995) Signalling by the p60c-src family of protein-tyrosine kinases. Int J Biochem Cell Biol 27(6):551–563. Review. PubMed PMID: 7545532

Keleman K, Dickson BJ (2001) Short- and long-range repulsion by the Drosophila Unc5 netrin receptor. Neuron. 32(4):605–617. PubMed PMID: 11719202

Kellogg EH, Hejab NMA, Poepsel S, Downing KH, DiMaio F, Nogales E (2018) Near-atomic model of microtubule-tau interactions. Science 360(6394):1242–1246. https://doi.org/10.1126/science.aat1780. Epub 2018 May 10. PubMed PMID: 29748322

Kendrew JC (1959) Structure and function in myoglobin and other proteins. Fed Proc 18(2, Part 1):740–751. PubMed PMID: 136722

Kendrew JC, Bodo G, Dintzis HM, Parrish RG, Wyckoff H, Phillips DC (1958) A three-dimensional model of the myoglobin molecule obtained by x-ray analysis. Nature 181 (4610):662–666

Khanapur M, Alvala M, Prabhakar M, Shiva Kumar K, Edwin RK, Sri Saranya PS, Patel RK, Bulusu G, Misra P, Pal M (2017) Mycobacterium tuberculosis chorismate mutase: A potential target for TB. Bioorg Med Chem 25(6):1725–1736. https://doi.org/10.1016/j.bmc.2017.02.001. Epub 2017 Feb 4. Review. PubMed PMID: 28202315

Khaitlina SY (2014) Intracellular transport based on actin polymerization. Biochemistry (Mosc) 79 (9):917–927. https://doi.org/10.1134/S0006297914090089. Review. PubMed PMID: 25385019

Khoshouei M, Radjainia M, Baumeister W, Danev R (2017) Cryo-EM structure of haemoglobin at 3.2 Å determined with the volta phase plate. Nat Commun 8:16099 https://doi.org/10.1038/ncomms16099

Kikuchi N, Fujiwara K, Ikeguchi M (2018) ß-strand twisting/bending in soluble and transmembrane ß-barrel structures. Proteins. https://doi.org/10.1002/prot.25576. [Epub ahead of print] PubMed PMID: 30019770

Kim KJ, Park MC, Choi SJ, Oh YS, Choi EC, Cho HJ, Kim MH, Kim SH, Kim DW, Kim S, Kang BS (2008) Determination of three-dimensional structure and residues of the novel tumor suppressor AIMP3/p18 required for the interaction with ATM. J Biol Chem 283 (20):14032–14040. https://doi.org/10.1074/jbc.M800859200. Epub 2008 Mar 14. PubMed PMID: 18343821

Kim TH, Mehrabi P, Ren Z, Sljoka A, Ing C, Bezginov A, Ye L, Pomès R, Prosser RS, Pai EF (2017) The role of dimer asymmetry and protomer dynamics in enzyme catalysis. Science 355 (6322). pii: eaag2355. doi:10.1126/science.aag2355. PubMed PMID: 28104837

Kim DH, Park M-J, Gwon GH, Silkov A, Xu Z-Y, Yang EC, Song S, Song K, Kim Y, Yoon HS, Honig B, Cho W, Cho Y, Hwang I (2014) Chloroplast targeting factor AKR2 evolved from an ankyrin repeat domain coincidentally binds two chloroplast lipids. Dev Cell 30(5):598–609

Kim KX, Rutherford MA (2016) Maturation of NaV and KV channel topographies in the auditory nerve spike initiator before and after developmental onset of hearing function. J Neurosci 36

(7):2111–2118. https://doi.org/10.1523/JNEUROSCI.3437-15.2016. PubMed PMID: 26888923

Kitaigorodsky AI (1973) Molecular crystals and molecules. Academic Press, New York

Klenchin VA, Schmidt DM, Gerlt JA, Rayment I (2004) Evolution of enzymatic activities in the enolase superfamily: structure of a substrate-liganded complex of the L-ALA-D/L-GLU epimerase from *Bacillus subtilis*. Biochemistry 43:10370–10378

Knox JR, Moews PC (1991) Beta-lactamase of Bacillus licheniformis 749-C. Refinement at 2 Angstroms resolution and analysis of hydration. J Mol Biol 220:435–455

Ko TP, Lai SJ, Hsieh TJ, Yang CS, Chen Y (2018) The tetrameric structure of sialic acid-synthesizing UDP-GlcNAc 2-epimerase from Acinetobacter baumannii: A comparative study with human GNE. J Biol Chem. 293(26):10119–10127. doi: https://doi.org/10.1074/jbc.RA118.001971. Epub 2018 May 15. PubMed PMID: 29764940; PubMed Central PMCID: PMC6028959

Kobe B (1999) Autoinhibition by an internal nuclear localization signal revealed by the crystal structure of mammalian importin. Nat Str Biol *6*:388–397

Kobe B, Deisenhofer J (1996) Mechanism of ribonuclease inhibition by ribonuclease inhibitor protein based on the crystal structure of its complex with ribonuclease A. J Mol Biol *264*:1028–1043

Kohl A, Binz HK, Forrer P, Stumpp MT, Plőckthun A, Grőtter MG (2002) Designed to be stable: crystal structure of a consensus ankyrin repeat protein. Proc Natl Acad Sci *100*:1700–1705

Kohn WD, Mant CT, Hodges RS (1977) Helical protein assembly motifs. J Biol Chem 272:2583–2586

Kollman JM, Pandi L, Sawaya MR, Riley M, Doolittle RF (2009) Crystal structure of human fibrinogen. Biochemistry 48(18):3877–3886. https://doi.org/10.1021/bi802205g

König H-G, Schwamborn R, Andresen S, Kinsella S, Watters O, Fenner B, Prehn JHM (2017) NF-κB regulates neuronal ankyrin-G via a negative feedback loop. Sci Rep 7:42006

Konno K, Rangel M, Oliveira JS, Dos Santos Cabrera MP, Fontana R, Hirata IY, Hide I, Nakata Y, Mori K, Kawano M, Fuchino H, Sekita S, Neto JR (2017) Decoralin, a novel linear cationic alpha-helical peptide from the venom of the solitary eumenine wasp Oreumenes decoratus. Peptides 28(12):2320–2327. Epub 2007 Sep 29

Kontur WS, Bingman CA, Olmsted CN, Wassarman DR, Ulbrich A, Gall DL, Smith RW, Yusko LM, Fox BG, Noguera DR, Coon JJ, Donohue TJ (2018) Novosphingobium aromaticivorans uses a Nu-class glutathione S-transferase as a glutathione lyase in breaking the ß-aryl ether bond of lignin. J Biol Chem 293(14):4955–4968. doi: https://doi.org/10.1074/jbc.RA117.001268. Epub 2018 Feb 15. PubMed PMID: 29449375; PubMed Central PMCID: PMC5892560.

Koonin EV, Dolja VV (2018) Metaviromics: a tectonic shift in understanding virus evolution. Virus Res. 246:A1–A3. https://doi.org/10.1016/j.virusres.2018.02.001

Kozjak-Pavlovic V (2017) The MICOS complex of human mitochondria. Cell Tissue Res. 367 (1):83–93. doi: https://doi.org/10.1007/s00441-016-2433-7. Epub 2016 Jun 1. Review. PubMed PMID: 27245231

Kranz TM, Berns A, Shields J, Rothman K, Walsh-Messinger J, Goetz RR et al (2016) Phenotypically distinct subtypes of psychosis accompany novel or rare variants in four different signaling genes. EBioMedicine 6:206–214

Krupovic M, Koonin EV(2014Jun 18) Evolution of eukaryotic single-stranded DNA viruses of the Bidnaviridae family from genes of four other groups of widely different viruses. Sci Rep 4:5347. doi: https://doi.org/10.1038/srep05347. PubMed PMID: 24939392; PubMed Central PMCID: PMC4061559

Krylov D, Mikhailenko I, Vinson C (1994) A thermodynamic scale for leucine zipper stability and dimerization specificity: e and g interhelical interactions. *EMBO J. 13*:2849–2861

Krzywda S, Brzozowski AM, Higashitsuji H, Fujita J, Welchman R, Dawson S, Mayer RJ, Wilkinson AJ (2004) The Crystal Structure of Gankyrin, an Oncoprotein Found in Complexes with Cyclin-dependent Kinase 4, a 19 S Proteasomal ATPase Regulator, and the Tumor Suppressors Rb and p53. J Biol Chem *279*:1541–1545

Kuan SL, Fischer S, Hafner S, Wang T, Syrovets T, Liu W, Tokura Y, Ng DYW, Riegger A, Förtsch C, Jäger D, Barth TFE, Simmet T, Barth H, Weil T (2018) Boosting antitumor drug efficacy with chemically engineered multidomain proteins. Adv Sci (Weinh). 5(8):1701036. doi: 10.1002/advs.201701036. eCollection 2018 Aug. PubMed PMID: 30128225; PubMed Central PMCID: PMC6097141

Kuang Q, Purhonen P, Ålander J, Svensson R, Hoogland V, Winerdal J, Spahiu L, Ottosson-Wadlund A, Jegerschöld C, Morgenstern R, Hebert H (2017) Dead-end complex, lipid interactions and catalytic mechanism of microsomal glutathione transferase 1, an electron crystallography and mutagenesis investigation. Sci Rep 7(1):7897. https://doi.org/10.1038/s41598-017-07912-3. PubMed PMID: 28801553; PubMed Central PMCID: PMC5554250

Kubo M, Ito E (2004) Structural dynamics of an ionotropic glutamate receptor. Proteins 56:411–419

Kühnel K, Jarchau T, Wolf E, Schlichting I, Walter U, Wittinghofer A, Strelkov SV (2004) The VASP tetramerization domain is a right-handed coiled coil based on a 15-residue repeat. Proc Natl Acad Sci 101:17027–17032

Kuijpers M, van de Willige D, Freal A, Chazeau A, Franker MA, Hofenk J, Rodrigues RJ, Kapitein LC, Akhmanova A, Jaarsma D, Hoogenraad CC (2016) Dynein regulator NDEL1 controls polarized cargo transport at the axon initial segment. Neuron 89(3):461–471. https://doi.org/10.1016/j.neuron.2016.01.022

Kummel D, Muller JJ, Roske Y, Henke N, Heinemann U (2006) Structure of the Bet3-Tpc6b core of trapp: two Tpc6 paralogs form trimeric complexes with Bet3 and Mum2. Ref J Mol Biol 361:22–32

Kümmel D, Oeckinghaus A, Wanga C, Krappmann D, Heinemann U (2008) Distinct isocomplexes of the TRAPP trafficking factor coexist inside human cells. FEBS Lett 582:3729–3733

Kunishima N, Shimada Y, Tsuji Y, Sato T, Yamamoto M, Kumasaka T, Nakanishi S, Jingami H, Morikawa K (2000) Structural basis of glutamate recognition by a dimaric metabotropic glutamate receptor. Nature 407:971–977

Kurochkina N, Bhaskar M, Yadav SP, Pant HC (2018) Phosphorylation, Dephosphorylation, and multiprotein assemblies regulate dynamic behavior of neuronal cytoskeleton: a mini-review. Front Mol Neurosci 11:373. https://doi.org/10.3389/fnmol.2018.00373. eCollection 2018. PubMed PMID: 30349458; PubMed Central PMCID: PMC6186834

Kurochkina N, Privalov G (1998) Heterogeneity of packing: structural approach. Protein Sci 7 (4):897–905. Erratum in: Protein Sci 7(5):1267. PubMed PMID: 9568896; PubMed Central PMCID: PMC2143988

Kurochkina N (2007) J Theor Biol 247:110–121

Kurochkina N (2008) J Theor Biol 255:188–198

Kurochkina N (2010a) Helix-helix interactions and their impact on protein motifs and assemblies. J Theor Biol 284:585–592

Kurochkina N, Yardeni T, Huizing M (2010) Molecular modeling of the bifunctional enzyme UDP-GlcNAc 2-epimerase/ManNAc kinase and predictions of structural effects of mutations associated with HIBM and sialuria. Glycobiology 20:322

Kurochkina N, Choekyi T (2011) Helix-helix interfaces and ligand binding. J Theor Biol 283:92–102

Kurochkina N (2010b) Helix-helix interfaces and their impact on protein motifs and assemblies. J Theor Biol 264:585

Kurochkina N (2011) Common structural characteristics of fibrous and globular proteins. In: Haggerty LM (ed) Protein Structure. Nova Science Publishers, Inc, Hauppauge

Kurochkina N, Guha U (2013) SH3 domains: modules of protein-protein interactions. Biol Rev 5:29–39

Kurochkina N, Iadarola M (2015) Helical assemblies: structure determinants. J Theor Biol 369C:80–84

Kurochkina N. (2016) Approximation of the protein molecule with formula for spatial location of amino acids within secondary structure elements.

Domains SH (2015) In: Kurochkina (ed) Structure, mechanisms, and applications. Springer, New Delhi

Kursula I, Wierenga R (2003) Crystal structure of triosephosphate isomerase complexed with 2-Phosphoglycolate at 0.83-Å resolution. J Biol Chem 278:9544–9551

de Lange O, Binder A, Lahaye T (2014) From dead leaf, to new life: TAL effectors as tools for synthetic biology. Plant J 78:753–771

Ladner JE, Parsons JF, Rife CL, Gilliland GL, Armstrong RN (2004) Parallel evolutionary pathways for glutathione transferases: structure and mechanism of the mitochondrial class kappa enzyme rGSTK1-1. Biochemistry 43(2):352–361. PubMed PMID: 14717589

Lance E, Arnich N, Maignien T, Biré R (2018) Occurrence of β-N-methylamino-l-alanine (BMAA) and Isomers in Aquatic Environments and Aquatic Food Sources for Humans. Toxins (Basel) 10 (2):pii: E83. https://doi.org/10.3390/toxins10020083. Review.PubMed PMID: 29443939; PubMed Central PMCID: PMC5848184

Lapouge K, Smith SJM, Walker PA, Gamblin SJ, Smerdon SJ, Rittinger K (2000) Structure of the TPR Domain of p67phox in Complex with Rac·GTP. Molecula Cell 6:899–907

Laribee NR (2018) Transcriptional and epigenetic regulation by the mechanistic target of rapamycin complex 1 pathway. J Mol Biol. pii: S0022-2836(18)30970-7. doi: 10.1016/j.jmb.2018.10.008. [Epub ahead of print] Review. PubMed PMID: 30359581

Lasker MV, Gajjar MM, Nair SK (2005) Molecular Structure of the IL-1R-Associated Kinase-4 Death Domain and Its Implications for TLR signaling. J Immun 175:4175–4179

Latorre R, Zaelzer C, Brauchi S (2009a) Structure-functional intimacies of transient receptor potential channels. Q Rev Biophys 42:201–246

Latorre R, Zaelzer C, Brauchi S (2009b) Structure-functional intimacies of transient receptor potential channels. Q Rev Biophys 42:201–246

Laursen M, Yatime L, Nissen P, Fedosova NU (2013) Crystal structure of the high-affinity Na+K+-ATPase-ouabain complex with Mg2+ bound in the cation binding site. Proc Natl Acad Sci U S A 110(27):10958–10963. https://doi.org/10.1073/pnas.1222308110. Epub 2013 Jun 17. PubMed PMID: 23776223; PubMed Central PMCID: PMC3704003

Lebioda L, Stec B, Brewer JM (1989) The structure of yeast enolase at 2.25-Å resolution. J Biol Chem 264:3685–2693

Lee G, Abdi K, Jiang Y, Michaely P, Bennett V, Marszalek PE (2006) Nanospring behaviour of ankyrin repeats. Nature 440:246–249

Lee SH, Joo K, Jung EJ, Hong H, Seo J, Kim J (2018a) Export of membrane proteins from the Golgi complex to the primary cilium requires the kinesin motor, KIFC1. FASEB J 32(2):957–968. https://doi.org/10.1096/fj.201700563R. Epub 2018 Jan 4. PubMed PMID: 29042452

Lee C-H, Kim M-S, Chung BM, Leahy DJ, Coulombe PA (2012) Structural basis for heteromeric assembly and perinuclear organization of keratin filaments. Nat Struct Mol Biol 19(7). https://doi.org/10.1038/nsmb.2330

Lee SO, Lee MK, Ku B, Bae KH, Lee SC, Lim HM, Kim SJ, Chi SW (2016b) High-resolution crystal structure of the PDZ1 domain of human protein tyrosine phosphatase PTP-Bas. Biochem Biophys Res Commun 478:1205–1210

Lee MS, Lim K, Lee MK, Chi SW (2018b) Structural Basis for the Interaction between p53 Transactivation Domain and the Mediator Subunit MED25. Molecules 23(10):pii: E2726. https://doi.org/10.3390/molecules23102726. PubMed PMID: 30360415

Lee H, Noh H, Mun J, Gu C, Sever S, Park S (2016a) Anks1a regulates COPII-mediated anterograde transport of receptor tyrosine kinases critical for tumorigenesis. Nat Commun 7:12799

Lee B, Richards FM (1971) The interpretation of protein structures: estimation of static accessibility. J Mol Biol 55:379–400

Lee H, Song C, Hong YS, Kim MS, Cho HR, Kang T, Shin K, Choi SH, Hyeon T, Kim DH (2017) Wearable/disposable sweat-based glucose monitoring device with multistage transdermal drug delivery module. Sci Adv 3(3):e1601314. doi:10.1126/sciadv.1601314. eCollection 2017 Mar. PubMed PMID: 28345030; PubMed Central PMCID: PMC5342654

Lee LK, Stewart AG, Donohoe M, Bernal RA, Stock D (2010) The structure of the peripheral stalk of Thermus thermophilus H+-ATPase/synthase. Nat Struct Mol Biol. 17(3):373–378. https://doi.org/10.1038/nsmb.1761. Epub 2010 Feb 21. PubMed PMID: 20173764; PubMed Central PMCID: PMC2912985

Lee SJ, Imamoto N, Sakai H, Nakagawa A, Kose S, Koike M, Yamamoto M, Kumasaka T, Yoneda Y, Tsukihara T (2000) The Adoption of a Twisted Structure of Importin-b is Essential for the Protein-protein Interaction Required for Nuclear Transport. *J Mol Biol 302*:251–264

Leo JC, Lyskowski A, Hattula K, Hartmann MD, Schwarz H, Butcher SJ, Linke D, Lupas AN, Goldman A (2011) The structure of E. coli IgG-binding protein D suggests a general model for bending and binding in trimeric autotransporter adhesins. Structure 19(7):1021–1030. doi: https://doi.org/10.1016/j.str.2011.03.021. PubMed PMID: 21742268

Lepage PK, Lussier MP, Barajas-Martinez H, Bousquet SM, Blanchard AP, Francoeur N, Dumaine R, Boulay G (2006) Identification of two domains involved in the Assembly of Transient Receptor Potential Canonical Channels. J Biol Chem 2006(*281*):30356–30364

Leung K. (2010) (R)-3-[(18)F]Fluoro-2-methyl-2-N-(methylamino)propanoic acid. [updated 2010 Jul 29]. Molecular Imaging and Contrast Agent Database (MICAD) [Internet]. Bethesda (MD): National Center for Biotechnology Information (US); 2004-2013. Available from http://www.ncbi.nlm.nih.gov/books/NBK45198/ PubMed PMID: 20662134.

Levine ZG, Fan C, Melicher MS, Orman M, Benjamin T, Walker S (2018) O-GlcNAc transferase recognizes protein substrates using an asparagine ladder in the tetratricopeptide repeat (TPR) Superhelix. J Am Chem Soc 140(10):3510–3513. https://doi.org/10.1021/jacs.7b13546. Epub 2018 Mar 5. PubMed PMID: 29485866; PubMed Central PMCID: PMC5937710

Lewis TL, Mao T, Svoboda K, Arnold DB (2009) Myosin-dependent targeting of transmembrane proteins to neuronal dendrites. Nat Neurosci *12*(5):568–576. https://doi.org/10.1038/nn.2318

Li Y, Li D, Wang D (2006a) Trimeric structure of the wild soluble chloride intracellular ion channel CLIC4 observed in crystals. Biophys Biochem Research Comm 343:1272–1278

Li H, Hallows A, Punzi S, Marquez E, Carrell HL, Pankiewicz KW, Watanabe KA, Goldstein BM (1994) Crystallographic studies of two alcohol dehydrogenase-bound analogs of thiazole-4-carboxamide adenine dinucleotide (TAD), the active anabolite of the antitumor agent tiazofurin. Biochemistry 33:23–32

Li J, Mahajan A, Tsai M-D (2006b) Ankyrin repeat: a unique motif mediating protein-protein interactions. Biochemistry *45*:15168–15178

Li Y, Meng X, Xiang Y, Deng J (2010) Structure function studies of vaccinia virus host range protein K1 reveal a novel functional surface for Ankyrin repeat proteins. J Virol 84:3331

Li B-S, Sun M-K, Zhang L, Takahashi S, Ma W, Vinade L et al (2001) Regulation of NMDA receptors by cyclin-dependent kinase-5. Proc Natl Acad Sci USA 98:12742–12747. https://doi.org/10.1073/pnas.211428098

Li Y, Zhou M, Hu Q, Bai XC, Huang W, Scheres SH, Shi Y (2017) Mechanistic insights into caspase-9 activation by the structure of the apoptosome holoenzyme. Proc Natl Acad Sci U S A 114(7):1542–1547. https://doi.org/10.1073/pnas.1620626114. Epub 2017 Jan 31. PubMed PMID: 28143931; PubMed Central PMCID: PMC5320974

Li K, Zhou H, Zhan L, Shi Z, Sun W, Liu D, Liu L, Liang D, Tan Y, Xu W, Xu E (2018) Hypoxic preconditioning maintains GLT-1 against transient global cerebral ischemia through upregulating Cx43 and inhibiting c-Src. Front Mol Neurosci 11:344. https://doi.org/10.3389/fnmol.2018.00344. eCollection 2018. PubMed PMID: 30323740; PubMed Central PMCID: PMC6172853

Li P, Wu G (2018) Roles of dietary glycine, proline, and hydroxyproline in collagen synthesis and animal growth. Amino Acids 50(1):29–38. https://doi.org/10.1007/s00726-017-2490-6. Epub 2017 Sep 20. Review. PubMed PMID: 28929384

Li X, Song B, Chen X, Wang Z, Zeng M, Yu D, Hu D, Chen Z, Jin L, Yang S, Yang C, Chen B (2013) Crystal structure of a four-layer aggregate of engineered TMV CP implies the importance of terminal residues for oligomer assembly. PLoS One 8(11):e77717. https://doi.org/10.1371/

journal.pone.0077717. eCollection 2013. PubMed PMID: 24223721; PubMed Central PMCID: PMC3817195

Lilja J, Zacharchenko T, Georgiadou M, Jacquemet G, Franceschi N, Peuhu E, Hamidi H, Pouwels J, Martens V, Nia FH, Beifuss M, Boeckers T, Kreienkamp HJ, Barsukov IL, Ivaska J (2017) SHANK proteins limit integrin activation by directly interacting with Rap1 and R-Ras. Nat Cell Biol 19:292–305

Lim LH, Pervaiz S (2007) Annexin 1: the new face of an old molecule. Faseb J *21*:968–975

Lin S-C, Lo Y-C, Wu H (2010) Helical assembly in the MyD88–IRAK4–IRAK2 complex in TLR/IL-1R signaling. Nature 465:885–890

Lishko PV, Procko E, Jin X, Phelps CB, Gaudet R (2007) The ankyrin repeats of TRPV1 bind multiple ligands and modulate channel sensitivity. Neuron 54:905–918

Littler DR, Harrop SJ, Brown LJ et al (2008) Comparison of vertebrate and invertebrate CLIC proteins: the crystal structures of Caenorhabditis elegans EXC-4 and Drosophila melanogaster DmCLIC. Proteins 71:364–378

Littler DR, Harrop SJ, Fairlie WD, Brown LJ, Pankhurst GJ, Pankhurst S, DeMaere MZ, Campbell TJ, Bauskin AR, Tonini R, Mazzanti M, Breit SN, Curmi PMG (2004) The intracellular chloride ion channel protein CLIC1 undergoes a redox-controlled structural transition. J Biol Chem *279*:9298–9305

Liu C, Peng X, Hu C, Li C, Li Q, Xu X (2016) Developmental profiling of ASD-related *shank3* transcripts and their differential regulation by valproic acid in zebrafish. Dev Genes Evol 226:389–400. https://doi.org/10.1007/s00427-016-0561-4

Liu J, Deng Y, Dey AK, Moore JP, Lu M (2009) Structure of the HIV-1 gp41membrane-proximal ectodomain region in a putative prefusion conformation. Biochemistry 48(13):2915–2923. https://doi.org/10.1021/bi802303b. PubMed PMID: 19226163; PubMed Central PMCID: PMC2765501

Liu JJ, Lu L, Hu FQ, Yuan H, Xu Q, Qin YF, Gong JH (2018) Methylene blue attenuates renal ischemia-reperfusion injury by negative regulation of NLRP3 signaling pathway. Eur Rev Med Pharmacol Sci 22(9):2847–2853. https://doi.org/10.26355/eurrev_201805_14986. PubMed PMID: 29771438

Liu J, Taylor DW, Krementsova EB, Trybus KM, Taylor KA (2006a) Three-dimensional structure of the myosin V inhibited state by cryoelectron tomography. Nature 442(7099):208–211. Epub 2006 Apr 16.

Liu J, Deng Y, Zheng Q, Cheng CS, Kallenbach NR, Lu M (2006b) A parallel coiled-coil tetramer with offset helices. Biochemistry 45(51):15224–15231. Epub 2006b Nov 29. PubMed PMID: 17176044

Lolis T, Alber T, Davenport RC, Rose D, Hartman FC, Petsko GA (1990) Structure of yeast triosephosphate isomerase at 1.9 Angstrom resolution. Biochemistry 29:6609–6618

Lomelino CL, Andring JT, McKenna R (2018) Crystallography and Its Impact on Carbonic Anhydrase Research. Int J Med Chem 2018:9419521. https://doi.org/10.1155/2018/9419521. eCollection 2018. Review. PubMed PMID: 30302289; PubMed Central PMCID: PMC6158936

Long SB, Tao X, Campbell EB, MacKinnon R (2007) Atomic structure of a voltage-dependent K+ channel in a lipid membrane-like environment. Nature 450:376–382

López-Benito S, Lillo C, Hernández-Hernández A, Chao MV, Arévalo JC (2016) ARMS/Kidins220 and synembryn-B levels regulate NGF-mediated secretion. J Cell Sci 129:1866–1877

Lotan A, Fenckova M, Bralten J, Alttoa A, Dixson L, Williams RW, van der Voet M (2014) Neuroinformatic analyses of common and distinct genetic components associated with major neuropsychiatric disorders. Front Neurosci 8:331

Lovejoy B, Choe S, Cascio D, McRorie DK, DeGrado WF, Eisenberg D (1993) Crystal structure of a synthetic triple-stranded alpha-helical bundle. Science 259:1288–1293

Lovejoy B, Le TC, Luthy R, Cascio D, O'Neil KT, DeGrado WF, Eisenberg D (1992) X-ray grade crystals of a designed a-helical coiled coil. Protein Sci 1:956–957

Lu G, Dolgner SJ, Hall TM (2009) Understanding and engineering RNA sequence specificity of PUF proteins. Curr Opin Struct Biol 19(1):110–115. doi: https://doi.org/10.1016/j.sbi.2008.12.009. Epub 2009 Jan 29. PubMed PMID: 19186050; PubMed Central PMCID: PMC2748946

Lumpkin EA, Caterina MJ (2007) Mechanisms of sensory transduction in the skin. Nature 445:858–865

MacGillavry HD, Kerr JM, Kassner J, Frost NA, Blanpied TA (2016) Shank–cortactin interactions control actin dynamics to maintain flexibility of neuronal spines and synapses. Eur J Neurosci 43:179–193

Madhurantakam C, Varadamsetty G, Grütter MG, Plückthun A, Mittl PRE (2012) Structure-based optimization of designed Armadillo-repeat proteins. Protein Sci *21*:1015–1028

Madrazo J, Brown JH, Litvinovich S, Dominguez R, Yakovlev S, Medved L, Cohen C (2001) Crystal structure of the central region of bovine fibrinogen (E5 fragment) at 1.4 Angstrom resolution. Proc Natl Acad Sci U S A 98(21):11967–11972

Maeda S, Nakagawa S, Suga M, Yamashita E, Oshima A, Fujiyoshi Y, Tsukihara T (2009) Structure of the connexin 26 gap junction channel at 3.5 A resolution. Nature 458 (7238):597–602. https://doi.org/10.1038/nature07869. PubMed PMID: 19340074

Maekawa K, Imagawa N, Naito A, Harada S, Yoshie O, Takagi S (1999) Association of protein-tyrosine phosphatase PTP-BAS with the transcription-factor-inhibitory protein IkappaBalpha through interaction between the PDZ1 domain and ankyrin repeats. Biochem J 337 (Pt 2):179–184

Main ER, Xiong Y, Cocco MJ, D'Andrea L, Regan L (2003) Design of stable alpha-helical arrays from an idealized TPR motif. Structure 11(5):497–508. PubMed PMID: 12737816

Maity TK, Venugopalan A, Linnoila I, Cultraro CM, Giannakou A, Nemati R, Zhang X, Webster JD, Ritt D, Ghosal S, Hoschuetzky H, Simpson RM, Biswas R, Politi K, Morrison DK, Varmus HE, Guha U (2015) Loss of MIG6 accelerates initiation and progression of mutant epidermal growth factor receptor-driven lung adenocarcinoma. Cancer Discov 5(5):534–549. https://doi.org/10.1158/2159-8290.CD-14-0750. Epub 2015 Mar 3. PubMed PMID: 25735773; PubMed Central PMCID: PMC4560174

Mak AN-S, Bradley P, Cernadas RA, Bogdanove AJ, Stoddard BL (2012) The crystal structure of TAL effector PthXo1 bound to its DNA target. Science 335:716–719

Malashkevich VN, Schneider BJ, McNally ML, Milhollen MA, Pang JX, Kim PS (1999) Core structure of the envelope glycoprotein GP2 from Ebola virus at 1.9-A resolution. Proc Natl Acad Sci U S A. 96(6):2662–2667. PubMed PMID: 10077567; PubMed Central PMCID: PMC15825

Malaspina D, Kranz TM, Heguy A, Harroch S, Mazgaj R, Rothman K et al (2016) Prefrontal neuronal integrity predicts symptoms and cognition in schizophrenia and is sensitive to genetic heterogeneity. Schizophr Res 172:94–100

Malenczyk K, Girach F, Szodorai E, Storm P, Segerstolpe Å, Tortoriello G, Schnell R, Mulder J

Martin RW, Kelly JE, Kelz JI. Advances in instrumentation and methodology for solid-state NMR of biological assemblies. J Struct Biol. 2018. pii: S1047-8477(18)30253–30253. https://doi.org/10.1016/j.jsb.2018.09.003. [Epub ahead of print] PubMed PMID: 30205196

Mathew E, Zhu H, Connelly SM, Sullivan MA, Brewer MG, Piepenbrink MS, Kobie JJ, Dewhurst S, Dumont ME (2018) Display of the HIV envelope protein at the yeast cell surface for immunogen development. PLoS One 13(10):e0205756. https://doi.org/10.1371/journal.pone.0205756. eCollection 2018. PubMed PMID: 30335821

Malik HS, Eickbush TH, Goldfarb DS (1997) Evolutionary specialization of the nuclear targeting apparatus. Proc Natl Acad Sci *94*:13738–13742

Mameza MG, Dvoretskova E, Bamann M, Hönck H-H, Güler T, Boeckers TM et al (2013) SHANK3 gene mutations associated with autism facilitate ligand binding to the Shank3 ankyrin repeat region. J Biol Chem 288:26697–26708

Mancia F, Keep NH, Nakagawa A, Leadlay PF, Mcsweeney S, Rasmussen B, Bosecke P, Diat O, Evans PR (1996) How coenzyme B12 radicals are generated: the crystal structure of methylmalonyl-CoA mutase at 2 Angstrom resolution. Structure (London) 4:339–350

Mapelli M, Massimiliano L, Crovace C, Seeliger MA, Tsai L-H, Meijer L, Musacchio A (2005) Mechanism of CDK5/p25 binding by CDK inhibitors. J Med Chem 2005(48):671–679

Marshall GR, Bosshard HE, Angiotensin II (1972Sep) Studies on the biologically active conformation. Circ Res.;31(9):Suppl 2:143–150. PubMed PMID: 4341473

Martinez J, Nguyen LD, Hinderlich S, Zimmer R, Tauberger E, Reutter W et al (2012) Crystal structures of N-Acetylmannosamine Kinase provide insights into enzyme activity and inhibition. J Biol Chem 287(17):13656–13665. https://doi.org/10.1074/jbc.M111.318170

Martins DB, Vieira MR, Fadel V, Santana VAC, Guerra MER, Lima ML, Tempone AG, Dos Santos Cabrera MP (2017) Membrane targeting peptides toward antileishmanial activity: design, structural determination and mechanism of interaction. Biochim Biophys Acta pii: S0304-4165(17):30248–30249. https://doi.org/10.1016/j.bbagen.2017.08.003. [Epub ahead of print]

Masood R, Ullah K, Ali H, Ali I, Betzel C, Ullah A (2018) Spider's venom phospholipases D: A structural review. Int J Biol Macromol 107(Pt A):1054–1065. https://doi.org/10.1016/j.ijbiomac.2017.09.081. Epub 2017 Sep 23. Review. PubMed PMID: 28951301

Mathews FS (1985) The structure, function, and evolution of cytochromes. Prog Biophys Mol Biol 45:1–56

Matsuura Y, Lange A, Harreman MT, Corbett AH, Stewart M (2003) Structural basis for Nup2p function in cargo release and karyopherin recycling in nuclear import. EMBO J 22 (20):5358–5369. PubMed PMID: 14532109; PubMed Central PMCID: PMC213792

Mattei S, Glass B, Hagen WJ, Kräusslich HG, Briggs JA (2016) The structure and flexibility of conical HIV-1 capsids determined within intact virions. Science 354(6318):1434–1437. PubMed PMID: 27980210

Maxson ME, Grinstein S (2014) The vacuolar-type H?-ATPase at a glance - more than a proton pump. J Cell Sci 127(Pt 23):4987–4993. https://doi.org/10.1242/jcs.158550. Review. PubMed PMID: 25453113

Mayer BJ (2001) SH3 domains: complexity in moderation. J Cell Science 114:1253–1263

Mayer ML (2005) Glutamate receptor ion channels. Curr Opin Neurobiol 15:282–288

Mayer ML, Ghosal A, Colman NP, Jane DE (2006) Crystal structures of the kainate receptor GluR5 ligand binding core dimer with novel GluR5-selective antagonists. J. Neurosci 26:2852–2861

McDermott SM, Stuart K (2017) The essential functions of KREPB4 are developmentally distinct and required for endonuclease association with editosomes. RNA 23(11):1672–1684. doi: https://doi.org/10.1261/rna.062786.117. Epub 2017 Aug 11. PubMed PMID: 28802260; PubMed Central PMCID: PMC5648035.

McFeeters RL, Oswald RE (2004) Emerging structural explanations of ionotropic glutamate receptor function. Faseb J 18:428–438

Medved L, Nieuwenhuizen W (2003) Molecular mechanisms of initiation of fibrinolysis by fibrin. Thromb Haemost. 2003 Mar;89(3):409–419.

Meents JE, Fischer MJ, McNaughton PA (2016) Agonist-induced sensitisation of the irritant receptor ion channel TRPA1. J Physiol 594(22):6643–6660. https://doi.org/10.1113/JP272237. Epub 2016 Aug 4. PubMed PMID: 27307078; PubMed Central PMCID: PMC5108891

Meier T, Krah A, Bond PJ, Pogoryclov D, Diederichs K, Faraldo-Gómez JD (2009) Complete ion-coordination structure in the rotor ring of Na+-dependent F-ATP synthases. J Mol Biol. 391 (2):498–507. doi:https://doi.org/10.1016/j.jmb.2009.05.082. Epub 2009 Jun 3. PubMed PMID: 19500592

Menuz K, Stroud RM, Nicoll RA, Hays FA (2007) TARP auxiliary subunits switch AMPA receptor antagonists into partial agonists. Science 318:815–817

Meotti FC, Figueiredo CP, Manjavachi M, Calixto JB (2017) The transient receptor potential ankyrin-1 mediates mechanical hyperalgesia induced by the activation of B1 receptor in mice. Biochem Pharmacol 125:75–83. https://doi.org/10.1016/j.bcp.2016.11.003. Epub 2016 Nov 4

Mercurio FA, Marasco D, Pirone L, Pedone EM, Pellecchia M, Leone M (2012) Solution structure of the first sam domain of odin and binding studies with the epha2 receptor. Biochemistry 51:2136–2145

Michaely P, Tomchick DR, Machius M, Anderson RGW (2002) Crystal structure of a 12 ANK repeat stack from human ankyrinR. EMBO J 21:6387–6396

Milburn CC, Boudeau J, Deak M, Alessi DR, van Aalten DMF (2004) Crystal structure of MO25α in complex with the C terminus of the pseudo kinase STE20-related adaptor. Nat Str & Mol Biol *11*:193–200

Miles WO, Lepesant JM, Bourdeaux J, Texier M, Kerenyi MA, Nakakido M, Hamamoto R, Orkin SH, Dyson NJ, Di Stefano L (2015) The LSD1 Family of Histone Demethylases and the Pumilio Posttranscriptional Repressor Function in a Complex Regulatory Feedback Loop. Mol Cell Biol 35(24):4199–4211. https://doi.org/10.1128/MCB.00755-15. Epub 2015 Oct 5. PubMed PMID: 26438601; PubMed Central PMCID: PMC4648826

Mills RD, Mulhern TD, Cheng H-C, Culvenor JG (2012) Analysis of LRRK2 accessory repeat domains: prediction of repeat length, number and sites of Parkinson's disease mutations. Biochem Soc Trans 40:1086–1089

Mitchell RA, Luwor RB, Burgess AW. The Epidermal Growth Factor Receptor: Structure-Function Informing the Design of Anticancer Therapeutics. Exp Cell Res. 2018. pii: S0014-4827(18) 30635–30639. https://doi.org/10.1016/j.yexcr.2018.08.009. [Epub ahead of print] Review. PubMed PMID: 30098332.

Miyanari Y (2014) TAL effector-mediated genome visualization (TGV). Methods 69:198–204

Mizutani K, Yamamoto M, Suzuki K, Yamato I, Kakinuma Y, Shirouzu M, Walker JE, Yokoyama S, Iwata S, Murata T (2011) Structure of the rotor ring modified with N,N-′-dicyclohexylcarbodiimide of the Na+-transporting vacuolar ATPase. Proc Natl Acad Sci U S A 108(33):13474–13479. https://doi.org/10.1073/pnas.1103287108. Epub 2011 Aug 3. PubMed PMID: 21813759; PubMed Central PMCID: PMC3158168

Mohanty S, Oruganty K, Kwon A, Byrne DP, Ferries S, Ruan Z, Hanold LE, Katiyar S, Kennedy EJ, Eyers PA, Kannan N (2016) Hydrophobic core variations provide a structural framework for Tyrosine Kinase evolution and functional specialization. PLoS Genet 12(2):e1005885. https://doi.org/10.1371/journal.pgen.1005885. eCollection 2016 Feb. Erratum in: PLoS Genet. 2016 Aug;12(8):e1006265. PubMed PMID: 26925779; PubMed Central PMCID: PMC4771162

Moitra J, Szilak L, Krylov D, Vinson C (1997) Leucine is the most stabilizing aliphatic amino acid in the d position of a dimeric leucine zipper coiled coil. Biochemistry 36:12567–12573

Molineros JE, Yang W, Zhou XJ, Sun C, Okada Y, Zhang H, Chua K. H, Lau YL, Kochi Y, Suzuki A, Yamamoto K, Ma J, Bang SY, Lee HS, Kim K, Bae SC, Zhang H, Shen N, Looger LL, Nath SK (2017) Confirmation of five novel susceptibility loci for Systemic Lupus Erythematosus (SLE) and integrated network analysis of 82 SLE susceptibility loci. Hum Mol Genet pii: ddx026

Moll JR, Ruvinov SB, Pastan I, Vinson C (2001) Designed heterodimerizing leucine zippers with a ranger of pI's and stabilities up to 10–15 M. Protein Sci 10:649–655

Monecke T, Haselbach D, Vo B, Russek A, Neumann P, Thomson E, Hurt E, Zachariae U, Stark H, Grubmüller H, Dickmanns A, Ficner R (2012) Structural basis for cooperativity of CRM1 export complex formation. Proc Natl Acad Sci 110:960–965

Mosavi LK, Cammett TJ, Desrosiers DC, Peng Z (2004) The ankyrin repeat as molecular architecture for protein recognition. Protein Sci 13:1435–1448

Mosavi LK, Minor DL, Peng Z (2002) Consensus-derived structural determinants of the ankyrin repeat motif. Proc Natl Acad Sci 99:16031–16034

Mott HR, Owen D, Nietlispach D, Lowe PN, Manser E, Limk L, Laue ED (1999) Structure of the small G protein Cdc42 bound to the GTPase-binding domain of ACK. Nature 399:384–388

Mykhailiuk PK, Kubyshkin V, Bach T, Budisa N (2017) Peptidyl-prolyl model study: how does the electronic effect influence the amide bond Conformation? J Org Chem. https://doi.org/10.1021/acs.joc.7b00803. [Epub ahead of print]

Mynott AV, Harrop SJ, Brown LJ, Breit SN, Kobe B, Curmi PM (2011) Crystal structure of importin-a bound to a peptide bearing the nuclear localisation signal from chloride intracellular channel protein 4. FEBS J 278(10):1662–1675. https://doi.org/10.1111/j.1742-4658.2011.08086.x. Epub 2011 Mar 30. PubMed PMID: 21388519

Nagatomo K, Ishii H, Yamamoto T, Nakajo K, Kubo Y (2010) The Met268Pro mutation of mouse TRPA1 changes the effect of caffeine from activation to suppression. Biophys J 99:3609–3618

Namba K, Stubbs G (1986) Structure of tobacco mosaic virus at 3.6 Angstroms resolution. Implications for assembly. Science 231:1401–1406

Namba K, Pattanayek R, Stubbs G (1989) Visualization of protein-nucleic acid interactions in a virus. Structure of intact tobacco mosaic virus at 2.9 Angstrom resolution by X-ray fiber diffraction. J Mol Biol 208:307–325

Nanao MH, Green T, Stern-Bach Y, Heinemann SF, Choe S (2004) Structure of the kainate receptor subunit GluR6 agonist-binding domain complexed with domoic acid. Proc Natl Acad Sci 102:1708–1713

Nanavati D, Austin DR, Catapano LA, Luckenbaugh DA, Dosemeci A, Manji HK et al (2011) The effects of chronic treatment with mood stabilizers on the rat hippocampal postsynaptic density proteome. J Neurochem 119:617–629

Narayanan A, Kumar S, Evrard AN, Paul LN, Yernool DA (2014) Asymmetric hetero-domain interface stabilizes a response regulator-DNA complex. Nat Commun 5:3282. https://doi.org/10.1038/ncomms4282

Natrajan G, Lamers MH, Enzlin JH, Winterwerp HHK, Perrakis A, Sixma TK (2003) Structures of Escherichia coli DNA mismatch repair enzyme MutS in complex with different mismatches: a common recognition mode for diverse substrates. Nucleic Acids Res 31(16):4814–4821

Navarro E, Fenude E, Celda B (2002) Solution structure of a D, L-alternating oligonorleucine as a model of double-stranded antiparallel beta-helix. Biopolymers 64(4):198–209

Navarro E, Fenude E, Celda (2004) Conformational and structural analysis of the equilibrium between single- and double-strand beta-helix of a D,L-alternating oligonorleucine. Biopolymers 73(2):229–241

Navarro E, Tejero R, Fenude E, Celda B (2001) Solution NMR structure of a D,L-alternating oligonorleucine as a model of beta-helix. Biopolymers 59(2):110–119

Nawrotek A, Knossow M, Gigant B (2011) The determinants that govern microtubule assembly from the atomic structure of GTP-tubulin. J Mol Biol 412(1):35–42. https://doi.org/10.1016/j.jmb.2011.07.029

Nelson CA, Wilen CB, Dai YN, Orchard RC, Kim AS, Stegeman RA, Hsieh LL, Smith TJ, Virgin HW, Fremont DH (2018) Structural basis for murine norovirus engagement of bile acids and the CD300lf receptor. Proc Natl Acad Sci USA 115(39):E9201–E9210. https://doi.org/10.1073/pnas.1805797115. Epub 2018 Sep 7. PubMed PMID: 30194229; PubMed Central PMCID: PMC6166816

Neubrand VE, Thomas C, Schmidt S, Debant A, Schiavo G (2010) Kidins220/ARMS regulates Rac1-dependent neurite outgrowth by direct interaction with the RhoGEF Trio. J Cell Sci 123:2111–2123

Neubrand VE, Cesca F, Benfenati F, Schiavo G (2012) Kidins220/ARMS as a functional mediator of multiple receptor signalling pathways. J Cell Sci 125:1845–1854

Newberry RW, Raines RT (2017) 4-Fluoroprolines: conformational analysis and effects on the stability and folding of peptides and proteins. Top Heterocycl Chem 48:1–25. https://doi.org/10.1007/7081_2015_196

Nicholas S (2015) Crystal structure of the tripeptide N-(benzyl-oxycarbon-yl)glycylglycyl-l-norvaline. Acta Crystallogr Sect E Crystallogr Commun 71(Pt 3):o216–o217. https://doi.org/10.1107/S205698901500393X

Niedergang F, Grinstein S 2018 How to build a phagosome: new concepts for an old process. Curr Opin Cell Biol 50:57–63. https://doi.org/10.1016/j.ceb.2018.01.009. Epub 2018 Feb 20. Review. PubMed PMID: 29471269

Nilius N, Prenen J, Owsianik G (2011) Irritating channels: the case of TRPA1. J Physiol 589 (7):1543–1549

Noble ME, Endicott JA, Brown NR, Johnson LN (1997) The cyclin box fold: protein recognition in cell-cycle and transcription control. Trends Biochem Sci 22:482–487

Nozawa K, Schneider TR, Cramer P (2017) Core mediator structure at 3.4 Å extends model of transcription initiation complex. Nature. https://doi.org/10.1038/nature22328. [Epub ahead of print]

Numazaki M, Tominaga M (2003) Molecular mechanisms of nociception and thermosensation: structures, expressions and functions of capsaicin receptor and its homologues. Seikagaku 75 (5):359–371. Review. Japanese. PubMed PMID: 12822433

Oder E, Safo MK, Abdulmalik O, Kato GJ (2016) New developments in anti-sickling agents: can drugs directly prevent the polymerization of sickle haemoglobin in vivo? Br J Haematol 175 (1):24-30. https://doi.org/10.1111/bjh.14264. Epub 2016 Sep 8. Review. PubMed PMID: 27605087; PubMed Central PMCID: PMC5035193

Ohshima T, Ward JM, Huh CG, Longenecker G, Veeranna Pant HC, Brady RO, Martin LJ, Kulkarni AB (1996) Targeted disruption of the cyclin-dependent kinase 5 gene results in abnormal corticogenesis, neuronal pathology and perinatal death. Proc Natl Acad Sci USA 93:11173–11178

Oka Y, Butnaru M, von Buchholtz L, Ryba NJP, Zuker CS (2013) High salt recruits aversive taste pathways. Nature 494:472–475

Oshaben KM, Horne WS (2014) Tuning assembly size in Peptide-based supramolecular polymers by modulation of subunit association affinity. Biomacromolecules 15(4):1436–1442. https://doi.org/10.1021/bm5000423. Epub 2014 Mar 17

Oswald RE (2004) Ionotropic glutamate receptor recognition and activation. Adv Protein Chem 68:313–349

O'Shea EK, Klemm JD, Kim PS, Alber T (1991) Crystal structure of GCN4 leucine zipper, a two-stranded parallel coiled coil. Science 254:539

Oxenoid K, Chou JJ (2005) The structure of phospholamban pentamer reveals a channel-like architecture in membranes. PNAS 102:10870–10875

Padmanabhan B, Adachi N, Kataoka K, Horikoshi M (2004) Crystal structure of the homolog of the oncoprotein gankyrin, an interactor of Rb and CDK4/6. J Biol Chem 279:1546–1552

Paliakasis CD, Kokkinidis M (1991) The stability of the four-α-helix bundle motif in proteins. Protein Eng 4:849

Pancaroglu R, Van Petegem F (2018) Calcium channelopathies: structural insights into disorders of the muscle excitation-contraction complex. Annu Rev Genet. https://doi.org/10.1146/annurev-genet-120417-031311. [Epub ahead of print] PubMed PMID: 30208288

Pan X, Xu S, Wu J, Duan Y, Zheng Z, Wang J, Song X, Zhou M (2018) Ankyrin-like protein AnkB interacts with CatB, affects catalase activity, and enhances resistance of Xanthomonas oryzae pv. oryzae and Xanthomonas oryzae pv. oryzicola to Phenazine-1-Carboxylic Acid. Appl Environ Microbiol. 84(4). pii: e02145–17. https://doi.org/10.1128/AEM.02145-17. Print 2018 Feb 15. PubMed PMID: 29180371; PubMed Central PMCID: PMC5795068

Pareek TK, Keller J, Kesavapany S, Agarwal N, Kuner R, Pant H, Iadarola MJ, Brady RO, Kulkarni AB (2006) Cyclin-dependent kinase 5 modulates nociceptive signaling through direct phosphorylation of transient receptor potential vanilloid 1. Proc Natl Acad Sci 104:660–665

Park MH (2006) The post-translational synthesis of a polyamine-derived amino acid, Hypusine, in the eukaryotic translation initiation factor 5A (eIF5A). J Biochem 139:161–169

Park S (2016) Defective Anks1a disrupts the export of receptor tyrosine kinases from the endoplasmic reticulum. BMB Rep 49:651–652

Park E, Kim N, Ficarro SB, Zhang Y, Lee BI, Cho A, Kim K, Park AKJ, Park WY, Murray B, Meyerson M, Beroukhim R, Marto JA, Cho J, Eck MJ (2015) Structure and mechanism of activity-based inhibition of the EGF receptor by Mig6. Nat Struct Mol Biol 22(9):703–711. https://doi.org/10.1038/nsmb

Park HH, Logette E, Raunser S, Cuenin S, Walz T, Tschopp J, Wu H (2007) Death domain assembly mechanism revealed by crystal structure of the oligomeric PIDDosome core complex. Cell 128:533–546

Parry DAD, Fraser RD, Squire JM (2008) Fifty years of coiled-coils and α-helical bundles: a close relationship between sequence and structure. J Struct Biol 163:258–269

Parthasarathy S, Eaazhisai K, Balaram H, Balaram P (2003) Structure of *Plasmodium Falciparum* triose-phosphate isomerase-2-phosphoglycerate complex at 1.1-Å resolution. J Biol Chem 278:52461

Passon DM, Lee M, Rackham O, Stanley WA, Sadowska A, Filipovska A, Fox AH, Bond CS (2012) Structure of the heterodimer of human NONO and paraspeckle protein component 1 and analysis of its role in subnuclear body formation. Proc Natl Acad Sci USA 109(13):4846–4850. https://doi.org/10.1073/pnas.1120792109. Epub 2012 Mar 13. PubMed PMID: 22416126; PubMed Central PMCID: PMC3324020

Paternain AV, Cohen A, Stern-Bach V, Lerma J (2003) A role for extracellular Na+ in the channel gating of native and recombinant kainate receptors. J Neurosci 23:8641–8648

Pauling L, Corey RB, Branson HR (1951) The structure of proteins: two hydrogen-bonded helical configurations of the polypeptide chain. Proc Natl Acad Sci USA 37:205

Paulson AR, Tong L (2012) Crystal structure of the Rna14-Rna15 complex. RNA Jun;18 (6):1154–1162. https://doi.org/10.1261/rna.032524.112. Epub 2012 Apr 18. PubMed PMID: 22513198; PubMed Central PMCID: PMC3358638

Pechik I, Madrazo J, Mosesson MW, Hernandez I, Gilliland GL, Medved L (2004) Crystal structure of the complex between thrombin and the central "E" region of fibrin. Proc Natl Acad Sci USA 101(9):2718–2723. https://doi.org/10.1073/pnas.0303440101

Pecqueur L, Duellberg C, Dreier B, Jiang Q, Wang C, Plückthun A et al (2012) A designed ankyrin repeat protein selected to bind to tubulin caps the microtubule plus end. Proc Natl Acad Sci USA 109(30):12011–12016. https://doi.org/10.1073/pnas.1204129109

Perez-Riba A, Synakewicz M, Itzhaki LS (2018) Folding cooperativity and allosteric function in the tandem-repeat protein class. Philos Trans R Soc Lond B Biol Sci. 373(1749). pii: 20170188. https://doi.org/10.1098/rstb.2017.0188. Review. PubMed PMID: 29735741; PubMed Central PMCID: PMC5941182

Perry AJ, Hulett JM, Likić VA, Lithgow T, Gooley PR (2006) Convergent evolution of receptors for protein import into mitochondria. Curr Biol 16:221–229

Perutz MF, Rossmann MG, Cullis AF, Muirhead H, Will G, North AC (1960) Structure of haemoglobin: a three-dimensional Fourier synthesis at 5.5-A. resolution, obtained by X-ray analysis. Nature 185(4711):416–422. PubMed PMID: 18990801

Pesti S, Balázs A, Udupa R, Szabó B, Fekete A, Bőgel G, Buday L (2012) Complex formation of EphB1/Nck/Caskin1 leads to tyrosine phosphorylation and structural changes of the Caskin1 SH3 domain. Cell Commun Signal 10:36. https://doi.org/10.1186/1478-811X-10-36

Petralia RS 2012 Distribution of extrasynaptic NMDA receptors on neurons. Sci World J 2012:267120. https://doi.org/10.1100/2012/267120. Epub 2012 Apr 30. Review. PubMed PMID: 22654580; PubMed Central PMCID: PMC3361219

Pfaender S, Sauer AK, Hagmeyer S, Mangus K, Linta L, Liebau S et al (2017) Zinc deficiency and low enterocyte zinc transporter expression in human patients with autism related mutations in SHANK3. Sci Rep 7:45190

Pflug A, Lukarska M, Resa-Infante P, Reich S, Cusack S (2017) Structural insights into RNA synthesis by the influenza virus transcription-replication machine. Virus Res 20. pii: S0168–1702(16)30782–1. https://doi.org/10.1016/j.virusres.2017.01.013. [Epub ahead of print]

Phelps CB, Huang RJ, Lishko PV, Wang RR, Gaudet R (2008) Structural analyses of the Ankyrin Repeat Domain of TRPV6 and related TRPV ion channels. Biochemistry 47:2476–2484

Phillips L, Trivedi JR (2018) Skeletal muscle channelopathies. Neurotherapeutics https://doi.org/10.1007/s13311-018-00678-0. [Epub ahead of print] Review. PubMed PMID: 30341599

Pielak RM, Schnell JR, Chou JJ (2009) Mechanism of drug inhibition and drug resistance of influenza A M2 channel. Proc Natl Acad Sci USA 106:7379–7384

Pitman KA, Borgland SL, MacLeod B, Puil E (2015) Isovaline does not activate GABA (B) receptor-coupled potassium currents in GABA(B) expressing AtT-20 cells and cultured rat hippocampal neurons. PLoS One 10(2):e0118497. https://doi.org/10.1371/journal.pone.0118497. eCollection 2015. PubMed PMID: 25706125; PubMed Central PMCID: PMC4337901

Plecs JJ, Harbury PB, Kim PS, Alber T (2004) Structural test of the parameterized-backbone method for protein design. J Mol Biol 342(1):289–297. PubMed PMID: 15313624.

Plested AJ, Mayer ML (2007) Structure and mechanism of kainate receptor modulation by anions. Neuron 53:829–841

Plested AJ, Vijauan R, Biggin PC, Mayer ML (2008) Molecular basis of kainate receptor modulation by sodium. Neuron 58:720–735

Poblete H, Oyarzún I, Olivero P, Comer J, Zuñiga M, Sepulveda RV, Báez-Nieto D, González Leon C, González-Nilo F, Latorre R (2015) Molecular determinants of phosphatidylinositol 4,5-bisphosphate (PI(4,5)P2) binding to transient receptor potential V1 (TRPV1) channels. J Biol Chem 290(4):2086–2098. https://doi.org/10.1074/jbc.M114.613620. Epub 2014 Nov 25. PubMed PMID: 25425643; PubMed Central PMCID: PMC4303662

Ponder JW, Richards FM (1987) Tertiary templates for proteins. J Mol Biol 193:775

Pospich S, Kumpula EP, von der Ecken J, Vahokoski J, Kursula I, Raunser S (2017) Near-atomic structure of jasplakinolide-stabilized malaria parasite F-actin reveals the structural basis of filament instability. Proc Natl Acad Sci USA 114(40):10636–10641. https://doi.org/10.1073/pnas.1707506114. Epub 2017 Sep 18. PubMed PMID: 28923924; PubMed Central PMCID: PMC5635891

Potoyan DA, Zheng W, Ferreiro DU, Wolynes PG, Komives EA (2016) PEST control of molecular stripping of NFκB from DNA transcription sites. J Phys Chem B 120(33):8532–8538. https://doi.org/10.1021/acs.jpcb.6b02359. Epub 2016 May 3. PubMed PMID: 27098223; PubMed Central PMCID: PMC5389414

Praditwongwan W, Chuankhayan P, Saoin S, Wisitponchai T, Lee VS, Nangola S, Hong SS, Minard P, Boulanger P, Chen CJ, Tayapiwatana C (2017) Crystal structure of an antiviral ankyrin targeting the HIV-1 capsid and molecular modeling of the ankyrin-capsid complex. J Comput Aided Mol Des 2014 Aug;28(8):869–884. https://doi.org/10.1007/s10822-014-9772-9. Epub 2014 Jul 5.

Preiss L, Yildiz O, Hicks DB, Krulwich TA, Meier T (2010) A new type of proton coordination in an F(1)F(o)-ATP synthase rotor ring. PLoS Biol 8(8):e1000443. https://doi.org/10.1371/journal.pbio.1000443. PubMed PMID: 20689804; PubMed Central PMCID: PMC2914638.

Privalov PL, Gill SJ (1988) Stability of protein structure and hydrophobic interaction. Adv Protein Chem 39:191–234

Prota AE, Bargsten K, Zurwerra D, Field JJ, Díaz JF, Altmann KH, Steinmetz MO (2013) Molecular mechanism of action of microtubule-stabilizing anticancer agents. Science 339 (6119):587–590. https://doi.org/10.1126/science.1230582. Epub 2013 Jan 3. PubMed PMID: 23287720

Pupo A, Baez-Nieto D, Martínez A, Latorre R, González C (2014) Proton channel models filling the gap between experimental data and the structural rationale. Channels (Austin) 8(3):180–192. Review. PubMed PMID: 24755912; PubMed Central PMCID: PMC4203746

Pylypenko O, Rak A, Durek T, Kushnir S, Dursina BE, Thomae NH, Constantinescu AT, Brunsveld L, Watzke A, Waldmann H, Goody RS, Alexandrov K (2006) Structure of doubly prenylated Ypt1:GDI complex and the mechanism of GDI-mediated Rab recycling. EMBO J 25 (1):13–23. Epub 2006 Jan 5. PubMed PMID: 16395334; PubMed Central PMCID: PMC1356364

Qi S, Pang Y, Hu Q, Liu Q, Li H, Zhou Y, He T, Liang Q, Liu Y, Yuan X, Luo G, Li H, Wang J, Yan N, Shi Y (2010) Crystal structure of the Caenorhabditis elegans apoptosome reveals an octameric assembly of CED-4. Cell 141(3):446–457. https://doi.org/10.1016/j.cell.2010.03.017. PubMed PMID: 20434985

Qin H, Srinivasula SM, Wu G, Fernandez-Alnemri T, Alnemri ES, Shi Y (1999) Structural basis of procaspase-9 recruitment by the apoptotic protease-activating factor 1. Nature 399:549

Raghavender US, Kantharaju Aravinda S, Shamala N, Balaram P (2010) Hydrophobic peptide channels and encapsulated water wires. J Am Chem Soc 132(3):1075–1086. https://doi.org/10.1021/ja9083978

Rak A, Pylypenko O, Durek T, Watzke A, Kushnir S, Brunsveld L, Waldmann H, Goody RS, Alexandrov K (2003) Structure of Rab GDP-dissociation inhibitor in complex with prenylated YPT1 GTPase. Science 302:646

Ramachandran GN, Ramakrishnan C, Sasisekharan V (1963) Stereochemistry of polypeptide chain configurations. J Mol Biol 7:95–99

Ramkumar K, Samanta S, Kyani A, Yang S, Tamura S, Ziemke E, Stuckey JA, Li S, Chinnaswamy K, Otake H, Debnath B, Yarovenko V, Sebolt-Leopold JS, Ljungman M, Neamati N (2016) Mechanistic evaluation and transcriptional signature of a glutathione S-transferase omega 1 inhibitor. Nat Commun 7:13084. https://doi.org/10.1038/ncomms13084. PubMed PMID: 27703239; PubMed Central PMCID: PMC5059489

Rao ST, Rossmann MG (1973) Comparison of super-secondary structure in proteins. J Mol Biol 76:241–256

Rappoport, JZ Reubold TF, Faelber K, Plattner N, Posor Y, Ketel K, Curth U, Schlegel J, Anand R, Manstein DJ, Noé F, Haucke V, Daumke O, Eschenburg S (2018) Crystal structure of the dynamin tetramer. Nature 2015 Sep 17;525(7569):404–408. https://doi.org/10.1038/nature14880. Epub 2015 Aug 24. PubMed PMID: 26302298

Reinert ZE, Lengyel GA, Horne WS (2013) Protein-like tertiary folding behavior from heterogeneous backbones.J Am Chem Soc 135(34):12528–12531. https://doi.org/10.1021/ja405422v. Epub 2013 Aug 15

Reinke SO, Hinderlich S (2007) Prediction of three different isoforms of the human UDP-N-acetylglucosamine 2-epimerase/N-acetylmannosamine kinase. FEBS Lett 581:3327

Reinke SO, Eidenschink C, Jay CM, Hinderlich S (2009) Biochemical characterization of human and murine isoforms of UDP-N-acetylglucosamine 2-epimerase/N-acetylmannosamine kinase (GNE). Glycoconj J 26:415

Richards FM (1977) Areas, volumes, packing, and protein structure. Ann Rev Biophys Bioeng 6:151–176

Riedl SJ, Li W, Chao Y, Schwarzenbacher R, Shi Y (2005) Structure of the apoptotic protease-activating factor bound to ADP. Nature 434:926

Ringel R, Sologub M, Morozov YI, Litonin D, Cramer P, Temiakov D (2011) Structure of human mitochondrial RNA polymerase. Nature 478(7368):269–273. https://doi.org/10.1038/nature10435. PubMed PMID: 21947009

Rogowski A, Briggs JA, Mortimer JC, Tryfona T, Terrapon N, Lowe EC, Baslé A, Morland C, Day AM, Zheng H, Rogers TE, Thompson P, Hawkins AR, Yadav MP, Henrissat B, Martens EC, Dupree P, Gilbert HJ, Bolam DN (2015) Glycan complexity dictates microbial resource allocation in the large intestine. Nat Commun. 6:7481. https://doi.org/10.1038/ncomms8481. Erratum in: Nat Commun. 2016;7:10705. PubMed PMID: 26112186; PubMed Central PMCID: PMC4491172

Romney DK, Murciano-Calles J, Wehrmüller JE, Arnold FH (2017) Unlocking reactivity of TrpB: a general biocatalytic platform for synthesis of tryptophan analogues. J Am Chem Soc 139 (31):10769–10776. https://doi.org/10.1021/jacs.7b05007. Epub 2017 Jul 28. PubMed PMID: 28708383; PubMed Central PMCID: PMC5589443

Romero-Hernandez A, Simorowski N, Karakas E, Furukawa H (2016) Molecular basis for subtype specificity and high-affinity Zinc inhibition in the GluN1-GluN2A NMDA receptor amino-terminal domain. Neuron 92(6):1324–1336. https://doi.org/10.1016/j.neuron.2016.11.006. Epub 2016 Dec 1. PubMed PMID: 27916457; PubMed Central PMCID: PMC5182123 5tq2

Roncalli V, Cieslak MC, Passamaneck Y, Christie AE, Lenz PH (2015) Glutathione S-Transferase (GST) gene diversity in the crustacean calanus finmarchicus—contributors to cellular detoxification. PLoS One 10(5):e0123322. https://doi.org/10.1371/journal.pone.0123322. eCollection 2015. PubMed PMID: 25945801; PubMed Central PMCID: PMC4422733

Rosenbaum T, Gordon-Shaag A, Munari M, Gordon SE (2004) Ca2+/calmodulin modulates TRPV1 activation by capsaicin. J Gen Physiol 123(1):53–62. PubMed PMID: 14699077; PubMed Central PMCID: PMC2217413

Rosengarth A, Gerke V, Luecke H (2001) X-ray structure of full-length annexin 1 and implications for membrane aggregation. J Mol Biol 306:489–498

Roy P (2017) Bluetongue virus structure and assembly. Curr Opin Virol 24:115–123. https://doi.org/10.1016/j.coviro.2017.05.003. Epub 2017 Jun 10. Review. PubMed PMID: 28609677

Ruba A, Yang W (2016) O-GlcNAc-ylation in the nuclear pore complex. Cell Mol Bioeng 9 (2):227–233. https://doi.org/10.1007/s12195-016-0440-0. Epub 2016 Apr 26. PubMed PMID: 28638491; PubMed Central PMCID: PMC5475274

Rubin DL, Lewis SE, Mungall CJ, Misra S, Westerfield M, Ashburner M, Sim I, Chute CG, Solbrig H, Storey MA, Smith B, Richter JD, Noy NF, Musen MA (2006) The national center for biomedical ontology: advancing biomedicine through structured organization of scientific knowledge. OMICS 10:185–198

Rubinson EH, Gowda AS, Spratt TE, Gold B, Eichman BF (2010) An unprecedented nucleic acid capture mechanism for excision of DNA damage. Nature 468:406–411

Rudolph MG, Wittinghofer A, Vetter IR (1999) Nucleotide binding to the G12V-mutant of Cdc42 investigated by X-ray diffraction and fluorescence spectroscopy: two different nucleotide states in one crystal. Protein Sci 8:778–787

Rudrabhatla P, Albers W, Pant HC (2009) Peptidyl-Prolyl Isomerase 1 regulates protein phosphatase 2A-mediated topographic phosphorylation of neurofilament proteins. J Neurosci 29 (47):14869–14880. https://doi.org/10.1523/JNEUROSCI.4469-09.2009

Le Rumeur E, Hubert JF, Winder SJ (2012) A new twist to coiled coil. FEBS Lett 586 (17):2717–2722. https://doi.org/10.1016/j.febslet.2012.05.004. Epub 2012 May 11. Review. PubMed PMID: 22584055

Sainlos M, Tigaret C, Poujol C, Olivier NB, Bard L, Breillat C, Thiolon K, Choquet D, Imperiali B (2011) Biomimetic divalent ligands for the acute disruption of synaptic AMPAR stabilization. Nat Chem Biol 7:81–91

Sakai H (2017) Overview of potential clinical applications of Hemoglobin Vesicles (HbV) as artificial red cells, evidenced by preclinical studies of the academic research consortium. J Funct Biomater 8(1):10. https://doi.org/10.3390/jfb8010010

Saleh T, Kalodimos CG (2017) Enzymes at work are enzymes in motion. Science 355 (6322):247–248. https://doi.org/10.1126/science.aal4632. PubMed PMID: 28104853

Salnikov ES, Anantharamaiah GM, Bechinger B (2018) Supramolecular organization of apolipoprotein-A-I-derived peptides within disc-like arrangements. Biophys J 115(3):467–477. https://doi.org/10.1016/j.bpj.2018.06.026. Epub 2018 Jul 11. PubMed PMID: 30054032; PubMed Central PMCID: PMC6085177

Samatey FA, Imada K, Nagashima S, Vonderviszt F, Kumasaka T, Yamamoto M, Namba K (2001) Structure of the bacterial flagellar protofilament and implications for a switch for supercoiling. Nature 410:331–337

Sarowar T, Grabrucker AM (2016) Actin-dependent alterations of dendritic spine morphology in shankopathies. Neural Plast 8051861

Sanders SS, Mui KKN, Sutton LM, Hayden MR (2014) Identification of binding sites in Huntingtin for the Huntingtin Interacting Proteins HIP14 and HIP14L. PlosOne 28:e90669

Sawyer N, Chen J, Regan L (2013) All repeats are not equal: a module-based approach to guide repeat protein design. J Mol Biol 425:1826–1838

Sayer C, Martinez-Torres RJ, Richter N, Isupov MN, Hailes HC, Littlechild JA, Ward JM (2014) The substrate specificity, enantioselectivity and structure of the (R)-selective amine: pyruvate transaminase from Nectria haematococca. FEBS J 281(9):2240–2253. https://doi.org/10.1111/febs.12778. Epub 2014 Apr 7. PubMed PMID: 24618038; PubMed Central PMCID: PMC4255305

Scalise M, Pochini L, Console L, Losso MA, Indiveri C (2018) The human SLC1A5 (ASCT2) amino acid transporter: from function to structure and role in cell biology. Front Cell Dev Biol 6: 96. https://doi.org/10.3389/fcell.2018.00096. eCollection 2018. Review. PubMed PMID: 30234109; PubMed Central PMCID: PMC6131531

Schaefer KN, Bonello TT, Zhang S, Williams CE, Roberts DM, McKay DJ, Peifer M (2018) Supramolecular assembly of the beta-catenin destruction complex and the effect of Wnt signaling on its localization, molecular size, and activity in vivo. PLoS Genet 14(4): e1007339. https://doi.org/10.1371/journal.pgen.1007339. eCollection 2018 Apr. PubMed PMID: 29641560; PubMed Central PMCID: PMC5912785

Scharfman HE, Chao MV (2013) The entorhinal cortex and neurotrophin signaling in Alzheimer's disease and other disorders. Cogn Neurosci 4:3–4. https://doi.org/10.1080/17588928.2013.826184

Scheraga HA, Chou KC, Nemethy G (1982) In: Srinivasan R, Sarma RH (eds) Conformation in biology. Adenine Press, Gilderland

Schmidt H, Carter AP (2016) Review: structure and mechanism of the dynein motor ATPase. Biopolymers 105(8):557–567. https://doi.org/10.1002/bip.22856

Schmidt H, Hansen G, Singh S, Hanuszkiewicz A, Lindner B, Fukase K et al (2012) Structural and mechanistic analysis of the membrane-embedded glycosyltransferase WaaA required for lipopolysaccharide synthesis. Proc Natl Acad Sci USA 109(16):6253–6258. https://doi.org/10.1073/pnas.1119894109

Schmieg N, Menendez G, Schiavo G, Terenzio M (2014) Signalling endosomes in axonal transport: travel updates on the molecular highway. Semin Cell Dev Biol 27:32–43

Schnell JR, Zhou G-P, Zweckstetter M, Rigby AC, Chou JJ (2005) Rapid and accurate determination of coiled-coil domains using NMR dipolar couplings: application to cGMP-dependent protein kinase Ia. Protein Sci 14:142421

Schnell JR, Chou JJ (2008) Structure and mechanism of the M2 proton channel of influenza A virus. Nature 451:591–595

Schneider NFZ, Cerella C, Lee JY, Mazumder A, Kim KR, de Carvalho A, Munkert J, Pádua RM, Kreis W, Kim KW, Christov C, Dicato M, Kim HJ, Han BW, Braga FC, Simões CMO, Diederich M (2018) Cardiac glycoside glucoevatromonoside induces cancer type-specific cell death. Front Pharmacol 9:70. https://doi.org/10.3389/fphar.2018.00070. eCollection 2018. PubMed PMID: 29545747; PubMed Central PMCID: PMC5838923

Scholze H, Boch J (2011) TAL effectors are remote controls for gene activation. Curr Opin Microbiol 14:47–53

Schrauber H, Eisenhaber F, Argos P (1993) Rotamers: to be or not to be? An analysis of amino acid side-chain conformations in globular proteins. J Mol Biol 230(2):592–612

Schulman (2000) 1fqv 1fs1

Schulze-Gahmen U, Kim S-H (2002) Structural basis for CDK6 activation by a virus-encoded cyclin. Nature Str Biol 9:177–181

Schulz GE, Schirmer RH (1982) Principles of protein structure. Ed: Cantor, C. R.,

Scott FL, Stec B, Pop C, Dobaczewska MK, Lee JJ, Monosov E, Robinson H, Salvesen GS, Schwarzenbacher R, Riedl SJ (2009) The Fas/FADD death domain complex structure unravels signaling by receptor clustering. Nature 457:1019–1022

Semela D (2017) Hepatitis C – Diagnostik und Therapie. Ther Umsch 74(3):101–108. https://doi.org/10.1024/0040-5930/a000892. German. PubMed PMID: 28777053

De Seranno S, Benaud C, Assard N, Khediri S, Gerke V, Baudier J, Delphin C (2006) Identification of an AHNAK binding motif specific for the Annexin2/S100A10 tetramer. J Biol Chem 281:35030–35038

Shaanan B (1983) Structure of human oxyhaemoglobin at 2.1 A resolution. J Mol Biol 171 (1):31–59

Shah RB, Thompson R, Sidi S (2015) A mitosis-sensing caspase activation platform? New insights into the PIDDosome. Mol Cell Oncol 3(3):e1059921. https://doi.org/10.1080/23723556.2015.

1059921. eCollection 2016 May. PubMed PMID: 27314076; PubMed Central PMCID: PMC4909455

Shao C, Zhang F, Kemp MM, Linhardt RJ, Waisman DM, Head JF, Seaton BA (2006) Crystallographic analysis of calcium-dependent heparin binding to Annexin A2. J Biol Chem 281:31689–31695

Sharma K, Dhillon A, Goyal A (2018) Insights into structure and reaction mechanism of β-mannanases. Curr Protein Pept Sci 19(1):34–47. https://doi.org/10.2174/1389203717666161013115724. Review. PubMed PMID: 27739373

Sharma M, Pandey GK (2016) Expansion and function of repeat domain proteins during stress and development in plants. Front Plant Sci 6:1218. https://doi.org/10.3389/fpls.2015.01218. eCollection 2015. Review. PubMed PMID: 26793205; PubMed Central PMCID: PMC4707873

Sheehan D, Meade G, Foley VM, Dowd CA (2001) Structure, function and evolution of glutathione transferases: implications for classification of non-mammalian members of an ancient enzyme superfamily. Biochem J 360:1–16

Shelby SJ, Colwill K, Dhe-Paganon S, Pawson T, Thompson DA (2013) MERTK interactions with SH2-domain proteins in the retinal pigment epithelium. PLoS One 8(2):e53964. https://doi.org/10.1371/journal.pone.0053964. Epub 2013 Feb 4. PubMed PMID: 23390493; PubMed Central PMCID: PMC3563642

Sheng M, Hoogenraad CC (2007) The postsynaptic architecture of excitatory synapses: a more quantitative view. Annu Rev Biochem 76:823–847

Chen S-C, Huang C-H, Lai S-J, Yang CS, Hsiao T-H, Lin C-H, Fu P-K, Ko T-P, Chen Y (2016) Mechanism and inhibition of human UDP-GlcNAc 2-epimerase, the key enzyme in sialic acid biosynthesis. Sci Rep 6:23274. https://doi.org/10.1038/srep23274

Sherr CJ, Roberts JM (1999) CDK inhibitors: positive and negative regulators of G1-phase progression. Genes Dev 13(12):1501–1512. Review. PubMed PMID: 10385618

Sheriff S, Hendrickson WA, Smith JL (1987) Structure of myohemerythrin in the azidomet state at 1.7/1.3 angstroms resolution. J Mol Biol 197:273–296

Shimizu T (2013) Structural basis for ß-galactosidase associated with lysosomal disease. Yakugaku Zasshi 133(5):509–517. Review. Japanese. PubMed PMID: 23649392

Shin H, Renatus M, Eckelman BP, Nunes VA, Sampaio CA, Salvesen GS (2005) The BIR domain of IAP-like protein 2 is conformationally unstable: implications for caspase inhibition. Biochem J 385(Pt 1):1–10. PubMed PMID: 15485395; PubMed Central PMCID: PMC1134667

Shirai Y, Li W, Suzuki T (2017) Role of splice variants of Gtf2i, a transcription factor localizing at postsynaptic sites, and its relation to neuropsychiatric diseases. Int J Mol Sci 18:411

Shoulders MD, Raines RT (2009) Collagen structure and stability. Annu Rev Biochem 78:929–958. https://doi.org/10.1146/annurev.biochem.77.032207.120833

Sicheri F, Moarefi I, Kuriyan J (1997) Crystal structure of the Src family tyrosine kinase Hck. Nature 385:602–609

Singh AK, Nguyen TH, Vidovszky MZ, Harrach B, Benko M, Kirwan A, Joshi L, Kilcoyne M, Berbis MÁ, Cañada FJ, Jiménez-Barbero J, Menéndez M, Wilson SS, Bromme BA, Smith JG, van Raaij MJ (2018) Structure and N-acetylglucosamine binding of the distal domain of mouse adenovirus 2 fibre. J Gen Virol 99(11):1494–1508. https://doi.org/10.1099/jgv.0.001145. Epub 2018 Oct 2. PubMed PMID: 30277856

Richa S, Shams K, Almiahoub Mohamed Y, Almutairi Mohammed M, Tomás G-M, Alvarez-Leefmans Francisco J, Di Mauricio F (2017) Plasma membrane targeting of endogenous NKCC2 in COS7 cells bypasses functional golgi cisternae and complex N-Glycosylation. Front Cell Dev Biol 4. https://doi.org/10.3389/fcell.2016.00150

Singh VB, Wooten AK, Jackson JW, Maggirwar SB, Kiebala M (2015) Investigating the role of ankyrin-rich membrane spanning protein in Human Immunodeficiency Virus Type-1 Tat-induced microglia activation. J Neurovirol 21:186–198

Sladky V, Schuler F, Fava LL, Villunger A (2017) The resurrection of the PIDDosome – emerging roles in the DNA-damage response and centrosome surveillance. J Cell Sci 130(22):3779–3787. https://doi.org/10.1242/jcs.203448. Review. PubMed PMID: 29142064

Smith DF (2004) Tetratricopeptide repeat cochaperones in steroid receptor complexes. Cell Stress Chaperones 9:109–121

Smith CV, Huang C-C, Miczak A, Russell DG, Sacchettini JC, Honer zu Bentrup K (2003) Biochemical and structural studies of malate synthase from *Mycobacterium tuberculosis*. J Biol Chem 278:1735–1743

Smith AI, Lew RA, Shrimpton CN, Evans RG, Abbenante G (2000) A novel stable inhibitor of endopeptidases EC 3.4.24.15 and 3.4.24.16 potentiates bradykinin-induced hypotension. Hypertension 35(2):626–630. PubMed PMID: 10679508

Sobhy H (2016) A review of functional motifs utilized by viruses. Proteomes 4(1) pii: E3. https://doi.org/10.3390/proteomes4010003. Review. PubMed PMID: 28248213; PubMed Central PMCID: PMC5217368

Soldati-Favre D (2008) Molecular dissection of host cell invasion by the apicomplexans: the glideosome. Parasite 15:197–205

Song X, Jensen MØ, Jogini V, Stein RA, Lee CH, Mchaourab HS, Shaw DE, Gouaux E (2018) Mechanism of NMDA receptor channel block by MK-801 and memantine. Nature 556 (7702):515–519. https://doi.org/10.1038/s41586-018-0039-9. Epub 2018 Apr 18. PubMed PMID: 29670280; PubMed Central PMCID: PMC5962351

Song X, Zhang Y, Wang Y (2011) Antimicrobial peptides peptaibols from trichoderma—a review. Wei Sheng Wu Xue Bao 51(4):438–444. Review. Chinese. PubMed PMID: 21796977

Sparks SE, Ciccone C, Lalor M, Orvisky E, Klootwijk R, Savelkoul PJ, Dalakas MC, Krasnewich DM, Gahl WA, Huizing M (2005) Use of a cell-free system to determine UDP-N-acetylglucosamine 2-epimerase and N-acetylmannoseamine kinase activities in human hereditary inclusion body myopathy. Glycobiology 15:1102

Spraggon G, Everse SJ, Doolittle RF (1997) Crystal structures of fragment D from human fibrinogen and its crosslinked counterpart from fibrin. Nature 389:455

Spudich JA (2011) Biochemistry. Molecular motors, beauty in complexity. Science 331 (6021):1143–1144. https://doi.org/10.1126/science.1203978. PubMed PMID: 21385703; PubMed Central PMCID: PMC4799654

Spyroulias GA, Papazacharias S, Pairas G, Cordopatis P (2002) Monitoring the structural consequences of Phe12—>D-Phe and Leu15—>Aib substitution in human/rat corticotropin releasing hormone. Implications for design of CRH antagonists. Eur J Biochem 269(24):6009–6019. PubMed PMID: 12473096

Squire JM, Paul DM, Morris EP (2017) Myosin and actin filaments in muscle: structures and interactions. Subcell Biochem 82:319–371. https://doi.org/10.1007/978-3-319-49674-0_11

Stahlberg H, Müller DJ, Suda K, Fotiadis D, Engel A, Meier T, Matthey U, Dimroth P (2001) Bacterial Na(+)-ATP synthase has an undecameric rotor. EMBO Rep 2(3):229–233. PubMed PMID: 11266365; PubMed Central PMCID: PMC1083843

Stanfield R, Cabezas E, Satterthwait A, Stura E, Profy A, Wilson I (1999) Dual conformations for the HIV-1 gp120 V3 loop in complexes with different neutralizing fabs. Structure 7(2):131–142. PubMed PMID: 10368281

Stella S, Molina R, López-Méndez B, Juillerat A, Bertonati C, Daboussi F, Campos-Olivas R, Duchateau P, Montoya G (2014) BuD, a helix-loop-helix DNA-binding domain for genome modification. Acta Crystallogr D Biol Crystallogr 70(Pt 7):2042–2052. https://doi.org/10.1107/S1399004714011183. Epub e2014 Jun 29. PubMed PMID: 25004980; PubMed Central PMCID: PMC4089491

Sternberg, Shio. Curvature in mathematics and physics

Stephens MJ (2014) Materials science. How shape affects microtubule and nanoparticle assembly. Science 343:981–982

Stetefeld J, Jenny M, Schulthess T, Landwehr R, Engel J, Kammerer RA (2000) Crystal structure of a naturally occurring parallel right-handed coiled coil tetramer. Nat Str Biol 7:772–776

Stevens J, Corper AL, Basler CF, Taubenberger JK, Palese P, Wilson IA (2004) Structure of the uncleaved human H1 hemagglutinin from the extinct 1918 influenza virus. Science 303:1866–1870

St-Jean M, Sygusch J (2007) Stereospecific proton transfer by a mobile catalyst in mammalian fructose-1,6-bisphosphate aldolase. J Biol Chem 282(42):31028-31037. Epub 2007 Aug 29. PubMed PMID: 17728250

Stoll VS, Kimber MS, Pai EF (1996) Insights into substrate binding by D-2-ketoacid dehydrogenases from the structure of Lactobacillus pentosus D-lactate dehydrogenase. Structure 4 (4):437–447. PubMed PMID: 8740366

Stone D, Kiem H-P, Jerome KR (2013) Targeted gene disruption to cure HIV. Curr Opin HIV AIDS 8:217–223

Strelkov SV, Herrmann H, Geisler N, Wedig T, Zimbelmann R, Aebi U, Burkhard P (2002) Conserved segments 1A and 2B of the intermediate filament dimer: their atomic structures and role in filament assembly. EMBO J 21(6):1255–1266. PubMed PMID: 11889032; PubMed Central PMCID: PMC125921

Streckfus CF, Arreola D, Edwards C, Bigler L (2012) Salivary protein profiles among HER2/neu-receptor-positive and -negative breast cancer patients: support for using salivary protein profiles for modeling breast cancer progression. J Oncol 2012:413256. https://doi.org/10.1155/2012/413256. Epub 2012 Apr 10. PubMed PMID: 22570650; PubMed Central PMCID: PMC3335259

Stumpp MT, Binz HK, Amstutz P (2008) DARPins: a new generation of protein therapeutics. Drug Discov Today 13:695–701

Subbalakshmi C, Basak P, Nagaraj R (2017) Self-assembly of t-Butyloxycarbonyl protected dipeptide methyl esters composed of leucine, isoleucine and valine into highly organized structures from alcohol and aqueous alcohol mixtures. Biopolymers https://doi.org/10.1002/bip.23033. [Epub ahead of print]

Sudha G, Singh P, Swapna LS, Srinivasan N (2015) Weak conservation of structural features in the interfaces of homologous transient protein–protein complexes. Protein Sci A Publ Protein Soc 24(11):1856–1873. https://doi.org/10.1002/pro.2792

Sullivan JM, Zimanyi CM, Aisenberg W, Bears B, Chen D-H, Day JW et al (2015) Novel mutations highlight the key role of the ankyrin repeat domain in TRPV4-mediated neuropathy. Neurology: Genetics 1:e29

Sundaram JR, Poore CP, Sulaimee NHB, Pareek T, Asad ABMA, Rajkumar R, Cheong WF, Wenk MR, Dawe GS, Chuang K-H, Pant HC, Kesavapan S (2013) Specific inhibition of p25/Cdk5 activity by the Cdk5 inhibitory peptide reduces neurodegeneration *in vivo*. J Neurosci 33 (1):334–343

Sun Y, Olson R, Horning M, Armstrong N, Mayer M, Gouaux E (2002) Mechanism of glutamate receptor desensitization. Nature 417:245–253

Sun X. Y. Wu, M. Gu, Y. Zhang (2014) Cell Rep 30 264

Sutton RB, Fasshauer F, Jahn R, Brunger AT (1998) Crystal structure of a SNARE complex involved in synaptic exocytosis at 2.4 Å resolution. Nature 395:347–353

Suzuki T, Li W, Zhang JP, Tian QB, Sakagami H, Usuda N, Kondo H, Fujii T, Endo S (2005) A novel scaffold protein, TANC, possibly a rat homolog of Drosophila rolling pebbles (rols), forms a multiprotein complex with various postsynaptic density proteins. Eur J Neurosci 21:339–350

Tabuchi K, Biederer T, Butz S, Sudhof TC (2002) CASK participates in alternative tripartite complexes in which Mint 1 competes for binding with caskin 1, a novel CASK-binding protein. J Neurosci 22:4264–4273

Tajima Y, Uyama E, Go S, Sato C, Tao N, Kotani M, Hino H, Suzuki A, Sanai Y, Kitajima K, Sakuraba H (2005) Distal myopathy with rimmed vacuoles: impaired O-glycan formation in muscular glycoproteins. Am J Pathol 166:1121

Takáts S, Glatz G, Szenci G, Boda A, Horváth GV, Hegedus K, Kovács AL, Juhász G (2018) Non-canonical role of the SNARE protein Ykt6 in autophagosome-lysosome fusion. PLoS Genet. 14(4): e1007359. https://doi.org/10.1371/journal.pgen.1007359. eCollection 2018 Apr. PubMed PMID: 29694367; PubMed Central PMCID: PMC5937789

Tamura Y, Kawano S, Endo T (2018) Organelle contact zones as sites for lipid transfer. J Biochem https://doi.org/10.1093/jb/mvy088. [Epub ahead of print] PubMed PMID: 30371789

Tao-Cheng J-H, Yang Y, Reese TS, Dosemeci A (2015) Differential Distribution of Shank and GKAP at the Postsynaptic Density. PLoS ONE 10:e0118750

Tarricone C, Dhavan R, Peng J, Areces LB, Tsai L-H, Musacchio A (2001) Structure and regulation of the CDK5-p25nck5a complex. Mol Cell 8:657–669

Tchesnokov EP, Raeisimakiani P, Ngure M, Marchant D, Götte M (2018) Recombinant RNA-dependent RNA polymerase complex of Ebola Virus. Sci Rep 8(1):3970. https://doi.org/10.1038/s41598-018-22328-3. PubMed PMID: 29507309; PubMed Central PMCID: PMC5838098

Taylor KC, Buvoli M, Korkmaz EN, Buvoli A, Zheng Y, Heinze NT, Cui Q, Leinwand LA, Rayment I (2015) Skip residues modulate the structural properties of the myosin rod and guide thick filament assembly. Proc Natl Acad Sci USA 112(29):E3806–E3815. doi: https://doi.org/10.1073/pnas.1505813112. Epub 2015 Jul 6. PubMed PMID: 26150528; PubMed Central PMCID: PMC4517226

Teramoto T, Balasubramanian A, Choi KH, Padmanabhan R (2017) Serotype-specific interactions among functional domains of dengue virus 2 nonstructural proteins (NS) 5 and NS3 are crucial for viral RNA replication. J Biol Chem 292(23):9465–9479. https://doi.org/10.1074/jbc.M117.775643. Epub 2017 Apr 10. PubMed PMID: 28396347; PubMed Central PMCID: PMC5465476

Teramoto T, Boonyasuppayakorn S, Handley M, Choi KH, Padmanabhan R (2014) Substitution of NS5 N-terminal domain of dengue virus type 2 RNA with type 4 domain caused impaired replication and emergence of adaptive mutants with enhanced fitness. J Biol Chem 289 (32):22385–22400. https://doi.org/10.1074/jbc.M114.584466. Epub 2014 Jun 5. PubMed PMID: 24904061; PubMed Central PMCID: PMC4139246

Terzyan SS, Cook PF, Heroux A, Hanigan M (2017) Structure of 6-diazo-5-oxo-norleucine-bound human gamma-glutamyl transpeptidase 1, a novel mechanism of inactivation Protein Sci. 26 (6):1196–1205. https://doi.org/10.1002/pro.3172. Epub 2017 Apr 17

Tewary SK, Liang L, Lin Z, Lynn A, Cotmore SF, Tattersall P et al (2015) Structures of minute virus of mice replication initiator protein N-terminal domain: insights into DNA nicking and origin binding. Virology 476:61–71. https://doi.org/10.1016/j.virol.2014.11.022

Teweri R, E Bailes, KA Bunting, JC Coates (2010) Trends Cell. Biol 20:470

Thomas VL, Golemi-Kotra D, Kim C, Vakulenko SB, Mobashery S, Shoichet BK (2005) Structural consequences of the inhibitor-resistant Ser130Gly substitution in TEM beta-lactamase. Biochemistry 44:9330–9338

Thomas SE, Mendes V, Kim SY, Malhotra S, Ochoa-Montaño B, Blaszczyk M, Blundell TL (2017) Structural biology and the design of new therapeutics: from HIV and cancer to mycobacterial infections: a paper dedicated to John Kendrew. J Mol Biol 2017 Jun 23. pii: S0022-2836(17) 30315–7. https://doi.org/10.1016/j.jmb.2017.06.014. [Epub ahead of print]

Tindi JO, Chávez AE, Cvejic S, Calvo-Ochoa E, Castillo PE, Jordan BA (2015) *ANKS1B* gene product AIDA-1 controls hippocampal synaptic transmission by regulating GluN2B subunit localization. J Neurosci 35:8986–8996. https://doi.org/10.1523/JNEUROSCI.4029-14.2015

Tomasetti C, Iasevoli F, Buonaguro EF, De Berardis D, Fornaro M, Fiengo ALC et al (2017) Treating the synapse in major psychiatric disorders: the role of postsynaptic density network in dopamine-glutamate interplay and psychopharmacologic drugs molecular actions. Int J Mol Sci 18:135

Toro-Roman A, Wu T, Stock AM (2005) A common dimerization interface in bacterial response regulators KdpE and TorR. Protein Science?: A Publication of the Protein Society 14 (12):3077–3088. https://doi.org/10.1110/ps.051722805

Torres MT, Pedron CN, da Silva Lima JA, da Silva PI da Silva FD, Oliveira VX (2017) Antimicrobial activity of leucine-substituted decoralin analogs with lower hemolytic activity. J Pept Sci https://doi.org/10.1002/psc.3029. [Epub ahead of print]

Tóth-Petróczy A, Tawfik DS (2014) The robustness and innovability of protein folds. Curr Opin Struct Biol 26:131–138. https://doi.org/10.1016/j.sbi.2014.06.007. Epub 2014 Jul 17. Review. PubMed PMID: 25038399

Touw DS, Nordman CE, Stuckey JA, Pecorano VL (2007) Identifying important structural characteristics of arsenic resistance proteins by using designed three-stranded coiled-coils. Proc Natl Acad Sci 104:11969–11974

Tseng T-S, Wang S-H, Chang T-W, Wei H-M, Wang Y-J, Tsai K-C et al (2016) Sarkosyl-induced helical structure of an antimicrobial peptide GW-Q6 plays an essential role in the binding of surface receptor OprI in Pseudomonas aeruginosa. PLoS ONE 11(10):e0164597. https://doi.org/10.1371/journal.pone.0164597

Tseng WC, Jenkins PM, Tanaka M, Mooney R, Bennett V (2015) Giant ankyrin-G stabilizes somatodendritic GABAergic synapses through opposing endocytosis of $GABA_A$ receptors. Proc Natl Acad Sci USA 112:1214–1219

Tu D, Li W, Ye Y, Brunger AT (2007) Structure and function of the yeast U-box-containing ubiquitin ligase Ufd2p. Proc Natl Acad Sci *104*:15599–15606

Uddin MS, Mamun AA, Labu ZK, Hidalgo-Lanussa O, Barreto GE, Ashraf GM (2018) Autophagic dysfunction in Alzheimer's disease: cellular and molecular mechanistic approaches to halt Alzheimer's pathogenesis. J Cell Physiol https://doi.org/10.1002/jcp.27588. [Epub ahead of print] Review. PubMed PMID: 30362531

Ueno H, Yasunaga T, Shingyoji C, Hirose K (2008) Dynein pulls microtubules without rotating its stalk. Proc Natl Acad Sci U S A 105(50):19702–19707. https://doi.org/10.1073/pnas.0808194105. Epub 2008 Dec 8. PubMed PMID: 19064920; PubMed Central PMCID: PMC2604933

Urosev D, Ferrer-Navarro M, Pastorello I, Cartocci E, Costenaro L, Zhulenkovs D, Maréchal JD, Leonchiks A, Reverter D, Serino L, Soriani M, Daura X (2013) Crystal structure of c5321: a protective antigen present in uropathogenic Escherichia coli strains displaying an SLR fold. BMC Struct Biol. 13:19. https://doi.org/10.1186/1472-6807-13-19. PubMed PMID: 24099525; PubMed Central PMCID: PMC3851747

Utreras E, Prochaskova M, Terse A, Gross J, Keller J, Iadarola MJ, Kulkarni AB (2013) TGF-β1 sensitizes TRPV1 through Cdk5 signaling in odontoblast-like cells. Mol Pain 9:24–38

Vajjhala PR, Lu A, Brown DL, Pang SW, Sagulenko V, Sester DP, Cridland SO, Hill JM, Schroder K, Stow JL, Wu H, Stacey KJ (2015) The inflammasome adaptor ASC induces procaspase-8 death effector domain filaments. J Biol Chem 290(49):29217–29230. https://doi.org/10.1074/jbc.M115.687731. Epub 2015 Oct 14. PubMed PMID: 26468282; PubMed Central PMCID: PMC4705927

Valenzuela-Chavira I, Contreras-Vergara CA, Arvizu-Flores AA, Serrano-Posada H, Lopez-Zavala AA, García-Orozco KD, Hernandez-Paredes J, Rudiño-Piñera E, Stojanoff V, Sotelo-Mundo RR, Islas-Osuna MA (2017) Insights into ligand binding to a glutathione S-transferase from mango: Structure, thermodynamics and kinetics. Biochimie 135:35–45. https://doi.org/10.1016/j.biochi.2017.01.005. Epub 2017 Jan 17. PubMed PMID: 28104507; PubMed Central PMCID: PMC5346462.

Valm AM, Cohen S, Legant WR, Melunis J, Hershberg U, Wait E, Cohen AR, Davidson MW, Betzig E, Lippincott-Schwartz J (2017) Applying systems-level spectral imaging and analysis to reveal the organelle interactome. Nature 546(7656):162–167. https://doi.org/10.1038/nature22369. Epub 2017 May 24. PubMed PMID: 28538724; PubMed Central PMCID: PMC5536967

van den Heuvel MG, Dekker C (2007) Motor proteins at work for nanotechnology. Science 317 (5836):333–6. Review. PubMed PMID: 17641191.PMID: 17641191

Vasilescu C, Alexianu M, Dan A (1984) Neuronal type of Charcot-Marie-Tooth disease with a syndrome of continuous motor unit activity. J Neurol Sci 63(1):11–25. PubMed PMID: 6699650

Vasudev PG, Ananda K, Chatterjee S, Aravinda S, Shamala N, Balaram P (2007) Hybrid peptide design. Hydrogen bonded conformations in peptides containing the stereochemically constrained gamma-amino acid residue, gabapentin. J Am Chem Soc 129(13):4039–4048. Epub 2007 Mar 10

Vasudev PG, Chatterjee S, Shamala N, Balaram P (2009) Gabapentin: a stereochemically constrained gamma amino acid residue in hybrid peptide design. Acc Chem Res 42 (10):1628–1639. https://doi.org/10.1021/ar9001153. Review. PubMed PMID: 19572698

Vaughn DE, Rodriguez J, Lazebnik Y, Joshua-Tor L (1999) Crystal structure of apaf-1 caspase recruitment domain: an α-helical greek key fold for apoptotic signaling. J Mol Biol 293:439–447

Venanzi M, Gatto E, Formaggio F, Toniolo C (2017) The importance of being Aib. Aggregation and self-assembly studies on conformationally constrained oligopeptides. J Pept Sci 23(2):104-116. https://doi.org/10.1002/psc.2956. Epub 2017 Jan 5. Review. PubMed PMID: 28054413

Venkataramani RN, MacLachlan TK, Chai X, El-Deiry WS, Marmorstein R (2002) Structure-based Design of p18INK4cProteins with Increased Thermodynamic Stability and Cell Cycle Inhibitory Activity. J Biol Chem 277:48827–48833

Verdon G, Oh S, Serio RN, Boudker O (2014)Coupled ion binding and structural transitions along the transport cycle of glutamate transporters. Elife 3:e02283. https://doi.org/10.7554/eLife.02283. PubMed PMID: 24842876; PubMed Central PMCID: PMC4051121

Verkerk AO, Amin AS, Remme CA (2018 Oct 1) Disease Modifiers of Inherited SCN5A Channelopathy. Front Cardiovasc Med. 5:137. doi: https://doi.org/10.3389/fcvm.2018.00137. eCollection 2018. Review. PubMed PMID: 30327767; PubMed Central PMCID: PMC6174200

Very N, Vercoutter-Edouart AS, Lefebvre T, Hardivillé S, El Yazidi-Belkoura I (2018 Oct 9) Cross-Dysregulation of O-GlcNAcylation and PI3K/AKT/mTOR Axis in Human Chronic Diseases. Front Endocrinol (Lausanne). 9:602. doi: 10.3389/fendo.2018.00602. eCollection 2018. Review. PubMed PMID: 30356686; PubMed Central PMCID: PMC6189293

Vitagliano L, Berisio R, Mazzarella L, Zagari A (2001) Structural bases of collagen stabilization induced by proline hydroxylation. Biopolymers 58(5):459–464

Vogensen SB, Frydenvang K, Greenwood JR, Postorino G, Nielsen B, Pickering DS, Ebert B, Bølcho U, Egebjerg J, Gajhede M, Kastrup JS, Johansen TN, Clausen RP, Krogsgaard-Larsen P. A tetrazolyl-substituted subtype-selective AMPA receptor agonist. J Med Chem. 2007 May 17;50(10):2408–2414. Epub 2007 Apr 25. PubMed PMID: 17455929

Vollmar M, Schlieper D, Winn M, Büchner C, Groth G (2009) Structure of the c14 rotor ring of the proton translocating chloroplast ATP synthase. J Biol Chem 284(27):18228–18235. https://doi.org/10.1074/jbc.M109.006916. Epub 2009 May 7. PubMed PMID: 19423706; PubMed Central PMCID: PMC2709358

Vuilleumier S (1997) Minireview. Bacterial glutathione *S*-transferases: What are they good for? J Bact 179:1431–1441

Yadav MK, Redman JE, Leman LJ, Alvarez-Gutiérrez JM, Zhang Y, Stout CD, Ghadiri MR (2005) Structure-based engineering of internal cavities in coiled-coil peptides. Biochemistry 44 (28):9723–9732. PubMed PMID: 16008357; PubMed Central PMCID: PMC1779508

Yakovlev S, Makogonenko E, Kurochkina N, Nieuwenhuizen W, Ingham K, Medved L (2000) Conversion of fibrinogen to fibrin: mechanism of exposure for tPA-plasminogen binding site. Biochemistry 39:15730

Yang J, Ma YQ, Page RC, Misra S, Plow EF, Qin J (2009) Structure of an integrin alphaIIb beta3 transmembrane-cytoplasmic heterocomplex provides insight into integrin activation. Proc Natl Acad Sci USA 106(42):17729–17734. https://doi.org/10.1073/pnas.0909589106. Epub 2009 Oct 1. PubMed PMID: 19805198; PubMed Central PMCID: PMC2764936

Yang B, Xie S, Berardi MJ, Zhao XM, Dev J, Yu W, Sun B, Chou JJ (2013a) Unusual architecture of the p7 channel from hepatitis C virus. Nature 498:521–525

Yang J, Zhang M, Wang X (2018a) Crystal structure of the chloroplast RNA editing factor MORF2. Biochem Biophys Res Commun 495(2):2038–2043. https://doi.org/10.1016/j.bbrc.2017.12.044. Epub 2017 Dec 8. PubMed PMID: 29229384

Yan J, Zhang Q, Guan Z, Wang Q, Li L, Ruan F, Lin R, Zou T, Yin P (2017) MORF9 increases the RNA-binding activity of PLS-type pentatricopeptide repeat protein in plastid RNA editing. Nat Plants 3:17037. https://doi.org/10.1038/nplants.2017.37. PubMed PMID: 28394309

Yardeni T, Jacobs K, Niethamer TK, Ciccone C, Anikster Y, Kurochkina N, Gahl WA, Huizing M (2013a) Murine isoforms of UDP-GlcNAc 2-epimerase/ManNAc kinase: secondary structures, expression profiles, and response to ManNAc therapy. Glycoconj J 30(6):609–618. https://doi.org/10.1007/s10719-012-9459-1. Epub 2012 Dec 25. PubMed PMID: 23266873; PubMed Central PMCID: PMC3622838

Young SC (2018) A systematic review of antiamyloidogenic and metal-chelating peptoids: two structural motifs for the treatment of Alzheimer's disease. Molecules 23(2). pii: E296. https://doi.org/10.3390/molecules23020296. Review. PubMed PMID: 29385058

Yuan S, Yu X, Topf M, Ludtke SJ, Wang X, Akey CW (2010a) Structure of an apoptosome-procaspase-9 CARD complex. Structure 18:571–583

Wada SI, Takesada A, Nagamura Y, Sogabe E, Ohki R, Hayashi J, Urata H (2017) Structure-activity relationship study of Aib-containing amphipathic helical peptide-cyclic RGD conjugates as carriers for siRNA delivery. Bioorg Med Chem Lett;27(24):5378–5381. https://doi.org/10.1016/j.bmcl.2017.11.018. Epub 2017 Nov 10. PubMed PMID: 29157863

Wagschal K, Lavigna P, Mant C, Hodges RS (1999) The role of position a in determining the stability and oligomerization state of a-helical coiled coils: 20 amino acid stability coefficients in the hydrophobic core of proteins. Protein Sci 8:2312–2329

Wang M, Ogé L, Perez-Garcia MD, Hamama L, Sakr S (2018) The PUF protein family: overview on PUF RNA targets, biological functions, and post transcriptional regulation. Int J Mol Sci 19 (2). pii: E410. https://doi.org/10.3390/ijms19020410. Review. PubMed PMID: 29385744; PubMed Central PMCID: PMC5855632

Wang L, Yang JK, Kabaleeswaran V, Rice AJ, Cruz AC, Park AY, Yin Q (2010a) The Fas-FADD death domain complex structure reveals the basis of DISC assembly and disease mutations. Nat Struct Mol Biol, in press

Wang C, Wei Z, Chen K, Ye F, Yu C, Bennett V, Zhang M (2014) Structural basis of diverse membrane target recognitions by ankyrins. eLife 3:e04353

Wang R, Wei Z, Jin H, Wu H, Yu C, Wen W, Chan LN, Wen Z, Zhang M (2009a) Autoinhibition of UNC5b revealed by the cytoplasmic domain structure of the receptor. Mol Cell 33(6):692–703. https://doi.org/10.1016/j.molcel.2009.02.016. PubMed PMID: 19328064

Wang L, Yang JK, Kabaleeswaran V, Rice AJ, Cruz AC, Park AY, Yin Q, Damko E, Jang SB, Raunser S, Robinson CV, Siegel RM, Walz T, Wu H (2010b) The Fas-FADD death domain complex structure reveals the basis of DISC assembly and disease mutations. Nat Struct Mol Biol 17(11):1324–1329. https://doi.org/10.1038/nsmb.1920. Epub 2010 Oct 10. PubMed PMID: 20935634; PubMed Central PMCID: PMC2988912

Wang C, Shang Y, Yu J, Zhang M (2012) Substrate recognition mechanism of atypical protein Kinase Cs revealed by the structure of PKCi in complex with a substrate peptide from Par-3. Structure 20:791–801

Watt IN, Montgomery MG, Runswick MJ, Leslie AG, Walker JE (2010a) Bioenergetic cost of making an adenosine triphosphate molecule in animal mitochondria. Proc Natl Acad Sci USA 107(39):16823–16827. https://doi.org/10.1073/pnas.1011099107. Epub 2010 Sep 16. PubMed PMID: 20847295; PubMed Central PMCID: PMC2947889

Wang R, Wei Z, Jin H, Wu H, Yu C, Wen W, Chan LN, Wen Z, Zhang M (2009b) Autoinhibition of UNC5b revealed by the cytoplasmic domain structure of the receptor. Mol Cell 33 (6):692–703. https://doi.org/10.1016/j.molcel.2009.02.016. PubMed PMID: 19328064

Wang X, McLachlan J, Zamore PD, Hall TM (2002) Modular recognition of RNA by a human pumilio-homology domain. Cell;110(4):501–512. PubMed PMID: 12202039

Warren JJ, Pohlhaus TJ, Changela A, Iyer RR, Modrich PL, Beese LS (2007) Structure of the human MutSa DNA lesion recognition complex. Mol Cell 26:579–592

Watt IN, Montgomery MG, Runswick MJ, Leslie AG, Walker JE (2010b) Bioenergetic cost of making an adenosine triphosphate molecule in animal mitochondria. Proc Natl Acad Sci USA

107(39):16823–16827. https://doi.org/10.1073/pnas.1011099107. Epub 2010 Sep 16. PubMed PMID: 20847295; PubMed Central PMCID: PMC2947889

Weber CH, Vincenz C (2001) The death domain superfamily: a tale of two interfaces? TIBS 26:475–481

Welbourn EM, Wilson MT, Yusof A, Metodiev MV, Cooper CE (2017) The mechanism of formation, structure and physiological relevance of covalent hemoglobin attachment to the erythrocyte membrane. Free Radic Biol Med 103:95–106. https://doi.org/10.1016/j.freeradbiomed.2016.12.024. Epub 2016 Dec 20. PubMed PMID: 28007575; PubMed Central PMCID: PMC5282401

Wells WW, Yang Y, Deits TL, Gan ZR (1993) Thioltransferases. Adv Enzymol Relat Areas Mol Biol;66:149–201. Review. PubMed PMID: 8430514

Weston MC, Schuck NP, Ghosal A, Rosenmund C, Mayer ML (2006) Conformational restriction blocks glutamate receptor desensitization. Nat Struct Mol Biol 13:1120–1127

Wiemuth D, van de Sandt L, Herr R, Gründer S (2012) Transient receptor potential N (TRPN1) from Xenopus interacts with the penta-EF-hand protein peflin. FEBS Lett;586(24):4276–4281. https://doi.org/10.1016/j.febslet.2012.10.046. Epub 2012 Nov 15. PubMed PMID: 23159937

Wierenga RK, Noble MEM, Vriend G, Naughe S, Hol WGJ (1991) Refined 1.83 Angstrom structure of trypanosomal triosephospate isomerase, crystallized in presence of 2.4M- Ammonium sulphate. A comparison with the structure of the trypanosomal triosephosphate isomerase-glycerol-3-phosphate complex. J Mol Biol 220:995–1015

Wierenga RK (2001) The TIM-barrel fold: a versatile framework for efficient enzymes. FEBS Lett 492:193–198

Willhoft O, Kerr R, Patel D, Zhang W, Al-Jassar C, Daviter T et al (2017) The crystal structure of the Sgt1-Skp1 complex: the link between Hsp90 and both SCF E3 ubiquitin ligases and kinetochores. Sci Rep 7:41626. https://doi.org/10.1038/srep41626

Willis BC, Ponce-Balbuena D, Jalife J (2015) Protein assemblies of sodium and inward rectifier potassium channels control cardiac excitability and arrhythmogenesis. Am J Physiol Heart and Circ Physiol 308:H1463–H1473

deWit J, Hong W, Luo L, Ghosh A (2011) Role of leucine-rich repeat proteins in the development and function of neural circuits. Annu Rev Cell Dev Biol 27:697–729

Won S, Levy JM, Nicoll RA, Roche KW (2017) MAGUKs: multifaceted synaptic organizers. Curr Opin Neurobiol 43:94–101

Wong AYC, Maclean DM, Bowie D (2007) Na+/Cl- dipole couples agonist binding to kainate receptor activation. J Neurosci 27:6800–6809

Wong E, Ng F-M, Yu C-Y, Lim P, Lim L-H, Traynelis SF, Low CM (2005) Expression and characterization of soluble amino-terminal domain of NR2B subunit of N-methyl-D-aspartate receptor. Prot Sci 14:2275–2283

Wong K, Perpich JD, Kozlov G, Cygler M, Abu Kwaik Y, Gehring K (2017) Structural mimicry by a bacterial F box effector hijacks the host ubiquitin-proteasome system. Structure 25 (2):376–383. https://doi.org/10.1016/j.str.2016.12.015

Wood MW, VanDongen HM, VanDongen AM (1995) Structural conservation of ion conduction pathways in K channels and glutamate receptors. Proc Natl Acad Sci USA 92(11):4882–4886. PubMed PMID: 7761417; PubMed Central PMCID: PMC41811

Woolfson DN (2017) Coiled-coil design: updated and upgraded. Subcell Biochem 82:35–61. https://doi.org/10.1007/978-3-319-49674-0_2

Woolfson DN (2010) Building fibrous biomaterials from -helical and collagen-like coiled-coil peptides. Biopolymers 94:118–127

Wu D, Xu C, Cederbaum A (2009) Role of nitric oxide and nuclear factor-kappaB in the CYP2E1 potentiation of tumor necrosis factor alpha hepatotoxicity in mice. Free Radic Biol Med 46 (4):480–491. https://doi.org/10.1016/j.freeradbiomed.2008.11.001. Epub 2008 Nov 19. PubMed PMID: 19063961

Wu H, Maciejewski MW, Marintchev A, Benashski SE, Mullen GP, King SM (2000a) Solution structure of a dynein motor domain associated light chain. Nat Struct Biol 7(7):575–579. PubMed PMID: 10876244

Wu H, Zeng H, Lam R, Tempel W, Kerr ID, Min J (2015) Structure of the human MLH1 N-terminus: implications for predisposition to Lynch syndrome. Acta Crystallogr Sect F Struct Biol Commun 71(Pt 8):981–985. https://doi.org/10.1107/S2053230X15010183

Wu G, Chai J, Suber TL, Wu JW, Du C, Wang X, Shi Y (2000b) Structural basis of IAP recognition by Smac/DIABLO. Nature 408(6815): 1008–1012. PubMed PMID: 11140638

Wybenga GG, Crismaru CG, Janssen DB, Dijkstra BW (2012) Structural determinants of the ß-selectivity of a bacterial aminotransferase. J Biol Chem;287(34):28495–28502. https://doi.org/10.1074/jbc.M112.375238. Epub 2012 Jun 28. PubMed PMID: 22745123; PubMed Central PMCID: PMC3436509

Xie W, Nangle LA, Zhang W, Schimmel P, Yang XL (2007) Long-range structural effects of a Charcot-Marie-Tooth disease-causing mutation in human glycyl-tRNA synthetase. Proc Natl Acad Sci USA 104(24):9976–9981. Epub 2007 Jun 1. PubMed PMID: 17545306; PubMed Central PMCID: PMC1891255

Xie L, Yang S (2016) Brain globins in physiology and pathology. Med Gas Res 6(3):154–163. https://doi.org/10.4103/2045-9912.191361

Xu M, Cooper EC (2015) An ankyrin-G N-terminal gate and protein kinase CK2 dually regulate binding of voltage-gated Sodium and KCNQ2/3 Potassium channels. J Biol Chem 290:16619–16632

Yan N, Chai J, Lee ES, Gu L, Liu Q, He J, Wu JW, Kokel D, Li H, Hao Q, Xue D, Shi Y (2005) Structure of the CED-4-CED-9 complex provides insights into programmed cell death in Caenorhabditis elegans. Nature 437(7060):831–837. PubMed PMID: 16208361

Yang Y, Zhang Y, Wei Z (2013b) Supramolecular helices: chirality transfer from conjugated molecules to structures. Adv Mater 25(42):6039–6049. https://doi.org/10.1002/adma.201302448. Epub 2013 Aug 21. PubMed PMID: 23966134

Yang H, Jiang X, Li B, Yang HJ, Miller M, Yang A, Dhar A, Pavletich NP (2017) Mechanisms of mTORC1 activation by RHEB and inhibition by PRAS40. Nature 552(7685): 368–373. https://doi.org/10.1038/nature25023. Epub 2017 Dec 13. PubMed PMID: 29236692; PubMed Central PMCID: PMC5750076

Yang J, Zhao Y, Li P, Yang Y, Zhang E, Zhong M, Li Y, Zhou D, Cao Y, Lu M, Shao F, Yan H (2018b) Sequence determinants of specific pattern-recognition of bacterial ligands by the NAIP-NLRC4 inflammasome. Cell Discov 4:22. https://doi.org/10.1038/s41421-018-0018-1. eCollection 2018. PubMed PMID: 29760946; PubMed Central PMCID: PMC5938239

Yang J, Cron P, Good VM, Thompson V, Hemmings BA, Barford D (2002) Crystal structure of an activated Akt/protein kinase B ternary complex with GSK3-peptide and AMP-PNP. Nat Struct Biol 9(12):940–944. PubMed PMID: 12434148

Yao I, Hata Y, Hirao K, Deguchi M, Ide N, Takeuchi M, Takai Y (1999) Synamon, a novel neuronal protein interacting with synapse-associated protein 90/postsynaptic density-95-associated protein. J Biol Chem 274:27463–27466

Yardeni T, Jacobs K, Niethamer TK, Ciccone C, Anikster Y, Kurochkina N et al (2013b) Murine isoforms of UDP-GlcNAc 2-epimerase/ManNAc kinase: secondary structures, expression profiles, and response to ManNAc therapy. Glycoconj J 30:609–618. https://doi.org/10.1007/s10719-012-9459-1

Yeo HJ, Yokoyama T, Walkiewicz K, Kim Y, Grass S, Geme JW (2007) The structure of the Haemophilus influenzae HMW1 pro-piece reveals a structural domain essential for bacterial two-partner secretion. J Biol Chem 282(42):31076–31084. Epub 2007 Aug 14. PubMed PMID: 17699157

Yin Z, Shi K, Banerjee S, Pandey KK, Bera S, Grandgenett DP, Aihara H (2016) Crystal structure of the Rous sarcoma virus intasome. Nature 530(7590):362–366. https://doi.org/10.1038/nature16950. PubMed PMID: 26887497; PubMed Central PMCID: PMC4881392

Yoshimura T, Stevens SR, Leterrier C, Stankewich MC, Rasband MN (2016) Developmental changes in expression of βIV spectrin splice variants at axon initial segments and nodes of Ranvier. Front Cell Neurosci 10:304

Yu X, Acehan D, Ménétret JF, Booth CR, Ludtke SJ, Riedl SJ, Shi Y, Wang X, Akey CW (2005) A structure of the human apoptosome at 12.8 A resolution provides insights into this cell death platform. Structure 13(11):1725–1735. PubMed PMID: 16271896

Yuan C, Li J, Selby TL (1999) Byeon I-J L and Tsai M-D (1999) tumor suppressor INK4: comparisons of conformational properties between p16INK4A and p18INK4C J. Mol Biol 294:201–211

Yuan T, Liu L, Zhang Y, Wei L, Zhao S, Zheng X, Huang X, Boulanger J, Gueudry C, Lu J, Xie L, Du W, Zong W, Yang L, Salamero J, Liu Y, Chen L (2015) Diacylglycerol guides the hopping of clathrin-coated pits along microtubules for exo-endocytosis coupling. Dev Cell 35(1):120-130. https://doi.org/10.1016/j.devcel.2015.09.004. Epub 2015 Oct 1. PubMed PMID: 26439397

Yuan S, Yu X, Topf M, Ludtke SJ, Wang X, Akey CW (2010b) Structure of an apoptosome-procaspase-9 CARD complex. Structure 18:571–583

Zamani MR, Aslani S, Salmaninejad A, Javan MR, Rezaei N (2016) PD-1/PD-L and autoimmunity: a growing relationship. Cell Immunol 10:27–41

Zhang X, Bennett V (1996) Identification of O-linked N-acetylglucosamine modification of ankyrinG isoforms targeted to nodes of Ranvier. J Biol Chem 271(49):31391–31398

Zhang C, Caj J, Zhang J, Li Z, Guo Z, Zhang X, Lu W, Zhang Y, Yuan A, Yu S, Fang Y (2014) Pharacogenetic effects of dopamine transporter gene polymorphisms on response to chlorpromazine and clozapine and on extrapyramidal syndrome in schizophrenia. Prog Neuropsychopharmacol Biol Physciatry 50:110

Zhang W, Cho Y, Lolis E, Howe JR (2008) Structural and single channel results indicate that the rates of ligand binding domain closing and opening directly impact AMPA receptor gating. J Neurosci 28:932–943

Zhang Q, Harding R, Hou F, Dong A, Walker JR, Bteich J, Tong Y (2016) Structural basis of the recruitment of ubiquitin-specific protease USP15 by spliceosome recycling factor SART3. J Biol Chem 291(33):17283–17292. https://doi.org/10.1074/jbc.M116.740787. Epub 2016 Jun 2. PubMed PMID: 27255711; PubMed Central PMCID: PMC5016127

Zhang Y (2017) Energetics, kinetics, and pathway of SNARE folding and assembly revealed by optical tweezers. Protein Sci;26(7):1252–1265. https://doi.org/10.1002/pro.3116. Epub 2017 Mar 8. Review. PubMed PMID: 28097727; PubMed Central PMCID: PMC5477538.

Zhang Z, Devarajan P, Dorfman AL, Morrow JS (1998) Structure of the ankyrin-binding domain of alpha-Na,K-ATPase. J Biol Chem;273(30):18681–18684. PubMed PMID: 9668035

Zhang Z, Roe SM, Diogon M, Kong E, El Alaoui H, Barford D (2010a) Molecular structure of the N-terminal domain of the APC/C subunit Cdc27 reveals a homo-dimeric tetratricopeptide repeat architecture. J Mol Biol 397(5):1316–1328. https://doi.org/10.1016/j.jmb.2010.02.045. Epub 2010 Mar 3. PubMed PMID: 20206185

Zhang X, Boyce M, Bhattacharya B, Zhang X, Schein S, Roy P, Zhou ZH (2010b) Bluetongue virus coat protein VP2 contains sialic acid-binding domains, and VP5 resembles enveloped virus fusion proteins. Proc Natl Acad Sci USA 107(14):6292-6297. https://doi.org/10.1073/pnas.0913403107. Epub 2010 Mar 23. PubMed PMID: 20332209; PubMed Central PMCID: PMC2852009

Zhang L, Chen S, Ruan J, Wu J, Tong AB, Yin Q, Li Y, David L, Lu A, Wang WL, Marks C, Ouyang Q, Zhang X, Mao Y, Wu H (2015) Cryo-EM structure of the activated NAIP2-NLRC4 inflammasome reveals nucleated polymerization. Science 350(6259):404–409. https://doi.org/10.1126/science.aac5789. Epub 2015 Oct 8. PubMed PMID: 26449474; PubMed Central PMCID: PMC4640189

Zhao H, Qin X, Yang D, Jiang Y, Zheng W, Wang D, Tian Y, Liu Q, Xu N, Li Z (2017) The development of activatable lytic peptides for targeting triple negative breast cancer. Cell Death Discov 3:17037. https://doi.org/10.1038/cddiscovery.2017.37. eCollection 2017. PubMed PMID: 29263848; PubMed Central PMCID: PMC5629628

Zhavoronkov A, Izumchenko E, Kanherkar RR, Teka M, Cantor C, Manaye K et al (2016) Pro-fibrotic pathway activation in trabecular meshwork and lamina cribrosa is the main driving force of glaucoma. Cell Cycle 15(12):1643–1652. https://doi.org/10.1080/15384101.2016.1170261

Zheng C-Y, Petralia RS, Wang Y-X, Kachar B, Wenthold RJ (2010) SAP102 is a highly mobile MAGUK in spines. J Neurosci 30:4757–4766

Zheng X, Kuang Y, Lv W, Cao D, Sun Z, Sun X (2016) Genome-wide association study for muscle fat content and abdominal fat traits in common carp (*Cyprinus carpio*). PLoS ONE 11:e0169127

Zheng Y-L, Li B-S, Rudrabhatla P, Shukla V, Amin ND, Maric D, Kesavapany S, Kanungo J, Pareek TK, Takahashi S, Grant P, Kulkarni AB, Pant HC (2010a) Phosphorylation of p27Kip1 at Thr187 by Cyclin-dependent Kinase 5 modulates neural stem cell differentiation. Mol Biol Cell 21:3601–3614

Zheng Y-L, Amin ND, Hu Y-F, Rudrabhatla P, Shukla V, Kanungo J, Kesavapany S, Grant P, Albers W, Pant HC (2010b) A 24-residue Peptide (p5), derived from p35, the Cdk5 neuronal activator, specifically inhibits Cdk5-p25 hyperactivity and tau hyperphosphorylation. J Biol Chem 285(44):34202–34212

Zhmurov A, Protopopova AD, Litvinov RI, Zhukov P, Mukhitov AR, Weisel JW, Barsegov V (2016) Structural basis of interfacial flexibility in fibrin oligomers. Structure 24(11):1907–1917. https://doi.org/10.1016/j.str.2016.08.009. Epub 2016 Sep 29. PubMed PMID: 27692965; PubMed Central PMCID: PMC5240993

Zhou M, Li Y, Hu Q, Bai XC, Huang W, Yan C, Scheres SH, Shi Y (2015a) Atomic structure of the apoptosome: mechanism of cytochrome c- and dATP-mediated activation of Apaf-1. Genes Dev;29(22):2349–2361. https://doi.org/10.1101/gad.272278.115. Epub 2015 Nov 5. PubMed PMID: 26543158; PubMed Central PMCID: PMC4691890

Zhou GP (2011) The structural determinations of the leucine zipper coiled-coil domains of the cGMP-dependent protein kinase I alpha and its interaction with the myosin binding subunit of the myosin light chains phosphase. Proteins Pept Lett 18:966–978

Zhou GP, Huang RB (2013) The pH-Triggered Conversion of the PrP(c) to PrP(sc.). Curr Top Med Chem 13:1152–1163

Zhou P, Chou J, Sanchez O, Yuan J, Wagner G (1999) Solution structure of Apaf-1 CARD and its interaction with caspase-9 CARD: A structural basis for specific adaptorycaspase interaction. Proc Natl Acad Sci 96:11265–11270

Zhou M, Li Y, Hu Q, Bai XC, Huang W, Yan C, Scheres SH, Shi Y (2015b) Atomic structure of the apoptosome: mechanism of cytochrome c- and dATP-mediated activation of Apaf-1. Genes Dev;29(22):2349–2361. https://doi.org/10.1101/gad.272278.115. Epub 2015 Nov 5. PubMed PMID: 26543158; PubMed Central PMCID: PMC4691890.

Zsombok A, Derbenev AV (2016) TRP Channels as Therapeutic Targets in Diabetes and Obesity. Pharmaceuticals (Basel) 9(3). pii: E50. https://doi.org/10.3390/ph9030050. Review. PubMed PMID: 27548188; PubMed Central PMCID: PMC5039503

Zynda ER, Grimm MJ, Yuan M, Zhong L, Mace TA, Capitano M et al (2015) A role for the thermal environment in defining co-stimulation requirements for $CD4^+$ T cell activation. Cell Cycle 14:2340–2354

Zweifel ME, Leahy DJ, Hughson FM, Barrick D (2003) Structure and stability of the ankyrin domain of the *Drosophila* Notch receptor. Protein Sci 12:2622–2632

Zukerberg LR, Patrick GN, Nikolic M, Humbert S, Wu CL, Lanier LM, Gertler FB, Vidal M, Van Etten RA, Tsai LH (2000) Cables links Cdk5 and c-Abl and facilitates Cdk5 tyrosine phosphorylation, kinase upregulation, and neurite outgrowth. Neuron 26(3):633–646. PubMed PMID: 10896159

Index

N. Kurochkina, *Protein Structure and Modeling*,
https://doi.org/10.1007/978-981-13-6601-7

MIX
Papier aus verantwortungsvollen Quellen
Paper from responsible sources
FSC® C105338

If you have any concerns about our products,
you can contact us on
ProductSafety@springernature.com

In case Publisher is established outside the EU,
the EU authorized representative is:
Springer Nature Customer Service Center GmbH
Europaplatz 3, 69115 Heidelberg, Germany

Printed by Libri Plureos GmbH
in Hamburg, Germany